Communications
in Computer and Information Science 1941

Rationale

The CCIS series is devoted to the publication of proceedings of computer science conferences. Its aim is to efficiently disseminate original research results in informatics in printed and electronic form. While the focus is on publication of peer-reviewed full papers presenting mature work, inclusion of reviewed short papers reporting on work in progress is welcome, too. Besides globally relevant meetings with internationally representative program committees guaranteeing a strict peer-reviewing and paper selection process, conferences run by societies or of high regional or national relevance are also considered for publication.

Topics

The topical scope of CCIS spans the entire spectrum of informatics ranging from foundational topics in the theory of computing to information and communications science and technology and a broad variety of interdisciplinary application fields.

Information for Volume Editors and Authors

Publication in CCIS is free of charge. No royalties are paid, however, we offer registered conference participants temporary free access to the online version of the conference proceedings on SpringerLink (http://link.springer.com) by means of an http referrer from the conference website and/or a number of complimentary printed copies, as specified in the official acceptance email of the event.

CCIS proceedings can be published in time for distribution at conferences or as post-proceedings, and delivered in the form of printed books and/or electronically as USBs and/or e-content licenses for accessing proceedings at SpringerLink. Furthermore, CCIS proceedings are included in the CCIS electronic book series hosted in the SpringerLink digital library at http://link.springer.com/bookseries/7899. Conferences publishing in CCIS are allowed to use Online Conference Service (OCS) for managing the whole proceedings lifecycle (from submission and reviewing to preparing for publication) free of charge.

Publication process

The language of publication is exclusively English. Authors publishing in CCIS have to sign the Springer CCIS copyright transfer form, however, they are free to use their material published in CCIS for substantially changed, more elaborate subsequent publications elsewhere. For the preparation of the camera-ready papers/files, authors have to strictly adhere to the Springer CCIS Authors' Instructions and are strongly encouraged to use the CCIS LaTeX style files or templates.

Abstracting/Indexing

CCIS is abstracted/indexed in DBLP, Google Scholar, EI-Compendex, Mathematical Reviews, SCImago, Scopus. CCIS volumes are also submitted for the inclusion in ISI Proceedings.

How to start

To start the evaluation of your proposal for inclusion in the CCIS series, please send an e-mail to ccis@springer.com.

Akram Bennour · Ahmed Bouridane ·
Lotfi Chaari

Editors

Intelligent Systems and Pattern Recognition

Third International Conference, ISPR 2023
Hammamet, Tunisia, May 11–13, 2023
Revised Selected Papers, Part II

Springer

Editors
Akram Bennour 🆔
Larbi Tebessi University
Tebessa, Algeria

Ahmed Bouridane 🆔
Sharjah University
Sharjah, United Arab Emirates

Lotfi Chaari 🆔
University of Toulouse
Toulouse, France

ISSN 1865-0929   ISSN 1865-0937 (electronic)
Communications in Computer and Information Science
ISBN 978-3-031-46337-2   ISBN 978-3-031-46338-9 (eBook)
https://doi.org/10.1007/978-3-031-46338-9

This Springer imprint is published by the registered company Springer Nature Switzerland AG
The registered company address is: Gewerbestrasse 11, 6330 Cham, Switzerland

Paper in this product is recyclable.

Preface

We are delighted to present the proceedings of the ISPR 2023: The Third International Conference on Intelligent Systems and Pattern Recognition. This event was meticulously organized by the Artificial Intelligence and Knowledge Engineering Research Labs at Ain Sham University in collaboration with the MIRACL laboratory at Sfax University, Tunisia. The conference served as a dynamic platform for interdisciplinary discourse, facilitating the exchange of cutting-edge developments across various domains of artificial intelligence and pattern recognition. Supported by the esteemed International Association of Pattern Recognition (IAPR), the conference took place in the picturesque locale of Hammamet, Tunisia, from May 11–13, 2023.

Within this compilation of proceedings lies a collection of papers that have been meticulously vetted and showcased during the conference. The conference drew an impressive array of scholarly contributions, with a grand total of 129 papers submitted across diverse domains within pattern recognition and artificial intelligence. These submissions underwent rigorous evaluation, undertaken by esteemed researchers hailing from various corners of the globe, each an authority in their respective fields. Following a thorough and competitive at least 3 reviews per submission in a double-blind review, 49 outstanding papers emerged triumphant, earning the opportunity to grace the conference podium. Ultimately, 44 of these distinguished papers were registered and adeptly presented and discussed during the event, attaining a rate of 34%.

We extend our heartfelt gratitude to the diligent reviewers, whose invaluable time and dedication were instrumental in evaluating the papers and offering insightful feedback to the authors. Our appreciation also extends to the esteemed keynote speakers, authors, and participants who collectively contributed to the conference's success. We extend our recognition to authors whose submissions were not selected this time; they are encouraged to explore the potential inclusion of their research papers in the forthcoming edition of ISPR.

The relentless efforts of the organizing committees have undoubtedly played a pivotal role in orchestrating this remarkable event, and for that, they deserve resounding praise.

A special acknowledgment goes to the International Association of Pattern Recognition (IAPR) and the event's sponsors, whose support has significantly enriched the conference's stature. We trust that ISPR 2023 fostered invaluable knowledge exchange and networking avenues for all participants. We eagerly anticipate your presence once again at the next edition.

August 2023

Akram Bennour
Ahmed Bouridane
Lotfi Chaari

Organization

General Chairs

Akram Bennour   Larbi Tebessi University, Algeria
Tolga Ensari   Arkansas Tech University, USA
Abdel-Badeeh Salem   Ain Shams University, Egypt

Steering Committee

Akram Bennour   Larbi Tebessi University, Algeria
Lotfi Chaari   INP-Toulouse, France
Moises Diaz   Universidad de Las Palmas de Gran Canaria, Spain
Bechir Alaya   Gabes University, Tunisia
Najib Ben Aoun   Al Baha University, Saudi Arabia
Abdel-Badeeh Salem   Ain Shams University, Egypt
Yahia Slimani   Manouba University, Tunisia
Atta ur Rehman Khan   Ajman University, UAE

International Advisory Board

Mohamed Elhoseny   University of Sharjah, UAE
K. C. Santosh   University of South Dakota, USA
José Ruiz-Shulcloper   University of Informatics Sciences, Cuba
Linas Petkevičius   Vilnius University, Lithuania
Mohammed al-Sarem   Tayba University, Saudi Arabia
Faiz Gargouri   Sfax University, Tunisia
Takashi Matsuhisa   Karelia Research Centre, Russian Academy of Science, Russia
Sabah Mohammed   Lakehead University, Canada
Mostafa M. Aref   Ain Shams University, Egypt

Program Chairs

Akram Bennour	Larbi Tebessi University, Algeria
Ahmed Bouridane	Sharjah University, UAE
Lotfi Chaari	INP-Toulouse, France

Publicity Chairs

Ahmed Cheikh Rouhou	National Engineering School of Sfax, Tunisia
Chintan M. Bhatt	Pandit Deendayal Energy University, India
Mohammed al-Chaabi	Tayba University, Saudi Arabia
Sean Eom	Southeast Missouri State University, USA
Ali Ismail Awad	United Arab University, Abu Dhabi, UAE
Majid Banaeyan	TU Wien, Austria
Mustafa Dagtekin	Istanbul University - Cerrahpaşa, Turkey
Mohamed Hammad	Menoufia University, Egypt
Mustafa Ali Abuzaraida	Utara University, Malaysia
Jinan Fiaidhi	Lakehead University, Canada

Sponsors and Exhibitions Chairs

Samir Tag	Larbi Tebessi University, Algeria
Hatem Haddad	Institut Supérieur des Arts Multimédia de la Manouba, Tunisia
Fedoua Drira	National Engineering School of Sfax, Tunisia
Gattel Abdejalil	Larbi Tebessi University, Algeria

Local Arrangement Committee

Bassem Bouaziz (LAC Chair)	University of Sfax, Tunisia
Adham Bennour	Constantine University, Algeria
Takwa Benaicha	Tunis University, Tunisia
Marzoug Soltan	Larbi Tebessi University, Algeria
Saméh Kchaou	University of Sfax, Tunisia
Walid Mahdi	University of Sfax, Tunisia
Yousra Bendaly Hlaoui	University of Tunis El Mana, Tunisia
Walid Barhoumi	University of Carthage, Tunisia

Publication Chairs

Tolga Ensari   Arkansas Tech University, USA
Mohamed Elhoseny   University of Sharjah, UAE
Mohammed al-Sarem   Tayba University, Saudi Arabia
Mohamed Hammad   Menoufia University, Egypt
Sajid Anwar   Institute of Management Sciences, Pakistan
Najib Ben Aoun   Al Baha University, Saudi Arabia
Varuna De Silva   Loughborough University, UK
Imad Rida   University of Technology of Compiègne, France

Program Committee

Niketa Gandhi   MIR Labs, USA
Souad Guessoum   Badji Mokhtar University, Annaba, Algeria
Takashi Matsuhisa   Institute of Applied Mathematical Research,
                              Karelia Research Centre, Russia
Ehlem Zigh   USTO, Algeria
Hatem Haddad   Institut Supérieur des Arts Multimédia de la
                              Manouba, Tunisia
Mousa Albashrawi   King Fahd University of Petroleum and Minerals,
                              KSA
Djeddar Afrah   Saad Dahleb University, Blida, Algeria
Yaâcoub Hannad   National Institute of Statistics and Applied
                              Economics, Morocco
Varuna De Silva   Loughborough University, UK
Mohamed Elhoseny   University of Sharjah, UAE
Mohamed Ben Halima   Institut Supérieur d'Informatique et de
                              Multimédia de Sfax, Tunisia
Karri Chiranjeevi   University of Porto, Portugal
Faiz Gargouri   University of Sfax, Tunisia
Linas Petkevičius   Vilnius University, Lithuania
Mohammed Benmohammed   Constantine University, Algeria
Faouzi Ghorbel   CRISTAL/ENSI, Tunisia
José Ruiz-Shulcloper   University of Informatics Sciences, Cuba
Mohammed al-Sarem   Tayba University, Saudi Arabia
Chintan M. Bhatt   Pandit Deendayal Energy University, India
K. C. Santosh   University of South Dakota, USA
Mohamed Hammad   Menoufia University, Egypt
Idris El-Feghi   SomaDetect, Canada
Bassem Bouaziz   University of Sfax, Tunisia

Najib Ben Aoun	Al Baha University, Saudi Arabia
Hemen Dutta	Gauhati University, India
Sirine Marrakchi	University of Sfax, Tunisia
Ali Abu Odeh	University of Technology, Bahrain
Salma Jamoussi	University of Sfax, Tunisia
Emna Fendri	Miracl/FSS, Tunisia
Ljubica Kazi	University of Novi Sad, Serbia
Ali Ismail Awad	United Arab Emirates University, UAE
Imran Siddiqi	Bahria University, Pakistan
Sabah Mohammed	Lakehead University, Canada
Sumaya Al Maadeed	Qatar University, Qatar
Azza Ouled Zaid	ISI, Tunisia
Jinan Fiaidhi	Lakehead University, Canada
Lotfi Chaari	INP-Toulouse, France
Sugata Gangopadhyay	Indian Institute of Technology Roorkee, India
Mohamed Hammami	Miracl/FSS, Tunisia
Ahmed Bouridane	Sharjah University, UAE
Imad Rida	University of Technology of Compiègne, France
Moises Diaz	Universidad de Las Palmas de Gran Canaria, Spain
Mustafa Ali Abuzaraida	Utara University, Malaysia
Yaâcoub Hannad	Mohammed V University of Rabat, Morocco
Mohammed al-Chaabi	Tayba University, Saudi Arabia
Xuewen Yang	InnoPeak Technology Inc., USA
Mustafa Dagtekin	Istanbul University - Cerrahpaşa, Turkey
Yahia Slimani	Manouba University, Tunisia
Hammad Afzal	National University of Sciences and Technology, Pakistan
Nemmour Hassiba	USTHB, Algeria
Vijayan K. Asari	University of Dayton, USA
Atta ur Rehman Khan	Ajman University, UAE
Slim M'Hiri	CRISTAL/ENSI, Tunisia
Sam Zaza	Middle Tennessee State University, USA
Bechir Alaya	Gabes University, Tunisia
Rachid Oumlil	École Nationale de Commerce et de Gestion, Morocco
Mahnane Lamia	Badji Mokhtar University, Annaba, Algeria
Tolga Ensari	Arkansas Tech University, USA
Fedoua Drira	National Engineering School of Sfax, Tunisia
Jerry Wood	Arkansas Tech University, USA
Toufik Sari	Badji Mokhtar University, Annaba, Algeria
Sean Eom	Southeast Missouri State University, USA

Walid Mahdi	University of Sfax, Tunisia
Anna-Maria Di Sciullo	UQAM, Canada
Bhaskar Ghosh	Arkansas Tech University, USA
Abdelkader Nasreddine Belkacem	Osaka University, Japan
Imed Riadh Ferah	RIADI/MSE, Tunisia
Faezeh Soleimani	Ball State University, USA
Ezeddine Zagrouba	ISI, Tunisia
Robin Ghosh	Arkansas Tech University, USA
Indira Dutta	Arkansas Tech University, USA
Suzan Anwar	Philander Smith College, USA
Nassima Bouchareb	Constantine University, Algeria
Shridhar Devamane	Global Academy of Technology, India
Hafidi Mohamed	Cadi Ayyad University, Morocco
Sid Ahmed Benabderrahmane	Inria, France
Nadia Zeghib	University of Constantine 2 - Abdelhamid Mehri, Algeria
Shahin Gelareh	University of Artois, France
Abdel-Badeeh Salem	Ain Shams University, Egypt
Nacereddine Zarour	Constantine 2 University, Algeria
Xinli Xiao	Arkansas Tech University, USA
Amel Benazza	Sup'Com, Tunisia
Weiru Chen	Arkansas Tech University, USA
Yudith Cardinale	Simón Bolívar University, Miranda, Venezuela
Mebarka Yahlali	Saida University, Algeria
Ankur Singh Bist	TowardsBlockchain, India
El-Sayed M. El-Horbaty	Ain Shams University, Egypt
Abdelghani Ghomari	University of Oran1, Algeria
Dragana Krstić	University of Niš, Serbia
Kechar Bouabdellah	University of Oran, Algeria
Krassimir Markov	Institute of Information Theories and Applications, Bulgaria
Dana Simian	Lucian Blaga University, Romania
Marina Nehrey	National University of Life and Environmental Sciences of Ukraine, Ukraine
Rossitsa Yalamova	University of Lethbridge, Canada
Ouassila Hioual	Constantine 2 University, Algeria
Vitalina Babenko	Kharkiv National University, Ukraine
El-Sayed A. El-Dahshan	Ain Shams University, Egypt
Mostafa M. Aref	Ain Shams University, Egypt
Vera Meister	Brandenburg University of Applied Sciences, Germany
Paata Kervalishvili	Georgian Technical University, Georgia

Walid Barhoumi	University of Carthage, Tunisia
Roumen Kountchev	Technical University of Sofia, Bulgaria
Nagwa Badr	Ain Shams University, Egypt
Rasha Ismail	Ain Shams University, Egypt
Natalya Shakhovska	Lviv Polytechnical National University, Ukraine
Anastasia Y. Nikitaeva	Southern Federal University, Russia
Francesco Sicurello	University of Milan-Bicocca, Italy
Maria Brojboiu	University of Craiova, Romania
Liliana Moga	Dunarea de Jos University of Galati, Romania
Nouhad Rizk	University of Houston, USA
Volodymyr Romanov	Glushkov Institute of Cybernetics of National Academy of Sciences of Ukraine, Ukraine
Romina Kountchev	Technical University of Sofia, Bulgaria
Cornelia Aida Bulucea	University of Craiova, Romania
Elena Nechita	Vasile Alecsandri University, Romania
Vicente Rodríguez Montequín	Universidad de Oviedo, Spain
Livia Bellina	MobileDiagnosis Onlus, Italy
Yukako Yagi	Harvard Medical School, USA
Qinghan Xiao	Defence R&D, Canada
Felix T. S. Chan	Macau University of Science and Technology, China
Redouane Tlemsani	University of Sciences and Technology of Oran, Algeria
Sami Saleh	Sains University, Malaysia
Djamel Samai	Ouargla University, Algeria
Sourour Ammar	Digital Research Center of Sfax, Tunisia
Yousra Bendaly Hlaoui	University of Tunis El Mana, Tunisia
Paulo Batista	University of Évora, Portugal
Baskar Arumugam	Amrita Vishwa Vidyapeetham University, India
Loukas Ilias	National Technical University of Athens, Greece
Ghalem Belalem	Université d'Oran, Algeria
Nabiha Azizi	Badji Mokhtar University, Annaba, Algeria
Mouna Rekik	Sousse University, Tunisia
Gattel Abdejalil	Larbi Tebessi University, Algeria
Leïla Boussaad	University of Batna, Algeria
Rebiha Zeghdane	University of Bordj Bou Arreridj, Algeria
Imran Mudassir	Air University Islamabad, Pakistan
Kais Khrouf	Jouf University, KSA
Nassima Aissani	University of Oran, Algeria
Abdellatif Rahmoun	ESI, Algeria
Djamila Mohdeb	University of Bordj Bou Arreridj, Algeria
Rudresh Dwivedi	Netaji Subhas University of Technology, India

Contents – Part II

Contents – Part I

Data Mining

Pattern Recognition

Simsiam Network Based Self-supervised Model for Sign Language Recognition

Deep R. Kothadiya[1], Chintan M. Bhatt[2]([✉]), and Imad Rida[3]

[1] U & P U. Patel Department of Computer Engineering, Chandubhai S. Patel
Institute of Technology (CSPIT), Faculty of Technology (FTE), Charotar University
of Science and Technology (CHARUSAT), Changa, India
deepkothadiya.ce@charusat.ac.in
[2] Department of Computer Science and Engineering, School of Engineering and
Technology, Pandit Deendayal Energy University, Gandhinagar, Gujarat, India
chintan.bhatt@sot.pdpu.ac.in
[3] Laboratoire Biomécanique et Bioingénierie UMR 7338, Centre de Recherches de
Royallieu, Université de Technologie de Compiègne, Compiègne, France
imad.rida@utc.fr

Abstract. To make a more accurate and robust deep learning model,
more labeled data is required. Unfortunately, in many areas, it's very difficult to manage properly labeled data. Sign language recognition is one
of the challenging areas of computer vision, to make a successful deep
learning model to recognize sign gestures in real-time, a huge amount of
labeled data is needed. Authors have proposed a self-supervised learning approach to address this problem. The proposed architecture used
Resnet50 v1 backbone-based simsiam encoder network to learn the similarity between two different images of the same class. Calculated cosine
similarity passes to MLP head for further classification. The proposed
study uses Indian and American Sign Language detests for simulation. The proposed methodology successfully achieve 74.59% of accuracy.
Authors have also demonstrated the impact of other self-supervised deep
learning models for sign language recognition.

Keywords: Sign Language · Deep Learning · Self-supervised
learning · Simsiam encoder

1 Introduction

The emergence of deep learning methods has constructive outcomes in feature
learning in visual representations, which are extremely useful for various tasks
using specific labels. However, since supervised learning itself is sparse, this requires the availability of a huge amount of labeled data to train deep learning
models. To reduce dependences on the large labeled datasets, recent methods
have explored the use of unsupervised or self-supervised learning techniques.
These techniques shine at automatically learning with useful input representations. Unsupervised or Self-supervised learning has been frequently deployed to

A. Bennour et al. (Eds.): ISPR 2023, CCIS 1941, pp. 3–13, 2024.
https://doi.org/10.1007/978-3-031-46338-9_1

design learning-rich language models [1–6]. Recently, these self-supervised models have been successfully leveraged in other domains, including visual classification and speech recognition. All of us have a distinctive way to communicate and navigate thoughts and emotions. There are so many people in the world having some level of hearing and speaking disability. The majority of people are limited to speech, and can't able to communicate with impaired people. Sign language is a non-verbal communication platform for connecting with the social environment by expressing emotions for hearing-impaired people. However, sign language is not a universal language, every country or region has its own singing language and gesture meaning, or we can also say a different structure of sign language varies from region to region. An independent gesture can be considered as sign normally called a symbol. In sign language, gestures are organized rhetorically [7]. The first educator for deaf people is "Charles-Michel, abbé de l'Epée." He has developed a structure to spell out the French word into signs. He has developed a French sign language, still in use in France. This structured concept is also used in the formation or design of American Sign Language (ASL) and British Sign Language (BSL) [8]. Nowadays, many well-structured sign languages are there like Ethiopian Sign Language (ESL), South African Sign Language (SASL), Indian Sign Language (ISL), and many more. Sign language is not just symbols or poses it involves gestures as well as facial emotion. The generation of gestures involves both hands and non-manual body parts such as direction of eyes, eyebrows, lips, orientation of the head, and other body parts like shoulder stomach, and many more. Mainly sign language divide into two categories static and dynamic. Static sign language involving gestures only may include one or two hands as per gesture requirement but ensure that the gesture has no movement, it's a steady pose. While in dynamic sign gestures might include non-manual body parts along with hand (single or both). Farther this categorization can be extended to isolated and continued sign language as a subset of dynamic sign language. This article aims to study the recognition of static signs form image datasets over 36 classes. Authors have approached self-supervised learning methodology for static sign recognition. Contribution of the article as i) Recognition of Gesture-based static sign language ii) proposed simsiam encoder based self-supervised methodology to recognize sign language. The rest of the paper is distributed as, section ii contains a study of related work done with self-supervised learning, section iii contains the architecture of the proposed methodology, and section iv contains experiments and result details.

2 Related Work

Shoaib Ahmed Siddiqui et al. [9] proposed a self-supervised methodology for document classification. The author used RVL-CDIP and Tobacco-3482 datasets to train and evaluate ResNet-50 encoder and pre-trained ImageNet. The proposed architecture has archived 86.75% and 88.52% accuracy for RVL-CDIP and Tobacco-3482 respectively. The author has also optimized the loss in the SimCLR and Barlow Twins methods in self-supervised training. The evaluation over both models represents the superiority of pretrained model over the design of the model from scratch.

Alan Preciado-Grijalva et al. [10] proposed a deep learning based self-supervised methodology for sonar image classification. The proposed methodology used RotNet, Denoising Autoencoders (DAE), and Jigsaw models to learn features from high-quality sonar images. The simulation of the proposed methodology was evaluated on pretrained and transfer learning methodology. The authors have considered SVM as a baseline model to compare proposed models, Baseline SVM gives an accuracy of 95.67%, while in RotNet and Jigsaw accuracy is improved by +1.6 and +1.42 compared to baseline SVM. On the other hand, DAE gets an accuracy of 92.22%, less than baseline SVM.

Yuan Gao et al. [11]. The proposed self-supervised pretrain model applies to the masked image modeling (MIM) method to remote sensing. The methodology used MIM to reconstruct masked patches and correlated with unseen parts in semantics. The authors have proposed a transformer-based transfer learning approach for remote sensing data. Authors have used NWPU-RESISC45 dataset (NWPU) [12], Aerial Image dataset (AID) [13], and UCMercedLand dataset (UCM) [14] open access datasets has archived 99.24 accuracies for 80% of the training dataset, 97.86% for 50% of training dataset. Bing Liu et al. [15] proposed an Ensemble Self-Supervised Feature Learning model for hyperspectral image classification. The proposed method uses automatic deep features conducive to classification without annotation. Authors used EfficientNet-B0 as backbone model.

Liangliang Song et al. [16]. Pro-posed Semi-Supervised Residual Network-assisted by Self-Supervised for Hyperspectral Image (HIS) Classification. The proposed methodology used a concatenation of semi-supervised and self-supervised approaches for high-spectrum image classification. The proposed semi-supervised residual network (SSRNet) assisted by a self-supervised model consists of two sub methodologies semi-supervised and self-supervised. The semi-supervised module improves performance by spectral feature shift, while the self-supervised module performs two tasks, masked band reconstruction, and spectral order forecast, to memorize features of HSI. The proposed methodology has achieved classification accuracy81.65%, 89.38%, 93.47%, and 83.93%, for Indian Pines, Pavia University, Salinas, and Houston2013 datasets respectively.

Yibo Zhao et al. [17] proposed a self-supervised knowledge distillation network (SSKDNet) for image classification. The backbone of the model and branch are optimized in knowledge distillation without independent pretraining. The authors have also proposed a feature fusion module to fuse feature maps dynamically. The proposed methodology finds accuracy as 99.62%, 95.96%, and 92.77% for UCM, AID, and NWPU public datasets respectively.

Fotini Patrona et al. [18]. Proposed gesture recognition with a self-supervised approach using CNN-LSTM architecture. The proposed architecture used pretrain CNN for a novel pretest task, which relayed on a spatiotemporal video frame. Authors have proposed Moving Interest Point (MIP) methodology for optical flow estimation, image frequency transform, and interest point detection.

The proposed methodology has archived 64.08% of the Correct Classification Rate over Unmanned Aerial Vehicle (UAV) dataset. Table 1 represent the comparative analysis of related work done for action and gesture recognition with self-supervised methodology.

Table 1. Comparative analytic of related work done for action and gesture recognition with self-supervised methodology.

Author	Year	Method	Dataset	Accuracy
F. Patrona et al. [18]	2022	MIP	UCF	64.08
S. Gidaris et al. [19]	2018	RotNet	UCF	60.00
Ying Wu and Thomas S. Huang [20]	2000	MLP		39.60
Dietz, A [21]	2018	InceptionV3	MFH	65.00
T. Chen et al. [22]	2020	SimCLR	Pets	69.30
R. Caramalau et al. [23]	2022	MoBYv2AL	MNIST	70.11

3   Methodology

3.1   Self-supervised Learning

Modern computer vision techniques have outstanding performance in object detection, recognition, classification, segmentation, and many more. However, supervised learning requires proper training on a large number of datasets, and those data must be labeled. This is how the success of deep learning based computer vision is based on the high availability of labeled data. Self-supervised learning is part of the machine learning technique to reduce dependencies of labeled data. Traditionally, the design of intelligent systems using machine learning methods has been majorly dependent on the quality of the labeled dataset. Self-supervised learning is part unsupervised learning technique of machine learning. The main reason behind self-supervised learning is to learn useful features of data from a huge set of unlabeled data. Finding good quality labeled data it's too time-consuming and expensive, especially for object detection, recognition, and segmentation. Practically it is also near to impossible to label every available data. Mostly unstructured data are not simple enough to label properly, data like defection detection from metal, surface detection, object identification from satellite images, and many more. Especially language-driven tasks are more difficult to recognize.

3.2   Visual Representation Learning

Recently self-supervised learning has archived promising results in the context of visual representation, the SSL technique designs pretest tasks, and the feature is formulated with unlabeled data but it requires a higher level of semantic understanding. To counter this issue SSL defines pretest to learn the representations of unstructured data, farther it can be used to solve other downstream tasks of visual representation. Sign language recognition is one of the most emerging areas of visual representation language. , which are used by impaired people for communication. Proposed methodology state that recognition of sign gestures forms an image.

SimSiam Architecture. Formally self-supervised approach needs a huge set of data for learning, to address this drawback, simsiam architecture used momentum encounter. Simsiam takes two random inputs from a set of augmented data buckets. The two input images are processed by an encoder network having backbone of ResNet50_V1, the encoder f shares weights between those two views, Eq. 1 denotes the output of encoder view. Negative cosine decay can be found in Eq. 2.

$$p_1 \triangleq h(f(x_1)) \ and \ z_2 \triangleq f(x_2) \tag{1}$$

$$D(p_1, z_2) = -\frac{P_1}{||P_1||_2} * \frac{Z_2}{||Z_2||_2} \tag{2}$$

Backbone ResNet 50_v1 was used for encoder f1 with 64 batch size. The projection head of MLP has used Batch Normalization (BN) in each intermediate layer and output layer as well. The input Z and P have size of 1024 with 4 hidden layers having h' (512) dimension. Stop-gradient method plays a key role in the proposed model to reduce gradient problem, Eq. 3 formulated a liner vector from both encoder networks with stopgard() method.

$$L_v = 1/2(D(p_2, stopgard(z_1) + D(p_1, stopgard(z_2))) \tag{3}$$

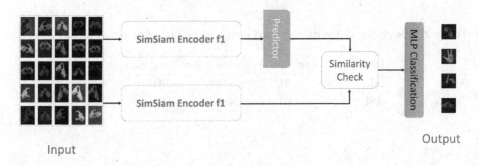

Fig. 1. Proposed SimSiam encoder based architecture for sign language recognition.

Figure 1 shows the proposed self-supervised based architecture for gesture-based sign recognition, The proposed network maximizes the similarity values between augmented images. The simsiam encoder performs down-streaming tasks to recognize the same gesture from different augmented images of that gesture sign. Initially, input images get augmented and then forwarded to ResNet-V1 for cosine decay learning. A calculated loss will be transferred to the simsiam encoder network which is completed by the linear classifier to identify a class of given gesture image.

4   Result and Experiments

4.1   Dataset

The Indian Sign Language and American sign language datasets have been used for simulation, Fig. 2 illustrates the sign gesture after augmentation, and simulation has used vertical flip, random rotation, and random color transformation to generate augmented input data. The Indian Sign Language (ISL) dataset consists of 36 classes and American Sign Language (ASL) consists of 35 classes [25]. each class contains average of 1000 sample images. Table 2 demonstrates the digital parameter of targeted dataset.

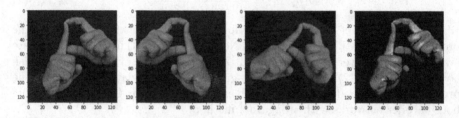

Fig. 2. Sample of augmented dataset used in proposed methodology.

Table 2. Description of the dataset used in the simulation of proposed methodology.

Dataset	Classes	dpi	Avg. Resolution	Avg. Sample
ISL	36	94	250×250	1200
ASL	35	96	400×400	840

4.2   Implementation Details

Proposes methodology used simsiam based encoder network to classify the class of sign gestures. Initially, an 80–20 train-test ratio was taken to train the proposed model. The proposed architecture used sign gesture images of 480×320 pixels with 3 channels. ResNet_v1 model used to generate cosine decay learning used to compare loss differences in the feed-forward network. An input image size is 32×32 has been taken for simulation. Different combinations in simulations were carried out with 100 epochs as a constant value.

Experiments with Different Train-Test Splits. Table 3 shows the accuracy of the proposed method over different values of train-test data. However, there is not much difference found for 80:20 to 70:30, but 60:40 make a huge difference in identifying the correct class of sign.

Table 3. Comparative analysis of simulation over different train-test split ratios.

Train-test Split	ISL Dataset	ASL Dataset
80:20	74.59%	77.41%
70:30	73.01%	75.84%
60:40	61.18%	69.62%

Experiment with Different Activation Functions. The proposed methodology was also simulated over different activation functions like ReLU and Softmax. Table 4 shows results for ISL and ASL databases over different activation. Pro- posed method performs high accuracy with softmax activation.

Table 4. Comparative analysis of simulation over different activation functions.

Activation Funcation	ISL Dataset	ASL Dataset
ReLLU	73.51%	75.09%
SoftMax	74.59%	77.41%

The authors have also demonstrated the impact of different classifiers on proposed methodology. Comparative analysis over different standard classifiers like Random Forest. Support Vector Machine (SVM), HMM, KNN, MLP. Table 5 demonstrate the accuracy and error rate of different classifier for the proposed model.

Table 5. Comparative analysis over different classifiers with SimSiam Encoder model.

Classifier	Params	Accuracy	Error Rate
Random Forest	4.1M	72.16%	27.84
SVM	4M	72.80%	27.20
HMM	2.4M	61.35%	38.65
KNN	4.4M	68.02%	31.98
MLP	4.2M	74.59%	25.41

The author has also demonstrated the SimSiam encoder over different deep learn- ing state-of-the-art models like Inception v2, and AlexNet. DenseNet121, MobileNet v2, and the proposed model. Figure 3 illustrates a comparative analysis with state of the art deep learning convolution model with simsiam encoder learning. Authors have also analyzed different self-supervised models like Sim- CLR [22], MoCoV2 [23], InfoMin [24], BYOL [25], and DINO [26] for sign language recognition images. Table 6 demonstrates the comparative analysis of self-supervised models.

Table 6. Comparative analysis over state-of-the-art self-supervised deep learning model for sign language datasets.

Model	Params	ISL (Acc)	ASL (Acc)
SimCLR	4.4M	74.08%	74.66%
MoCoV2	8M	70.21%	68.03%
InfoMin	4M	72.57%	72.69%
BYOL	1.8M	69.07%	68.47%
DINO	2.1M	68.12%	68.90%
SimSiam	4.2M	74.59%	77.41%

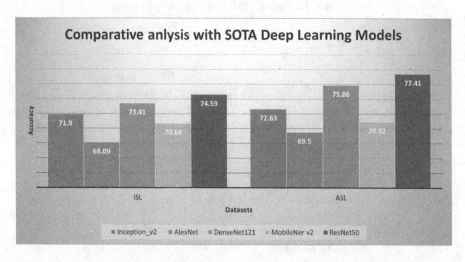

Fig. 3. Comparative analysis with state-of-the-art deep learning model as a backbone for Simsiam encoder network.

4.3 Discussion

The proposed methodology used a simsiam based encoder network to recognize sign gestures from images. Self-supervised based methodology archives 74.59% accuracy for Indian sign language (ISL) with 36 classes, the best accuracy among all different experiments. The proposed self-supervised based methodology used augmentation especially crop function to make model more generalized and robust model.

5 Conclusion

Recognition of sign language is an essential component of human commuter interaction and interactive medium to communicate with people having impaired. In this article, the authors proposed a self-supervised approach for Indian sign language recognition. The proposed methodology aims to reduce this communication barrier with a self-supervised technique. The proposed methodology achieves 74.59% of accuracy with a simsiam encoder network over Indian sign language. The methodology used augmentation and crop based images, which makes the model more generalized to identify signs from images. The results of the simulation show that the proposed methodology perform well on static Indian sign language (ISL) and American sign language dataset (ASL). Sign gesture recognition can be enhanced by cross domain based learning with self-supervised and semi-supervised models.

References

1. Ranzato, M., Szummer, M.: Semi-supervised learning of compact document representations with deep networks. In: Proceedings of the 25th international conference on Machine learning, pp. 792–799 (2008)
2. Brown, T., et al.: Language models are few-shot learners. Adv. Neural. Inf. Process. Syst. **33**, 1877–1901 (2020)
3. Mikolov, T., Sutskever, I., Chen, K., Corrado, G.S., Dean, J.: Distributed representations of words and phrases and their compositionality. In: Advances in Neural Information Processing Systems 26 (2013)
4. Bojanowski, P., Grave, E., Joulin, A., Mikolov, T.: Enriching word vectors with subword information. Trans. Assoc. Comput. Linguist. **5**, 135–146 (2017)
5. Devlin, J., Chang, M.-W., Lee, K., Toutanova, K.: BERT: pre-training of deep bidirectional transformers for language understanding. arXiv preprint arXiv:1810.04805 (2018)
6. Kothadiya, D., Chaudhari, A., Macwan, R., Patel, K., Bhatt, C.: The convergence of deep learning and computer vision: Smart city applications and research challenges. In: 3rd International Conference on Integrated Intelligent Computing Communication & Security (ICIIC 2021), pp. 14–22, Atlantis Press (2021)
7. Kothadiya, D., Bhatt, C., Sapariya, K., Patel, K., Gil-González, A.-B., Corchado, J. M.: Deepsign: sign language detection and recognition using deep learning. Electronics **11**(11) (2022)

8. de L'Epée, C.-M.: Institution des sourds et muets, par la voie des signes méthodiques: ouvrage qui contient le project d'une langue universelle, par l'entremise des signes naturels assujettis à une méthode, vol. 1. Chez Nyon l'ainé, 1776

9. Li, P., et al.: SelfDoc: self-supervised document representation learning. In: Proceedings of the IEEE/CVF Conference on Computer Vision and Pattern Recognition, pp. 5652–5660 (2021)

10. Preciado-Grijalva, A., Wehbe, B., Firvida, M.B., Valdenegro-Toro, M.: Self-supervised learning for sonar image classification. In: Proceedings of the IEEE/CVF Conference on Computer Vision and Pattern Recognition, pp. 1499–1508 (2022)

11. Gao, Y., Sun, X., Liu, C.: A general self-supervised framework for remote sensing image classification. Remote Sensing **14**(19), 4824 (2022)

12. Cheng, G., Han, J., Lu, X.: Remote sensing image scene classification: benchmark and state of the art. Proc. IEEE **105**(10), 1865–1883 (2017)

13. Xia, G.-S., et al.: Aid: A benchmark data set for performance evaluation of aerial scene classification. IEEE Trans. Geosci. Remote Sens. **55**(7), 3965–3981 (2017)

14. Zou, Q., Ni, L., Zhang, T., Wang, Q.: Deep learning based feature selection for remote sensing scene classification. IEEE Geosci. Remote Sens. Lett. **12**(11), 2321–2325 (2015)

15. Liu, B., Gao, K., Yu, A., Ding, L., Qiu, C., Li, J.: ES2FL: ensemble self-supervised feature learning for small sample classification of hyperspectral images. Remote Sensing **14**(17), 4236 (2022)

16. Song, L., Feng, Z., Yang, S., Zhang, X., Jiao, L.: Self-supervised assisted semi-supervised residual network for hyperspectral image classification. Remote Sensing **14**(13), 2997 (2022)

17. Zhao, Y., Liu, J., Yang, J., Wu, Z.: Remote sensing image scene classification via self-supervised learning and knowledge distillation. Remote Sensing **14**(19), 4813 (2022)

18. Patrona, F., Mademlis, I., Pitas, I.: Gesture recognition by self-supervised moving interest point completion for CNN-LSTMs. In: 2022 IEEE 14th Image, Video, and Multidimensional Signal Processing Workshop (IVMSP), pp. 1–5, IEEE (2022)

19. Gidaris, S., Singh, P., Komodakis, N.: Unsupervised representation learning by predicting image rotations. arXiv preprint arXiv:1803.07728 (2018)

20. Wu, Y., Huang, T.S.: Self-supervised learning for visual tracking and recognition of human hand. In: AAAI/IAAI, pp. 243–248 (2000)

21. Dietz, A., Pösch, A., Reithmeier, E.: Hand hygiene monitoring based on segmentation of interacting hands with convolutional networks. In: Medical Imaging 2018: Imaging Informatics for Healthcare, Research, and Applications, vol. 10579, pp. 273–278, SPIE (2018)

22. Chen, T., Kornblith, S., Norouzi, M., Hinton, G.: A simple framework for contrastive learning of visual representations. In: International Conference on Machine Learning, pp. 1597–1607, PMLR (2020)

23. Caramalau, R., Bhattarai, B., Stoyanov, D., Kim, T.-K.: MoBYv2AL: self-supervised active learning for image classification. arXiv preprint arXiv:2301.01531 (2023)

24. Tian, Y., Sun, C., Poole, B., Krishnan, D., Schmid, C., Isola, P.: What makes for good views for contrastive learning? Adv. Neural. Inf. Process. Syst. **33**, 6827–6839 (2020)

25. Grill, J.-B., et al.: Bootstrap your own latent-a new approach to self-supervised learning. Adv. Neural Inf. Process. Syst. **33**, 21271–21284 (2020)
26. Caron, M., et al.: Emerging properties in self-supervised vision transformers. In: Proceedings of the IEEE/CVF International Conference on Computer vision, pp. 9650–9660 (2021)

Study of Support Set Generation Techniques in LAD for Intrusion Detection

Sneha Chauhan[1,2]($\boxtimes$), Sugata Gangopadhyay[1]($\boxtimes$),
and Aditi Kar Gangopadhyay[1]($\boxtimes$)

[1] Indian Institute of Technology Roorkee, Roorkee, India
{schauhan1,sugata.gangopadhyay}@cs.iitr.ac.in,
aditi.gangopadhyay@ma.iitr.ac.in
[2] National Institute of Technology Uttarakhand, Uttarakhand, India

Abstract. Support Set generation is an essential process in the Logical Analysis of Data (LAD). The process of binarization results in an increase in the dimensions of the dataset, which can make the classification process more challenging. The support set generation step is performed to select the important features from the binarized dataset. In this paper, five techniques, namely Set covering problem, Mutual Information Greedy algorithm, Information Gain, Gain ratio, and Gini Index, are used to find the minimal support set for the classification of the Intrusion Detection dataset. LAD uses partially defined Boolean functions to generate positive and negative patterns from the historical observations, which are then transformed into rules for the classification of future observations. The LAD classifier is built using different techniques, and their performances on the NSL-KDD dataset are recorded.

Keywords: Logical Analysis of Data · Support Set generation · Set Covering problem

1 Introduction

The Logical Analysis of Data (LAD) is a data analysis technique used for the binary classification of data. It operates by identifying the structural characteristics of a dataset and generating patterns that can effectively differentiate between normal and abnormal traffic within a network. Unlike other machine learning techniques, LAD focuses on analyzing a subset of attributes to classify positive and negative observations and extract patterns using combinatorial techniques.

Due to an increase in the use of the Internet in today's scenario, security and privacy of data has become a major challenge in every field. A lot of research is being done in detecting vulnerabilities in the system and avoiding security violations by the attackers and intruders. Many machine learning and deep learning models have been utilized to develop intrusion detection systems that can detect abnormal activities in the network and take appropriate actions.

A. Bennour et al. (Eds.): ISPR 2023, CCIS 1941, pp. 14–28, 2024.
https://doi.org/10.1007/978-3-031-46338-9_2

This paper provides an introduction to the fundamental concepts of Logical Analysis of Data (LAD). We have applied the LAD technique to the NSL-KDD intrusion detection dataset in order to detect attacks by generating patterns. The paper discusses different combinatorial optimization models and explores methods to identify the minimal subset of attributes. The LAD classifier has an advantage over other techniques as it can give an explanation through its patterns, about the features that are involved in the attack. Thus, we can know about the system vulnerabilities using LAD patterns.

The paper consists of the following sections. Related work on intrusion detection system is described in Sect. 2. Section 3 introduces LAD methodology and its implementation. Section 3.2 describes various techniques to find the minimal support set. Section 4 of the paper presents a comprehensive comparative analysis of the performance of various LAD variations on the NSL-KDD dataset. Lastly, Sect. 5 provides concluding remarks and summarizes the findings and techniques demonstrated throughout the paper.

2 Related Work

In [17], a novel intrusion detection system (IDS) based on anomaly detection is presented. The system leverages Software Defined Network to gather network traffic features, which are subsequently fed into a voting system comprised of diverse machine learning algorithms for prediction. The KDDCup99 and NSL-KDD datasets were employed to evaluate the effectiveness of decision tree, random forest, SVM, XGboost, and DNN. Similarly, in another study [3], machine learning classifiers were utilized to monitor network traffic and identify malicious activities. The NSL-KDD dataset was employed to test various techniques such as random forest, decision tree, and XGBoost for attack detection. Notably, XGBoost demonstrated superior performance with an accuracy of 95%. Additionally, S.M. Kasongo explored the implementation of an IDS framework in his paper [16], utilizing different types of Recurrent Neural Networks (RNNs) including Simple RNN, Long-Short Term Memory (LSTM) and Gated Recurrent Unit (GRU). XGBoost was employed for feature selection, and NSL-KDD and UNSW-NB15 datasets were used to test the IDS model. The evaluation encompassed binary and multi-class classifications, with XGBoost-LSTM achieving the highest accuracy rates of 88.13% and 86.93%, respectively. In another research work [14], a hybrid IDS model was proposed where principal component analysis (PCA) was employed for dimensionality reduction, while an optimized support vector machine (SVM) served as the classification method for network traffic data. The performance evaluation of this approach was conducted using the NSL-KDD and KDDCup datasets. In the paper [18], machine learning algorithms, specifically decision tree and K-nearest neighbors (KNN), were implemented in an intrusion detection system. The system employed a univariate feature selection technique based on Analysis of Variance (ANOVA). The performance evaluation was carried out using the NSL-KDD dataset.

3   Logical Analysis of Data

The Logical Analysis of Data is a technique that analyzes data using combinatorial and optimization models to generate patterns that can classify positive and negative observations. The LAD technique was first introduced by Peter L. Hammer in 1986 [12]. This method was initially applied only on the binary data. The authors of the paper [5] extended LAD's approach for non-binary data. The main idea behind LAD is to derive logical explanations that can differentiate positive and negative observations. LAD focuses on the logical patterns and selecting features or a collection of features directly affecting the outcome. The patterns produced by LAD can explain the classification results by formal reasoning, which can be easily understood by human experts. In recent times, LAD can be applied to many areas varying from medical application to classification and financial ratings.

The dataset used in this context is divided into two subsets, denoted as Ω^+ and Ω^-, which represent positive and negative observations, respectively. Each observation is represented as an n-dimensional vector, with individual components referred to as attributes or features. Hence, we can say that our dataset has n features whose values distinguish a positive observation from a negative observation. Before studying LAD, let us first introduce some terminology that will be required for understanding LAD concepts.

A Boolean function of n variables refers to a mapping from the set $\{0,1\}^n$ to the set $\{0,1\}$, where n is a positive integer. In this context, a vector $X = (x_1, x_2, \ldots, x_n) \in \{0,1\}^n$ is considered a positive vector of the Boolean function f if $f(X) = 1$. Similarly, if $f(X) = 0$ then X is categorized as a negative vector of the Boolean function f. The Table 1 shows a Boolean function on 3 variables [10].

Table 1. Truth Table

x_1	x_2	x_3	$f(x_1, x_2, x_3)$
0	0	0	1
0	0	1	0
0	1	0	1
0	1	1	1
1	0	0	0
1	0	1	0
1	1	0	1
1	1	1	0

The dataset Ω, consisting of Ω^+ and Ω^-, can be expressed as a partially defined Boolean function (pdBf) ϕ. This function ϕ maps elements from Ω to $\{0,1\}$, where Ω is a subset of $\{0,1\}^n$. To illustrate, Table 2 shows a partially defined Boolean function with eight variables. There are only 7 points out of 2^8 points defined for the function. Few of the points are positive and few are negative. We can add new points to the table. In the same way, pdBf can also be extended by defining its value on the new observations [9]. A Boolean function that is defined for all 2^n vectors and which agrees to the pdBF ϕ where $\Omega \subset \{0,1\}^n$ then it is called an extension of ϕ. LAD aims to find such an extension, that can classify the unknown observations correctly. Due to the inherent difficulty in finding the exact extension, our approach is to seek an approximate extension, denoted as ϕ', for the pdBf ϕ. The basic components of LAD are described below [1, 13]:

- A technique called Binarization is used in LAD so that it can be applied to non binary data. This process is explained in detail in the Sect. 3.1.
- Binarization may lead to many redundant attributes in the dataset. To eliminate irrelevant and redundant attributes, we extract a minimal subset S from the feature set which is capable to distinguish positive and negative observations.
- Next step is to detect patterns from the projections Ω_S^+ and Ω_S^- of Ω^+ and Ω^- respectively.
- Patterns which cover all observations are selected to create a model.
- A classifier is developed that can classify the observations as positive or negative based on their underlying patterns..
- Testing of classifier is carried out to verify the efficiency of the model.

Table 2. Partially defined Boolean function for 8 variables

x_1	x_2	x_3	x_4	x_5	x_6	x_7	x_8	$f(x_1, x_2, x_3)$
0	0	0	0	0	0	0	0	1
0	0	1	0	1	0	0	1	0
0	1	0	0	1	0	1	1	1
0	1	1	1	0	0	0	0	0
1	0	0	0	1	0	1	0	1
1	0	1	0	0	1	0	0	0
1	1	0	1	0	0	0	0	1
1	1	1	0	1	1	0	1	0

3.1   Binarization

In real life applications, the data may not be always in binary form and numerical. To apply LAD on such problems, there is a need of a process that can transform such data to binary. In the paper [4], a procedure called as Binarization is described which transforms the real data to binary data.

One of the simple attributes is a nominal attribute. For example, the shape attribute can take values as circular, triangular, rectangular, etc. To convert these attributes into binary attributes, we can associate each value v_i of the attribute x with a corresponding Boolean variable $b(x, v_i)$, such as

$$b(x, v_i) = \begin{cases} 1, \text{if } x = v_i. \\ 0, \text{otherwise} \end{cases}$$

If the nominal attribute has only two values, only one Boolean variable is required with values 0 and 1.

The numerical attributes, like temperature, water level, duration, etc., can be denoted as normal and abnormal depending on whether their values are above or below a certain threshold. These attributes are binarized by comparing their values to a threshold or cutpoint value. A numerical attribute can be represented by two types of Boolean variables: Level variables and Interval variables. The values of these variables will depend on the cutpoints. The cutpoints are selected such that the partially defined Boolean function has an extension within the class of all Boolean functions C_{ALL} [4]. Let's consider a numerical attribute x which has $k + 1$ values, $x_0, x_1, x_2, \ldots, x_{k+1}$ arranged in the descending order. Each attribute will have finite values, so the number of cutpoints will also be finite. A cutpoint is introduced between two values x_i and x_{i+1} of attribute x if they belong to different classes. A cutpoint value is calculated by taking mean of the values i.e. $\frac{x_i + x_{i+1}}{2}$.

A level variable created for a cutpoint β is represented by a Boolean variable $b(x, \beta)$ such that

$$b(x, \beta) = \begin{cases} 1, \text{if } x \geq \beta \\ 0, \text{otherwise} \end{cases}$$

Similarly, an interval variable is produced for every two cutpoints β_1 and β_2 and is represented by a Boolean variable $b(x, \beta_1, \beta_2)$ such that

$$b(x, \beta_1, \beta_2) = \begin{cases} 1, \text{if } \beta_1 \leq x \leq \beta_2 \\ 0, \text{otherwise} \end{cases}$$

The binarization process for the example given in Table 3 is explained here. In this example, positive observations have the class label as 1 and negative ones have class label 0. The attribute x_2 is a nominal attribute which has 3 distinct values, black, blue and red. So, three Boolean variables will be generated for the variable x_2. The rest of the variables are numerical. In order to obtain cutpoints for the variable x_1, we sort the dataset over x_1 in descending order to get Table 4. Before calculating the cutpoints, we perform a preprocessing step as done in the paper [11]. For the consecutive observations having same value v_i for an attribute but different class label, a new class label is introduced and all such observations are removed except one. Now the dataset will have a new class label for this observation as shown in Table 5. The cutpoints obtained for the attribute x_1 using the formula mentioned above are 3.05, 2.45 and 1.65. Thus, we will have 3 level variables and 3 interval variables for the attribute x_1. Similarly, cutpoints for all the other attributes will be calculated. Finally, by utilizing the level variables (Table 6) and interval variables (Table 7), we obtain the binary dataset presented in Table 8.

Table 3. Original dataset

	x_1	x_2	x_3	x_4
Ω^+	3.5	black	3.8	31
	2.6	blue	1.6	29
	1.0	red	2.1	20
Ω^-	3.5	red	1.6	20
	2.3	black	2.1	14

Table 4. Single feature with class label

x_1	Class Label
3.5	1
3.5	0
2.6	1
2.3	0
1.0	1

Table 5. Feature with new class label

x_1	Class Label
3.5	2
2.6	1
2.3	0
1.0	1

Table 6. Level Variables

b_1	b_2	b_3	b_7	b_8	b_9	b_{10}	b_{11}	b_{13}	b_{14}
$x_1 \geq 3.05$	$x_1 \geq 2.45$	$x_1 \geq 1.65$	$x_2 = black$	$x_2 = blue$	$x_2 = red$	$x_3 \geq 2.95$	$x_3 \geq 1.85$	$x_4 \geq 24.5$	$x_4 \geq 17$

Table 7. Interval Variables

b_4	b_5	b_6	b_{12}	b_{15}
$1.65 \leq x_1 \leq 3.05$	$2.45 \leq x_1 \leq 3.05$	$1.65 \leq x_1 \leq 2.45$	$1.85 \leq x_3 \leq 2.95$	$17 \leq x_4 \leq 24.5$

Table 8. Binarized table

b_1	b_2	b_3	b_4	b_5	b_6	b_7	b_8	b_9	b_{10}	b_{11}	b_{12}	b_{13}	b_{14}	b_{15}	Class Label
1	1	1	0	0	0	1	0	0	1	1	0	1	1	0	1
0	1	1	1	1	0	0	1	0	0	0	0	1	1	0	1
0	0	0	0	0	0	0	0	1	0	1	1	0	1	1	1
1	1	1	0	0	0	0	0	1	0	0	0	0	1	1	0
0	0	1	1	0	1	1	0	0	0	1	1	0	0	0	0

3.2 Support Set Generation

A dataset may contain irrelevant and redundant attributes. Also, the Binarization process increases the number of attributes, some of which may be redundant. In order to eliminate such attributes, a combinatorial optimization model is required. Consider a dataset $(\Omega^+ \cup \Omega^-)$ obtained after binarization, where Ω^+ and Ω^- denote sets of true and false observations, respectively. When the dataset is divided into the set of true and false points, it is assumed that no observation belongs to both the sets simultaneously. A support set S is a set of attributes such that $(\Omega_S^+ \cap \Omega_S^-) = \phi$ where Ω_S^+ and Ω_S^- are projections of S on Ω^+ and Ω^- respectively. This means that no observation should belong to both the true and false set simultaneously. A support set is considered minimal when removal of any attribute results in $(\Omega_S^+ \cap \Omega_S^-) \neq \phi$. It is ensured that the support set generated must follow the basic property that an observation either belongs to positive or negative set only [9]. The problem of identifying a support set can be considered equivalent to set covering problem. A binary variable y_i is associated with each variable x_i such that $y_i = 1$, if x_i is part of the support set, otherwise

$y_i = 0$. Consider two observations $a = (a_1, a_2, \ldots, a_n)$ and $b = (b_1, b_2, \ldots, b_n)$. We can say that a and b are positive and negative observations respectively, only if they differ in at least one variable. This can be represented by following equation [9]

$$\sum_{i=1}^{n}(a_i \oplus b_i)y_i \geq 1$$

A minimal support set for a dataset can be obtained by solving the set covering problem given below:

$$\min \sum_{i=1}^{n} y_i \geq 1$$

s.t. $\sum_{i=1}^{n}(a_i \oplus b_i)y_i \geq 1$ for each pair $a \in \Omega^+$ and $b \in \Omega^- y \in \{0,1\}^n$

We have implemented the set covering problem to obtain the support set for our dataset. Apart from this, there are several implementations proposed to address the set covering problem in [5,6]. In our study, we have utilized the Mutual Information Greedy algorithm, which is one of the implementations proposed in [2], to address the set covering problem. The authors in [8] have used this algorithm and compared their LAD results with deep learning models on several datasets. In this algorithm, entropy score is used to select the best features.

The support set generation process is similar to process of feature selection from a dataset. In both the processes, redundant and unnecessary attributes are removed, thus speeding up the algorithm's work process. There are 3 techniques which are used in case of decision trees to select the most suitable attribute for tree splitting. Information gain, Gini Index and Information gain ratio are such techniques which can be used for feature selection.

Information gain of each feature is evaluated using its entropy. Entropy is the measure of uncertainty and information gain is related to how much uncertainty is reduced with respect to a feature. The feature with the highest Information Gain is selected to be part of the support set. The entropy formula for binary classification is given below.

$$E = -(P(0) * log_2(P(0)) + P(1) * log_2(P(1)))$$

The information gain of an attribute A can be calculated using the formula given below:

Info-Gain = Entropy(Dataset) − (Count (Group 1)/
Count (Dataset) * Entropy (Group 1)+
Count (Group 2)/ Count (Dataset) * Entropy (Group 2))

Here, attribute A has only two values 0 or 1. So Group 1 represents the observations for which value of A is 1 and Group 2 represents observations having value of A as 0. The Information Gain Ratio is an extension of information gain which overcomes the bias. The Gain ratio incorporates split information to normalize the information gain [15].

$$\text{SplitInfo(A)} = -\sum_{j=1}^{v} \frac{|D_j|}{|D|} * log_2 \frac{|D_j|}{|D|}$$

The above equation gives the splitting information for the dataset D partitioned into v subset based on the v values of the attribute A.
The gain ratio is defined as:

$$\text{Gain Ratio} = \text{Info-Gain(D,A)} / \text{SplitInfo(A)}$$

In the paper [7], the authors have used Information gain ratio to select support set for logical analysis of data. The LAD is applied on NSL-KDD and UNSW-NB15 datasets and the results show that this technique performed equivalently well in comparison to other machine learning methods. In this paper, the authors have used level and interval variables whereas for our experiment, we have considered only level variables in binarization step.

Gini Index or Gini impurity is the method used in a decision tree algorithm CART (Classification and Regression Tree) to create split points. The Gini index of the dataset D is given as

$$\text{Gini(D)} = 1 - \sum_{i=1}^{k} p_i^2$$

where p_i is the probability that an observation in the dataset D belongs to the class C_i.

Gini(D) = 0 is the minimum value which means that all observations belong to the same class. The Gini index uses binary split for each attribute. The dataset D is partitioned into D1 and D2 when a feature X has a binary split and its gini index is

$$\text{Gini}_X(D) = \frac{|D1|}{|D|}\text{Gini(D1)} + \frac{|D2|}{|D|}\text{Gini(D2)}$$

$$\Delta\text{Gini(X)} = \text{Gini(D)} - \text{Gini}_X(D)$$

The feature with minimum gini index is selected for the splitting.

3.3 Pattern Generation

After the binarization and support set generation process, main step of LAD is carried out which is pattern generation. The pattern generation step focuses on producing quality patterns that can cover maximum observations and also the best patterns should not be missed. Pattern generation approach is based on combinatorial enumeration techniques. Top-down and Bottom-up approach can be followed while generating patterns [5].

In top-down approach, each positive observation is associated by its characteristic term. A characteristic term is always a positive pattern and it is likely that even after a few literals are dropped, it will still be a pattern. The top-down approach removes literals from the term one by one until it gets a pattern.

The bottom-up approach begins with a single literal which covers a few positive observations. If no negative observations are covered by this term then it is a positive pattern. Otherwise, literals are added to the term one at a time until a pattern is obtained.

In our paper, a breadth first enumerative technique is used to generate patterns in which bottom up approach is applied. At each stage n, all the positive patterns are obtained along with candidate terms which are examined in the next stage $n + 1$. A candidate term is a term that covers both positive and negative observations. From these candidate terms, patterns are obtained in the next stage by eliminating some of its literals. The terms of degree n are partitioned in to three sets at each stage n:

- P_n is a set of terms that cover only positive observation and no negative observation.
- C_n is a set of terms that cover at least one positive and negative observation.
- G_n is a set of all remaining terms.

Thus, P_n is the set of positive prime patterns and C_n is the set of candidate terms to be examined in the next stage of the algorithm. The set G_n is not considered further in the process.

The elements of set C_n are examined in the lexicographic order given by

$$e_1 \prec \overline{e_1} \prec e_2 \prec \overline{e_2} \prec \ldots$$

To generate term of degree $n + 1$, a literal is added to a term $t \in C_n$ such that this literal is larger than all other literals in t. Consider literals in T be $l_1 < l_2 < \ldots < l_n$. Let a literal l which satisfies $l < l_n$ is added to this term to obtain a new term T'. Consider a term T" obtained from T' by dropping literal l_n. Both T and T" are of degree $n - 1$ where T" is smaller lexicographically from T. Hence, if T"$\in C_n$ then T' is already examined as T" is smaller than T'. Thus, it enforces a restriction that literal added to a term has to be larger than other literals. If the term T covers only positive observations and no negative observation then it is positive pattern and is added to the set P_n. The algorithm 1 illustrates the pattern generation process [7,11]. In the algorithm, we are using a threshold k which is the least number of positive observations, a pattern needs to cover to qualify as a positive pattern. This k value varies with the dataset and number of features in the support set. We obtain k value by performing experiments numerous times.

3.4   Design of Classifier

The patterns generated in the previous step provide a tool for classifying the unknown observations. These patterns are converted into rules, and the classifier

---

Algorithm 1 Pattern Generation Methodology

Input: $\Omega_S^+, \Omega_S^- \subset \{0,1\}^n$
D: maximum degree of generated patterns
n : no. of input variables
Result: P : Set of prime patterns.
$P = \phi$
$C_0 = \{\phi\}$
for $d = 1, \ldots, D$ **do**
    if $d < D$ **then**
        | $C_d = \phi$
    end
    for $T \in C_{d-1}$ **do**
        $p = $ maximum index of literal in T
        for $s = p + 1, \ldots, n$ **do**
            for $l_{new} \in \{l_s, \bar{l}_s\}$ **do**
               $T_1 = T \| l_{new}$
               for $i = 1, \ldots, d - 1$ **do**
                   $T_2 = $ remove i^{th} literal from T_1
                   if $T_2 \in C_{d-1}$ **then**
                   | Continue
                   else
                   | Break and continue with next term of C_{d-1}
                   end
               end
            end
            if $k \leq \sum_{v \in \Omega_S^+} T_1(v)$ **then**
               if $1 \notin T_1(\Omega_S^-)$ **then**
               | $P = P \cup \{T_1\}$
               | Delete the covered observations from the dataset
            else
               if $d < D$ **then**
               | $C_d = C_d \cup \{T_1\}$
               end
            end
        end
        end
    end
end

---

is constructed using these rules. A pattern obtained for the dataset shown in Table 8 is $b_2 b_{13}$. Here, b_2 means $x_1 \geq 2.45$ and b_{13} means $x_4 \geq 24.5$. Thus, the rule generated using this pattern is $(x_1 \geq 2.45) \wedge (x_4 \geq 24.5) \implies Label = 1$. The pseudo code can be given as follows:

$$\textbf{if } ((x_1 \geq 2.45) \wedge (x_4 \geq 24.5)) \textbf{ then}$$
$$Label = 1$$
$$\textbf{end if}$$

In a similar way, other patterns can also be converted into rules and can be combined using if else-if structure to build the classifier. The positive and negative rules can be used to develop hybrid classifiers.

4   Results

In order to perform our comparative study, we have used the dataset of Intrusion Detection Systems. NSL-KDD dataset is the most popular dataset for IDSs. This dataset consists of 42 attributes or features including a class label. Many attacks have been mentioned in the dataset. However, for our experiment, all the attacks are clubbed into a single label called 'attack'. Thus, the dataset has two labels, normal which is denoted by 1 and attack which is represented by 0. To build the LAD classifier, a part of the NSL-KDD dataset (KDDTrain_20percent dataset) consisting of 25000 observations, is used.

The NSL-KDD dataset contains real-valued data. Hence, the binarization step of LAD is carried out to generate binary values for the dataset. We have used the cutpoint technique to generate the binary variables. In this step, we have only produced level variables for all the numeric features in the dataset. Each cutpoint introduces a binary variable, and the total number of binary variables corresponds to the total number of cutpoints across all features. If a feature generates a large number of cutpoints, it results in a significant increase in the number of binary variables associated with that feature. Also, having more number of cutpoints indicate that the feature value varies a lot and may not effect the classification process. So we consider a threshold value to limit the count of binary variables in the dataset. This means that if the number of cutpoints is more than the threshold value, then such a feature is removed. Here, we have considered the threshold value as 175 by analysing the data. Thus, 1044 binary variables were generated for the dataset in this step. We have not incorporated interval variables for our experiment as the dataset size would increase too much.

After the binarization process, a support set is generated from the binarized dataset. We have used five different techniques to find the support set for the given dataset. These techniques have been described in the Sect. 3.2. These support sets, obtained by different support set generation techniques, are then used to generate patterns and build five different LAD classifiers. The support of a pattern is the number of observations covered by a pattern [8]. Its value is fixed after several rounds of analysis on the dataset. This value depends on the dataset used.

Since support set generation is similar to the set covering problem, we have implemented the set covering problem in Python language to find the support set for the NSL-KDD dataset. The support set obtained had 37 features. The pattern generation process was carried out on the support set to generate 19 patterns. The set cover problem took maximum time out of all the techniques to produce the support set.

The second technique used for generating support set is the Mutual Information Greedy algorithm. We get 17 features in the support set by applying this algorithm. The pattern generation process is carried out on the dataset using this support set of features. A total of 14 positive patterns are generated, where each pattern covers at least 30 observations of the dataset.

Using the Information Gain criteria, 39 features were selected out of 1044 binary variables for the support set. During the pattern generation step, 21 patterns were found, where each pattern covered at least 60 positive observations. Next method that we used to generate the minimal support set is Information Gain Ratio. A total of 40 features were obtained in the support set, and 13 patterns were produced, where each pattern is covering at least 100 positive observations. The support set generated using the Gini Index method consists of 38 features. Twenty-five positive patterns were generated in the third step. Each pattern generated covers at least 60 positive observations from the dataset.

The classifiers are built using the patterns generated from each support set technique. The NSL-KDD dataset consists of two test datasets: KDDTest$^+$ and KDDTest21. These two datasets are used to test the classifier. The KDDTest$^+$ consists of 22544 observations, and KDDTest21 has 11850 observations. The performance of different LAD classifiers developed using five different support sets have been compared using the parameters: accuracy, sensitivity, specificity, precision, and F1 score. The test results of all the techniques on both the datasets are presented in Tables 9 and 10, respectively.

Table 9. Experimental Results for KDDTest$^+$ Dataset

Algorithm	Accuracy	Sensitivity	Specificity	Precision	F1-score
Set Cover	0.921	0.892	0.942	0.921	0.906
Mutual Information Greedy	0.722	0.394	0.971	0.911	0.551
Information Gain	0.925	0.868	0.967	0.952	0.908
Information Gain Ratio	0.935	0.913	0.951	0.933	0.923
Gini Index	0.925	0.941	0.912	0.889	0.915

Table 10. Experimental Results for KDDTest21 Dataset

Algorithm	Accuracy	Sensitivity	Specificity	Precision	F1-score
Set Cover	0.876	0.657	0.924	0.658	0.658
Mutual Information Greedy	0.804	0.097	0.961	0.358	0.153
Information Gain	0.909	0.696	0.956	0.781	0.736
Information Gain Ratio	0.902	0.751	0.935	0.719	0.735
Gini Index	0.869	0.808	0.883	0.605	0.692

Table 10 shows the results of LAD using different support set techniques on the KDDTest21 Dataset. The results achieved by the LAD technique on this dataset are lower in comparison to that of the KDDTest$^+$ dataset, as the observations that can be easily classified are removed from the KDDTest21 dataset. Here, both Information Gain and Information Gain ratio have achieved similar accuracy of 90.9% and 90.2%, respectively. The specificity is the highest for the mutual information greedy algorithm and the lowest for the Gini index. A high specificity shows that LAD has correctly classified most of the attack observations. The information gain and information gain ratio for this dataset also worked well for support set generation. The rest of the techniques did have good accuracy, but precision and F1 score were not good.

Figure 1 shows the performance comparison of the existing classification techniques with our LAD model that uses Information gain ratio as support set generation technique. We have used Gain ratio based LAD model for comparison because it has the highest accuracy compared to other support set generation techniques as shown in Table 9. The techniques have been compared based on the accuracy. Simple RNN(83.7%), GRU(84.66%) and LSTM(88.13%) are implemented in the paper [16]. Paper [3] describes the XGBoost algorithm(95.55%). Random Forest(77.4%), SVM(73.22%), Decision Tree(82.72%) are taken from the paper [17]. From the figure, we can conclude that our LAD model has done well in comparison to other techniques and has second highest accuracy of 93.5% and XGBoost has the highest accuracy of 95.55%. Since our LAD model is a binary classifier it has not been compared with multiclass classifiers.

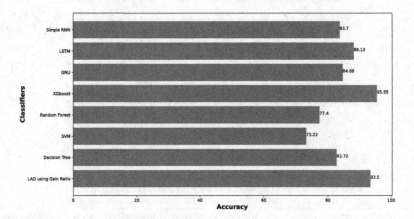

Fig. 1. Comparison of LAD using Information Gain Ratio with other classification techniques on KDDTest$^+$ dataset

The information gain ratio has been used with Logical Analysis of Data to develop an intrusion detection system in the study [7]. The authors have compared their IDS model with the existing classification techniques and it has performed equivalently well. They have used two datasets, NSL-KDD and UNSW-NB15 for validating their model. Another paper [8] has used Mutual information greedy algorithm to generate support set for LAD and their study show that the

LAD has worked well in comparison to the deep learning models like DNN and CNN. Here, they used UNSW-NB15 and CSE-CIC-IDS2018 datasets.

5 Conclusion

The paper studies the performance of the Logical Analysis of Data using different support set generation techniques on an intrusion detection dataset. The support set generation is an important step of LAD as it select the features that have higher impact on the classification. It is similar to feature selection which is applied in machine learning models. Five different methods namely, Set cover, Mutual information greedy algorithm, Information gain, Gain ratio and Gini index are used to generate the support set for binarized NSL-KDD dataset. LAD classifiers are developed using patterns generated from five different support sets of data. In terms of accuracy, LAD with the information gain ratio technique has done well compared to other techniques, with an accuracy of 93.5%. The result obtained using set cover technique can be further enhanced by improving its implementation. When trained on a larger dataset, the set cover problem can be a good solution for support set generation. For our study, only level variables are used to keep the dataset size small. Results may improve if both level and interval variables are considered in binarization. The LAD using the mutual information greedy algorithm has correctly classified the attack observations with specificity of 97.1%, but many normal observations have been detected as attack which resulted in lower accuracy. In all LAD classifiers, the specificity is higher than the sensitivity, indicating that the LAD classifier effectively distinguishes abnormal activity from normal activity. F1 score can be enhanced if hybrid patterns are used in the classifier so that the misclassification can be reduced. Through this study, we can conclude that LAD can be used along with other machine learning models in detection of attacks. The different support set generation methods can be tried for different datasets. Further, we can use LAD on larger datasets as in our study, a small part of the dataset is used and LAD has performed significantly well. Thus, LAD can perform in near real time and can give reasoning behind an abnormal activity.

References

1. Alexe, G., Alexe, S., Bonates, T.O., Kogan, A.: Logical analysis of data – the vision of Peter L. Hammer. Ann. Math. Artif. Intell. **49**(1–4), 265–312 (2007). https://doi.org/10.1007/s10472-007-9065-2
2. Almuallim, H., Dietterich, T.G.: Learning Boolean concepts in the presence of many irrelevant features. Artif. Intell. **69**(1–2), 279–305 (1994)
3. Alzahrani, A.O., Alenazi, M.J.F.: Designing a network intrusion detection system based on machine learning for software defined networks. Future Internet **13**(5), 111 (2021)
4. Boros, E., Hammer, P.L., Ibaraki, T., Kogan, A.: Logical analysis of numerical data. Math. Program. **79**(1–3), 163–190 (1997)

5. Boros, E., Hammer, P.L., Ibaraki, T., Kogan, A., Mayoraz, E., Muchnik, I.: An implementation of logical analysis of data. IEEE Trans. Knowl. Data Eng. **12**(2), 292–306 (2000)

6. Bruni, R.: Reformulation of the support set selection problem in the logical analysis of data. Ann. Oper. Res. **150**(1), 79–92 (2007)

7. Chauhan, S., Gangopadhyay, S.: Design of intrusion detection system based on logical analysis of data (LAD) using information gain ratio. In: Dolev, S., Katz, J., Meisels, A. (eds.) Cyber Security, Cryptology, and Machine Learning: 6th International Symposium, CSCML 2022, Be'er Sheva, Israel, June 30 – July 1, 2022, Proceedings, pp. 47–65. Springer, Cham (2022). https://doi.org/10.1007/978-3-031-07689-3_4

8. Chauhan, S., Mahmoud, L., Gangopadhyay, S., Gangopadhyay, A.K.: A comparative study of LAD, CNN and DNN for detecting intrusions. In: Yin, H., Camacho, D., Tino, P. (eds.) Intelligent Data Engineering and Automated Learning – IDEAL 2022: 23rd International Conference, IDEAL 2022, Manchester, UK, November 24–26, 2022, Proceedings, pp. 443–455. Springer, Cham (2022). https://doi.org/10.1007/978-3-031-21753-1_43

9. Chikalov, I., et al.: Logical analysis of data: theory, methodology and applications. In: Chikalov, I., et al. (eds.) Three Approaches to Data Analysis: Test Theory, Rough Sets and Logical Analysis of Data, pp. 147–192. Springer, Berlin, Heidelberg (2013). https://doi.org/10.1007/978-3-642-28667-4_3

10. Crama, Y., Hammer, P.L.: Boolean functions: Theory, algorithms, and applications. Cambridge University Press (2011)

11. Das, T.K., Gangopadhyay, S., Zhou, J.: SSIDS: semi-supervised intrusion detection system by extending the logical analysis of data. arXiv preprint arXiv:2007.10608 (2020)

12. Hammer, P.L.: Partially defined Boolean functions and cause-effect relationships. In: Proceedings of the International Conference on Multi-attribute Decision Making via OR-based Expert Systems. University of Passau (1986)

13. Hammer, P.L., Bonates, T.O.: Logical analysis of data-an overview: from combinatorial optimization to medical applications. Ann. Oper. Res. **148**(1), 203–225 (2006)

14. Ikram, S.T., Cherukuri, A.K.: Improving accuracy of intrusion detection model using PCA and optimized SVM. J. Comput. Inf. Technol. **24**(2), 133–148 (2016)

15. Karegowda, A.G., Manjunath, A., Jayaram, M.: Comparative study of attribute selection using gain ratio and correlation based feature selection. Int. J. Inf. Technol. Knowl. Manage. **2**(2), 271–277 (2010)

16. Kasongo, S.M.: A deep learning technique for intrusion detection system using a recurrent neural networks based framework. Comput. Commun. **199**, 113–125 (2023). https://doi.org/10.1016/j.comcom.2022.12.010

17. Oqbah Ghassan Abbas, Khaldoun Khorzom, M.A.: Machine learning based intrusion detection system for software defined networks. Int. J. Eng. Res. Technol. (IJERT) **09**(9) (2020)

18. Pathak, A., Pathak, S.: Study on decision tree and KNN algorithm for intrusion detection system. Int. J. Eng. Res. **9**(05) (2020). https://doi.org/10.17577/IJERTV9IS050303

Minimal Window Duration for Identifying Cognitive Decline Using Movement-Related Versus Rest-State EEG

Basma Jalloul[1]([✉]), Siwar Chaabene[1,2], and Bassem Bouaziz[1,2]

[1] Multimedia InfoRmation Systems and Advanced Computing (MIRACL), University of Sfax, Sfax, Tunisia
basma1707@gmail.com, bassem.bouaziz@isims.usf.tn
[2] Digital Research Center of Sfax (CRNS), B.P. 275, 3021 Sakiet Ezzit, Sfax, Tunisia

Abstract. Until recently, diagnosing people with neurophysiological disorders such as mild cognitive impairment (MCI) was challenging. The common diagnostic techniques used are invasive in nature and time-consuming due to their reliance on the intervention of an expert neuropsychologist and manual diagnosis. Therefore, the adoption of artificial intelligence (AI) and especially machine learning (ML) has proven most useful. It provided healthcare practitioners with an effective tool to diagnose patients faster with higher accuracy. In this paper, a method to separate MCI subjects from healthy controls using movement-related Raw electroencephalogram (EEG) is evaluated, and a new effective EEG segment length is discovered. A variety of binary classifiers are trained and our proposed segment length of 12 s with 50% overlap produces an accuracy of 97.27%.

Keywords: MCI classification · EEG · supervised ML · binary classification · movement-related

1 Introduction

Alzheimer's disease (AD) is one of the most common health ailments that can affect an aging person. Its effects can manifest in various ways [1,2]. Physiologically, it disrupts the affected individual's thinking and affects their behavior. Physically, it can cause motor disruption and violent outbursts, while psychologically it may result in severe depression and aggression. Moreover, AD shortens the affected individual's life span, which can be as little as 3.1 years in some extreme cases [3]. In the case of this disease, prevention is key, as once a diagnosis is made, it cannot be reversed. Therefore, continuous screening for its prodromal stage, mild cognitive impairment (MCI), is imperative. Unlike AD, MCI only affects the individual's cognitive capacities especially the memory while leaving the other faculties intact [4]. However, MCI is a risk state for dementia,

© The Author(s), under exclusive license to Springer Nature Switzerland AG 2024
A. Bennour et al. (Eds.): ISPR 2023, CCIS 1941, pp. 29–42, 2024.
https://doi.org/10.1007/978-3-031-46338-9_3

as more than half of the diagnosed individuals progress to dementia over 5 years [5].

MCI diagnosis used to rely on invasive and expensive techniques such as biomarker analysis [6–8] and neuroimaging [9–11]. In recent years, diagnostic efforts have gravitated toward more accessible techniques. Tools such as Electroencephalogram (EEG) [12], Magnetoencephalography (MEG) [13], functional Near-Infrared Spectroscopy (fNIRS) [14] and Heart Rate Variability (HRV) [15] have been considered as potential alternative diagnostic tools, and their contribution to MCI identification accuracy is being studied. Albeit these methods are easier to use and minimally invasive, they do not generate humanly readable data, instead, they generate a high volume of data that requires intense deciphering and interpretation. To make this process easier, artificial intelligence (AI) i.e. machine learning (ML) has been added to the equation. In the last twenty years, ML has been gradually minimizing the human intervention necessary for diagnosing neurological disorders, especially AD and MCI.

AD's symptoms are very noticeable as they manifest as pathological amyloid depositions [8], metabolic alterations [16] and structural atrophy [17]. On the other hand, MCI can be more subtle, as it presents very mild symptoms that can sometimes be easy to miss. Motor-related symptoms such as aimless movement, wandering [18] and balance and gait alteration [19] can be detected using postural tests, and movement-related data such as EEG and HRV can be captured. However, most of the studies exploring MCI's characteristics have been overlooking this type of data and focusing on the rest-state [20–23] instead, thus making the characterization incomplete. For this reason, further attention has to be awarded to studying the motor decline in MCI patients and its neurophysiological manifestations, as they can be key to early MCI identification.

To make detection an easier task, a wide variety of ML algorithms are used to derive and characterize the neurodegenerative patterns of MCI. The most popular algorithms are K-nearest neighbor classifiers (KNN) [24], boosting-based learning [25], and support vector machines (SVM) [26]. Remarkable results [27–29] comparable to that of an experienced neuroradiologist were reached which makes early intervention more achievable.

Recent studies have shown promising results in using EEG data to detect AD and MCI. For instance, the authors in [23] evaluated the optimal window length for detecting AD using EEG data over different brain regions. They found that longer window lengths provided better classification accuracy, and that certain brain regions were more informative than others. In another study [22], used EEG spectral features and segment lengths of (0.5 s, 1 s, 1.5 s, and 2 s) to automatically diagnose MCI. Their results showed high classification accuracy, indicating that EEG data could be a useful tool for early detection of cognitive decline.

In this article, we use a non-invasive EEG technique called $Emotiv\ Epoc_x$ headset to capture the neurophysiological response of an MCI patient during rest-state and a postural task. This headset has been used in various anomaly detection systems [30,31]. Based on our data, we train a variety of ML binary classifiers on different EEG segment lengths to try to achieve the best accuracy. Moreover, we explore the efficiency of using movement-related EEG in comparison to rest-state EEG by training the models on different segments of our dataset representing both. In follows, we outline the proposed method in Sect. 2 and we present our results in Sect. 3. We discuss our findings in Sect. 4. Finally, we conclude the paper by summarizing our key findings in Sect. 5.

2   Materials and Methods

In this work, we propose a new MCI detection method using EEG signals during rest-state and motor task. Figure 1 presents the pipeline of the proposed method. The subsections that follow detail the data acquisition protocol, the steps of data preparation, data classification, and the evaluation metrics used for experimental validation.

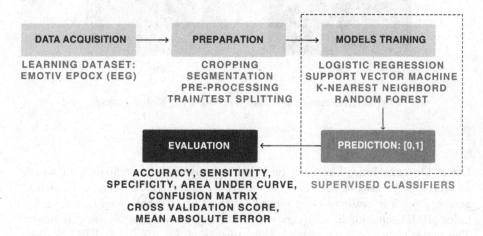

Fig. 1. Pipeline of the proposed method.

2.1   Data Acquisition Protocol

The data used in this work was acquired during the *Characterization of MCI Patients Based on EEG and HRV Signals During High Demanding Cognitive and Postural Tasks* study at *Otto von Guericke University Magdeburg, Germany* in 2021. The data acquisition protocol used in this work is described in detail below, along with a brief experiment summary.

Experiment Summary. The study consists of EEG and HRV recording during cognitive and motor tasks. As shown in Fig. 2, the experiment is split into two parts: cognitive testing and motor testing. At the start of the experiment, the participants receive a health evaluation that lasts for 10 min, then their EEG and HRV are recorded while they are in a rest state. Next, participants perform cognitive tasks using the Neurotrack program as shown in Fig. 2a that last for 30 min, followed by a 5 min rest. Participants then take the CERAD test which is a standardized test designed to measure primary cognitive deficits for 30 min followed by a 5 min rest. Finally, participants perform postural test for 10–15 minutes using the Balance Master system as shown in Fig. 2b to assess their motor control and balance. In this paper, we focus only on EEG signals during the Balance Master task.

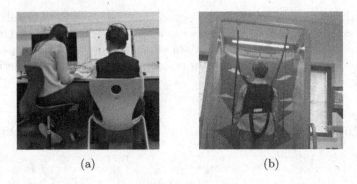

(a)   (b)

Fig. 2. Experiment: (a) Cognitive Tasks (b) Motor Tasks.

Data Description. The data included a total of 24 participants, of which 13 were MCI patients and 11 were healthy controls (HC). The participants' general health is evaluated and records of their age, weight, height, body mass index (BMI), smoking habits (cigarette numbers) and chronic diseases are made. The material used for data acquisition consists of *Emotiv Epoc_x* EEG headset illustrated in Fig. 3a that consists of 14 electrodes: AF3, AF4, F3, F4, F7, F8, FC5, FC6, T7, T8, O1, O2, P7 and P8 and 2 references electrodes: DRL and CMS placed as shown in Fig. 3b. The headset has a sampling frequency of 128 Hz and its raw output is exported via the EmotivPro software as EDF or CSV files.

2.2 Data Preparation

Before being used to train ML models, the raw data needs to be pre-processed and split into training/testing subsets. In the following sub-subsections, we outline the different steps for data pre-processing.

Cropping. We did not use the entirety of the data recorded during the experiment; we cropped the data to only preserve EEG recorded during the Balance Master postural tests to obtain our movement-related segments. Then, we cropped the rest-state related segments to compare with. Both segments are 10 min in length and have relatively the same number of samples.

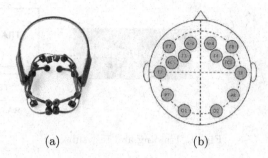

(a)   (b)

Fig. 3. EEG Acquisition: (a) *Emotiv Epoc_x EEG Headset* (b) Electrodes Placement.

Segmentation. According to the literature, the EEG segment length is usually set either arbitrarily or based on a literature survey. In [23], the authors conducted tests on a variety of segment lengths varying between 5 s and 12 s and came to the conclusion that a window size of 12 s is optimal. While in [22], the authors tested short segment sizes from 0.5 to 2 s with varying degrees of overlap and concluded that a 1 s window size with 50% overlap was the best choice. We tested both suggested window sizes on our dataset, and a 12 s segment size with 50% overlap yielded the best results, which combines the approaches from both studies.

Pre-processing. We cleaned the raw signals by passing them through 3rd-order median and low pass filters to eliminate noise and artifacts [32]. Afterward, the signals are converted from the time-domain to the frequency-domain using the fast Fourier transform (FFT). The Wavelet transformation was used in converting the signals to the time-frequency domain. We theorize that signals obtained in the time-frequency domain and extracted in the time-frequency domain result in getting a higher classification accuracy. The resulting signals were combined to form a hybrid feature set and used as is.

Train Test Splitting. Our data needs to be split into training and testing subsets per Fig. 4 as this is mandatory for the evaluations of our classifiers. The training subset (i.e. training dataset) is used to fit the models, while the testing subset is used to make predictions against the expected values. We have chosen to split our in 0.8:0.2 ratio, with 80% of the data being used for training and 20% for evaluation.

2.3   Supervised Classifiers

To look for optimal classification accuracy, a variety of supervised ML classifiers are trained in this work. Descriptions for each ML classifier are presented in follows.

Fig. 4. Training and test subsets.

Logistic Regression (LR) is a classification technique that uses a logistic function to model the dependent variable. The variable in question is dichotomous which means that there could only be two possible classes. As a result, this technique [33] is often used in a binary classification where the classes are either 0 or 1.

Support Vector Machine (SVM) represents a set of related supervised learning methods used mostly as a regression and classification tool. This classifier is used to make predictions using the machine learning theory to boost prediction accuracy and avoid overfitting at the same time. SVM uses the hypothesis space of a linear function in a high dimensional feature space and implements a learning bias derived from statistical learning theory [34]. For the parameters, we have set the classifier's kernel to Radial Basis Function (RBF) in this instance.

K-Nearest-Neighbours (KNN) is a popular and widely used classifier and is known for its simplicity and efficiency. Its training phase only consists of storing all training samples as a classifier and deferring the decision on generalizing beyond the training data until a new instance is encountered and for that, it has been called a "lazy learner". KNN classifies data by determining the class of an unclassified point by counting the majority class votes of its "k-nearest" neighbor points [35]. We implemented a method that looks for the optimal K value for each classifier and training is done using the best-found value.

Random Forest (RF) is a combination of decision trees such that each tree's prediction depends on a random vector's value sampled independently with the same distribution. The generalization error of a random forest converges to a limit as the size of the forest becomes larger thus depending on the strength of the individual trees and their correlation [36]. For our parameters, the same process described above is repeated to find the best X value for *random_state* and the value of *n_estimators* is set to 700.

2.4   Evaluation Metrics

We evaluate the performance of our models through accuracy, sensitivity, specificity, AUC, confusion matrix, cross-validation score and mean absolute error.

Sensitivity and specificity have been employed instead of recall and precision as they are best suited for a binary classification problem. Different calculations involving True Positives (TP), True Negatives (TN), False Positives (FP) and False Negatives (FN) are used to obtain these metrics [37]. This is further explained in the following subsection:

Accuracy (ACC). Also called fraction correct (FC) represents the ratio of correct classifications to the correct and incorrect classifications per Eq. 1.

$$ACC = \frac{TP + TN}{TP + TN + FP + FN} \tag{1}$$

Sensitivity (SEN). Also called Recall or True Positive Rate (TPR), is the ratio of correct positive classifications of all positive classifications made per Eq. 2. The higher the sensitivity the fewer actual positives go undetected.

$$SEN = \frac{TP}{TP + FN} \tag{2}$$

Specificity (SPC). Also called True Negative Rate (TNR) is the ratio of correct negative classifications that are actually negative to the total of real negative classifications per Eq. 3. With higher specificity, fewer positive classifications are labeled as negative.

$$SPC = \frac{TN}{TN + FP} \tag{3}$$

Area Under Curve (AUC). Represents the area under the ROC curve. The bigger this score is, the more likely that the classifier ranks the positive samples in comparison to the negative samples which makes for a better classification. AUC is widely used to evaluate the predictive capability of a classifier and is considered a better performance metric compared to accuracy [38].

Shuffle Split Cross Validation (CV). Also called *Monte Carlo cross-validation*, is a repeated random sub-sampling cross-validation method that splits the dataset randomly into training and validation. This method splits the dataset randomly in each iteration to evaluation the model on different data subsets each time.

Mean Absolute Error (MAE). Refers to the average difference between the prediction of an observation and the ground truth value of that observation in a group of predictions as a measurement of the magnitude of errors.

3   Results

Our binary classifiers produce [1,0] predictions for MCI and Healthy respectively. Based on the accuracy of these predictions, the models are evaluated. In the upcoming sections, we present the following results: i) Results obtained from models trained on 2 different segment lengths inspired by literature studying the classification of MCI and AD. ii) Results of models trained using our proposed segment length. iii) Results of our proposed method and segment length applied on the rest-state portion of our EEG signal. iv) Results of evaluating our method in comparison to the best scoring method in the literature.

3.1   Segment Length Evaluation

Segment Length of 1s with 50% Overlap. We trained our models using 1 s with a 50% overlap window size proposed by Kashefpoor et al. [22]. The best performing classifier (RF) achieved a classification accuracy of **95.17%** (sensitivity = 95.41%, specificity = 94.92%, AUC = 0.99, cross-validation score = 0.94 ±0.00) as shown in Table 1 along with the results of the rest of the models.

Table 1. Accuracy comparison of 1 s segments with 50% overlap classifiers.

Metrics	*LR*	*SVM*	*KNN*	*RF*
ACC (%)	68.04	74.87	95.11	**95.17**
SEN (%)	67.66	74.05	95.14	**95.41**
SPC (%)	69.79	86.82	95.08	**94.92**
AUC	0.77	0.89	0.95	**0.99**
CV	0.68 0.01	0.74 0.00	0.94 0.00	**0.94 0.00**
MAE	0.32	0.25	0.05	**0.05**

Segment Length of 12s with No Overlap. We trained our models on non overlapping 12 s window size proposed by Tzimourta et al. [23]. The best performing classifier (RF) achieved a classification accuracy of **95.52%** (sensitivity = 95.42%, specificity = 95.61%, AUC = 0.99, cross-validation score = 0.92 ±0.01) as shown in Table 2 along with the results of the rest of the models.

Table 2. Accuracy comparison of non overlapping 12 s segments classifiers.

Metrics	LR	SVM	KNN	RF
ACC (%)	77.13	71.30	93.27	**95.52**
SEN (%)	75.02	71.69	91.03	**95.42**
SPC (%)	85.44	88.45	96.03	**95.61**
AUC	0.87	0.85	0.93	**0.99**
CV	0.78 0.02	0.72 0.03	0.93 0.01	**0.92 0.01**
MAE	0.23	0.29	0.07	**0.04**

Proposed Segment Length Results. We trained our models using 12 s segments with a 50% overlap which combines two aspects from the previously tested lengths. We test the theory that the longer segment length with larger overlap can preserve more information thus achieving better results. Our theory is confirmed as the best performing classifier (RF) outperformed the previous classifiers. It achieved a classification accuracy of **97.27%** (sensitivity = 99.00%, specificity = 95.86%, AUC = 1.00, cross-validation score = 0.94 ±0.00) as shown in Table 3 along with the results of the rest of the models.

Table 3. Accuracy comparison of 12 s segments with 50% overlap classifiers.

Metrics	LR	SVM	KNN	RF
ACC (%)	77.95	78.64	96.36	**97.27**
SEN (%)	76.08	76.21	96.27	**99.00**
SPC (%)	81.36	91.50	96.45	**95.86**
AUC	0.89	0.89	0.96	**1.00**
CV	0.81 0.01	0.77 0.02	0.95 0.01	**0.94 0.00**
MAE	0.22	0.21	0.04	**0.03**

3.2   Movement-Related vs. Rest-State EEG Comparison

We train our models using our proposed segment length on the rest-state EEG portion of our dataset to compare the efficiency of the different types of EEG data. According to the results presented below in Table 4, these models did not score as highly as their previous counterparts. The highest score achieved was **92.81%** (sensitivity = 92.36%, specificity = 93.32%, AUC = 0.93, cross-validation score = 0.90 ±0.01) which is not as impressive.

Table 4. Accuracy obtained using the rest-state sub-dataset.

Metrics	LR	SVM	KNN	RF
ACC (%)	77.08	75.28	**92.81**	91.01
SEN (%)	76.35	74.06	**92.36**	89.39
SPC (%)	78.31	80.43	**93.32**	93.25
AUC	0.81	0.86	**0.93**	0.96
CV	0.73 0.02	0.96 0.01	**0.90 0.01**	0.88 0.01
MAE	0.23	0.25	**0.07**	0.09

3.3   Comparative Analysis

To explore new characteristics that can help with the early detection of MCI, we compare the efficiency of movement-related vs rest-state data by training the same models on both and comparing the accuracies acquired. Given the fact that movement-related EEG models outperfomed their counterparts by a significant margin as shown in Table 5, this direction seems like the correct path towards early MCI detection and with further fine-tuning, can become even more effective.

Table 5. Rest-state vs Movement-related results

Modalities	Classifiers' Accuracy (%)			
	LR	SVM	KNN	RF
Rest-state	77.08	75.28	92.81	91.01
Movement-related	77.95	78.64	96.36	97.27

Additionally, we compare our work with the most similar and recent method for MCI classification using EEG in the literature. In [39], the authors use both ML and EEG signals to obtain a binary classification of HC vs MCI. They propose a method that consists of extracting the multi-domain (time, frequency, and information theory) features from resting-state EEG signals before and after a cognitive task from 15 MCI participants and 15 age-matched HC. Afterwards, features are normalized in the range of (0,1) and principal component analysis (PCA) is used to perform feature selection. An SVM and a KNN model are trained using these features, and an accuracy of 98.57% is obtained in this work. We implement this method using our EEG data, and Table 6 showcases the accuracy values of the testing set using the competing method.

Table 6. Accuracy test comparison with related works.

Method	Accuracy	Classification Model
Li et al. [39]	85.63%	KNN
Proposed Method	**97.27%**	**KNN**

4   Discussion

4.1   Raw Data Aspect

Our main finding for this aspect is that training models on RAW movement-related EEG data can result in high accuracy scores and is a viable approach. Keeping the raw aspect of EEG data can keep the approach simple but effective, as preprocessing techniques such as normalization, data scaling, and PCA may cause information loss or skew the data in the wrong direction. By using raw data, we were able to avoid these issues and still achieve high accuracy scores.

4.2   Cognitive vs Motor Tasks Data

The advantage of movement-state EEG over rest-state EEG in MCI studies is that it allows for investigation of motor-related abnormalities that are often present in MCI patients, such as difficulty with balance and gait. Movement-state EEG also provides a more ecologically valid representation of brain activity, making it a valuable tool for studying neural mechanisms in a naturalistic context. Rest-state EEG, on the other hand, is better suited for investigating cognitive and emotional aspects of MCI.

4.3   Segment Length Impact on Accuracy

Regarding the segment length, we note that it performed differently from a model to another; while LR and RF models performed better with a longer segment length, SVM and KNN seem to have benefited from the smaller segment size. However, adding an overlap to the long segment length seems to have enhanced the classification results amongst all of the classifiers. This improvement is noted on the results obtained by all of the classifiers. This mostly goes back to the fact that more data is preserved and some segments can be crucial as they may contain MCI indicators.

4.4   Limitations and Assumptions

The use of different EEG segment lengths for MCI classifications has limitations. These include a trade-off between temporal and frequency resolution, sensitivity to data segmentation, computational complexity, and a risk of over-fitting. The choice of window size can significantly impact MCI classification results, and it is essential to consider these limitations when selecting a segment length. The main assumption we presented in this work is that the optimal EEG window size can be determined through the analysis of the extracted EEG features. This assumption assumes that the selected window size provides the best balance between temporal and frequency resolution and maximizes the separation between MCI and non-MCI groups.

5   Conclusion

In this paper, we used Raw EEG to train our binary classifiers. We also tested various EEG segment lengths proposed by the best works in literature and proposed our own. The high accuracy scores achieved by our method using the movement-related EEG sub-dataset in comparison to the rest-state EEG sub-dataset proved that it is more efficient and presents key MCI characteristics. To extend our work, a multimodal approach that combines multiple modalities will be implemented. Adding HRV as a complementary modality to EEG and re-training models on a hybrid feature set may be a worthwhile experiment. Additionally, we will explore longer EEG window sizes, which can help us study the trade-off between higher temporal resolution and higher frequency resolution and its effect on accuracy. Furthermore, we will add a feature extraction step to the data pre-processing to improve the interpretability, analysis, and classification of complex and voluminous EEG data. Once the multimodal approach has been developed, further clinical trials will be followed up in the future to validate our new methods.

Acknowledgment. This work is a part of the ICT-Rollator project. This project aims to assist seniors in daily life activities, provide physical and cognitive training, and foster the social inclusion of seniors.

References

1. Jewart, R.D., Green, J., Lu, C.J., Cellar, J., Tune, L.E.: Cognitive, behavioral, and physiological changes in Alzheimer disease patients as a function of incontinence medications. Am. J. Geriatric Psychiatry **13**, 324–328 (2005)
2. Mukaetova-Ladinska, E.B., Cerejeira, J., Lagarto, L.: Behavioral and psychological symptoms of dementia. Front. Neurol. **3**, 73 (2012)
3. Wolfson, C., et al.: A reevaluation of the duration of survival after the onset of dementia. New England J. Med. **344**, 1111–1116 (2001)
4. Petersen, R.C., Smith, G.E., Waring, S.C., Ivnik, R.J., Tangalos, E.G., Kokmen, E.: Mild cognitive impairment: clinical characterization and outcome. Arch. Neurol. **56**, 303–8 (1999)
5. Gauthier, S., et al.: Mild cognitive impairment. Lancet, **367**, 1262–1270 (2006)
6. Paola S., et al.: FDG-PET and CSF biomarker accuracy in prediction of conversion to different dementias in a large multicentre MCI cohort. NeuroImage Clin. **18**, 167–177 (2018)
7. Galluzzi, S., et al.: Supporting evidence for using biomarkers in the diagnosis of MCI due to ad. J. Neurol. **260**, 640–650 (2013)

8. Jack, C.R., et al.: Hypothetical model of dynamic biomarkers of the Alzheimer's pathological cascade. Lancet Neurol. **9**, 119–28 (2010)
9. Pihlajamaki, M., Jauhiainen, A., Soininen, H.: Structural and functional MRI in mild cognitive impairment. Current Alzheimer Res. **6**, 179–185 (2009)
10. Jack. C.R., rt al.: Prediction of ad with MRI-based hippocampal volume in mild cognitive impairment. Neurology **52**, 1397–1397 (1999)
11. Hojjati, S.H., Ebrahimzadeh, A., Khazaee, A., Babajani-Feremi, A.: Predicting conversion from mci to ad by integrating RS-FMRI and structural MRI. Comput. Bio. Med. **102**, 30–39 (2018)
12. Toural, J.E.S., Pedrón, A.M., Reyes, E.J.M.: A new method for classification of subjects with major cognitive disorder, Alzheimer type, based on electroencephalographic biomarkers. Inf. Med. Unlocked **23**, 100537 (2021)
13. Yang, S., Bornot, J.M.S., Wong-Lin, K., Prasad, G.: M/EEG-based bio-markers to predict the mci and Alzheimer's disease: a review from the ml perspective. IEEE Trans. Biomed. Eng. **66**, 2924–2935 (2019)
14. Grässler, B., et al.: Multimodal measurement approach to identify individuals with mild cognitive impairment: study protocol for a cross-sectional trial. BMJ Open, **11**, e046879 (2021)
15. Nicolini, P., et al.: Autonomic dysfunction in mild cognitive impairment: evidence from power spectral analysis of heart rate variability in a cross-sectional case-control study. PloS One **9**, e96656 (2014)
16. Nestor, P.J., Scheltens, P., Hodges, J.R.: Advances in the early detection of Alzheimer's disease. Nature Med. **10**, S34–41 (2004)
17. Jack C.R., et al.: Brain atrophy rates predict subsequent clinical conversion in normal elderly and amnestic MCI. Neurology **65**, 1227–31 (2005)
18. Khan, T., Jacobs, P.G.: Prediction of mild cognitive impairment using movement complexity. IEEE J. Biomed. Health Inf. **25**, 227–236 (2021)
19. Bahureksa, L., et al.: The impact of mild cognitive impairment on gait and balance: a systematic review and meta-analysis of studies using instrumented assessment. Gerontology **63**, 67–83 (2017)
20. Meghdadi, A.H., et al.: Resting state EEG biomarkers of cognitive decline associated with Alzheimer's disease and mild cognitive impairment. PloS One textbf16, e0244180 (2021)
21. Cecchetti, G., et al.: Resting-state electroencephalographic biomarkers of Alzheimer's disease. NeuroImage Clin. **31**, 102711 (2021)
22. Kashefpoor, M., Rabbani, H., Barekatain, M.: Automatic diagnosis of mild cognitive impairment using electroencephalogram spectral features. J. Med. Signals Sens. **6**, 25–32 (2005)
23. Tzimourta, K.D., et al.: EEG window length evaluation for the detection of Alzheimer's disease over different brain regions. Brain Sci. **9**, 81 (2019)
24. Ricciardi, C., et al.: Machine learning can detect the presence of mild cognitive impairment in patients affected by Parkinson's Disease, pp. 1–6. IEEE (2020)
25. Morra, J.H., Tu, Z., Apostolova, L.G., Green, A.E., Toga, A.W., Thompson, P.M.: Comparison of adaboost and support vector machines for detecting Alzheimer's disease through automated hippocampal segmentation. IEEE Tran. Med. Imaging **29**, 30–43 (2010)
26. Jiang, J., Kang, L., Huang, J., Zhang, T.: Deep learning based mild cognitive impairment diagnosis using structure MR images. Neurosci. Lett. **730**, 134971 (2020)

27. Davatzikos, C., Bhatt, P., Shaw, L.M., Batmanghelich, K.N., Trojanowski, J.Q.: Prediction of MCI to ad conversion, via MRI, CSF biomarkers, and pattern classification. Neurobio. aging **32**, 2322.e19–2322.e27 (2011)

28. Moradi, E., Pepe, A., Gaser, C., Huttunen, H., Tohka, J.: Machine learning framework for early MRI-based Alzheimer's conversion prediction in MCI subjects. NeuroImage **104**, 398–412 (2015)

29. Gao, F., Yoon, H., Xu, Y., Goradia, D., Luo, J., Wu, T., Su, Y.: AD-NET: age-adjust neural network for improved mci to ad conversion prediction. NeuroImage Clin. **27**, 102290 (2020)

30. Boudaya, A., Bouaziz, B., Chaabene, S., Chaari, L., Ammar, A., Hökelmann, A.: EEG- Based Hypo-vigilance Detection Using Convolutional Neural Network. In: International Conference On Smart Living and Public Health (ICOST), pp. 69–78 (2020)

31. Chaabene, S., Bouaziz, B., Boudaya, A., Hökelmann, A., Ammar, A., Chaari, L.: Convolutional neural network for drowsiness detection using EEG signals. Sensors **21**, 1734 (2021)

32. Attallah, O.: An effective mental stress state detection and evaluation system using minimum number of frontal brain electrodes. Diagnostics (Basel, Switzerland) **10**(5), 292 (2020)

33. Fernandes, A.A., Filho, D.B.F., da Rocha, E.C., Nascimento, W.d.S.: Read this paper if you want to learn logistic regression. Revista de Sociologia e Política **28** (2020)

34. Jakkula, V.: Tutorial on support vector machine (svm). Sch. EECS Wash. State Univ. **37**(2.5), 3 (2006)

35. Bzdok, D., Krzywinski, M., Altman, N.: Machine learning: supervised methods. Nature Methods **15**, 5–6 (2018)

36. Breiman, L.: Random forests. Mach. Learn.**45**(1), 5–32 (2001)

37. Verma, A., Srivastava, V.: Performance measures in machine learning classification: Accuracy, precision, recall, f1 score and confusion matrix. Int. J. Eng. Technol. **7**(4.28), 63–68 (2018)

38. Fawcett. T., An introduction to roc analysis. Pattern Recogn. Lett. **27**, 861–874 (2006)

39. Li, Y., Xiao, S., Li, Y., Li, Y., Yang, B.: Classification of mild cognitive impairment from multi-domain features of resting-state EEG, pp. 256–259. IEEE (2020)

Modeling Graphene Extraction Process Using Generative Diffusion Models

Modestas Grazys[✉]

Institute of Computer Science, Vilnius University, Didlaukis st. 47, 08303 Vilnius,
Lithuania
modestas.grazys@mif.stud.vu.lt

Abstract. Graphene, a two-dimensional material composed of carbon
atoms arranged in a hexagonal lattice, possess a unique array of properties that make it a highly sought-after material for a wide range of
applications. Its extraction process, a chemical reaction's result, is represented as an image that shows areas of the synthesized material. Know-
ing the initial conditions (oxidizer) the synthesis result could be modeled
by generating possible visual outcomes. A novel text2image pipeline to
generate experimental images from chemical oxidizers is proposed. Key
components of a such pipeline are a textual input encoder and a conditional generative model. In this work, the capabilities of certain text
model and generative diffusion model are investigated and some conclusions are drawn providing further suggestions for further full text2image
pipeline development

1 Introduction

Graphene is an exotic wonder material with many advantages. It is the thinnest
and strongest measured material in the universe. Graphene can maintain 10^6
times higher current density than copper, has record thermal conductivity and
stiffness, is impermeable to gases, and combines contradictory properties such as
brittleness and ductility [6]. It is a one-atom-thick material - a plane made up of
carbon atoms lined up in a hexagonal lattice, resembling a beehive. Graphene has
a very wide range of applications: manufacture of transistors [1]; production of
conductive plates with single-electron-transistor (SET) circuitry; digital memory
for quantum repeaters [27]; plasma displays [23] and much more.

Graphene could be synthesized using methods described in [20,21]. Yet, the
development of a digital tool capable of visualizing the potential output of a
graphene synthesis reaction using a selected oxidant could give an insight into
the expected result beforehand. Instead of complex chemical process modeling,
a such process could be modeled by analyzing images. The mathematical model
for image encoding in our case is a neural network, which uses the process of
automatic conversion of text into an image (*text2image* [19]), like Stable Diffusion
or DALL-E 2 [18,22]. Model input - chemical name (text) of the oxidant used

Supported by Vilnius University.

in graphene synthesis. Then the output of the model - a visual representation (picture) of the result of a possible reaction. However, this will be discussed in more detail in a later work, in the future. And in the current context, more attention is paid to the separate components of the model architecture: the component that prepares the input text embeddings that condition the output and the generative component that provides the visual result of the synthesis simulation (diffusion model selected).

For the component which prepares the input text embeddings (text encoder) the Contrastive Language-Image Pre-training model (CLIP) was selected [17]. The CLIP model can be conceptualized as a conventional image classifier with the ability to learn from the natural textual description of a photo. The usage of this model does not need complex and limiting data labeling methodologies. Also, the CLIP model can extract latent text embeddings, which can later influence the generative diffusion model in the process of conditional image generation. The simulation tool for the graphene synthesis reaction described in this paper specifically requires the text encoder component of the CLIP model for the latent text representations (text embeddings), precisely because of the influence on the diffusion model. The challenge here is to adapt appropriately this model to classify images of graphene synthesis correctly. As the images among oxidant classes look similar and the amount of data available is very limiting currently, because a larger amount of samples was not generated from the initial small dataset, provided by scientists working on graphene synthesis.

For the generative component a diffusion model [12, 24] was selected. In this context, it is a generative neural network, like generative adversarial networks (*GANs*) [7] or variational autoencoders (*VAEs*) [4] capable of generating images from the input vector by applying a denoising process. The text latents encoded by the CLIP text encoder can influence the input vector of the diffusion model, thus conditioning the output of the diffusion model. In this way, potential imaging results of the graphene synthesis reaction can be generated conditioned by the entered name of the oxidant. However, this paper presents the capabilities of the diffusion model to generate imaging results for the graphene synthesis reaction by discarding this conditional input application and leaving this task for the future. The challenge here is to inspect the capability of a diffuser to generate graphene synthesis images and draw further conclusions about the usage of a conditional diffuser for the general development objective.

The novelty of this work is the proposed usage of CLIP for digital components required for the subsequent modeling of the graphene synthesis simulation tool. This proposed solution allows the creation of a digital tool that can visually represent the possible output of the graphene synthesis reaction using the chosen oxidant. The tasks of the work include investigating the applicability of the CLIP model, the training process, and the ability of the diffusion-based generator (diffuser) to unconditionally generate images matching graphene synthesis reaction results.

2   Methodology

2.1   Overview

This text discusses digital tools that will be used to visualize the potential outcome of a graphene synthesis reaction using a selected oxidant. The two main tools used in this methodology are CLIP (Contrastive Language-Image Pretraining) and diffusion models. The CLIP model is used for image classification, learning to predict which oxidant corresponds to which image showing the result of a graphene synthesis reaction using an indicated oxidant. The CLIP text encoder can extract latent text embeddings from oxidant names. Text embeddings can then influence the input vectors of the diffusion model and thus condition the model output. Diffusion models, on the other hand, are used to create flexible and easily controlled generative models by systematically and slowly destroying image structure through an iterative diffusion process and learning the distribution of the added noise to reconstruct the image from the noise at inference. This section provides a brief summary of CLIP and diffusion model training and inference, including pseudocodes for the training and inference algorithms. The article concludes by highlighting the advantages of using CLIP and diffusion models in this context.

2.2   CLIP

The CLIP model [17] was introduced by the OpenAI company and is able to learn to predict which caption matches which image from the natural textual description of the photo. In this way, the limiting usability of computer vision systems is avoided, since conventional systems of this type require additional labeling information for any other visual concept. In implemented experimentation, the CLIP model is trained by providing (photo, text) data pairs, where text corresponds to the oxidant which was used to synthesize the result seen in the given photo. Conventional image classifiers simultaneously train a visual feature extractor and a linear classifier to predict the class label, while the CLIP model simultaneously trains a visual feature encoder and a text encoder to predict suitable (photo, text) pairs. Encoded oxidizer namings (latents) during training are encouraged to match synthesis result image latents as much as possible by minimizing pairs of their dot products. The minimized values are located on the main diagonal of this resulting matrix. The used architecture in our experimentation can be seen in Fig. 1.

During inference, the text encoder simulates a zero-shot linear classifier by embedding the oxidizer class names of the graphene images dataset and classifies the graphene synthesis results by their corresponding oxidizers. The image latent vector is scalar multiplied with each of the possible naming latent vectors and the smallest value of all obtained is selected. Then an oxidizer naming, which latent vector provided this smallest value in product with image latent, is selected and passed as an output. This process is illustrated by Fig. 2.

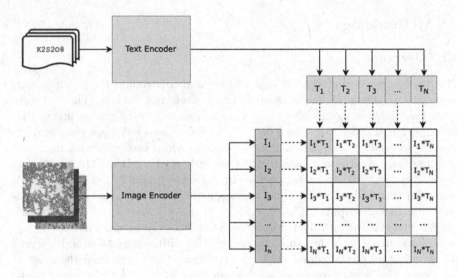

Fig. 1. CLIP training

Due to the property of learning from the natural textual description of an image, the CLIP model requires fewer data to achieve the same accuracy result compared to conventional computer vision classification models. These require a larger volume of training material corresponding to the distribution of use. This is relevant as the available amount of graphene data is very limited.

The CLIP model with its ability to embed textual input is also used in models for generating images from a text: DALL-E 2, Stable Diffusion [18,22]. It is this use case that is relevant to the purpose of this paper. Other use cases of the CLIP model: content moderation [5]; image captioning [2]; image search engine [14]; and much more.

2.3 Diffusion Model

The invention of the diffusion model was inspired by the idea of non-equilibrium thermodynamics: to systematically and slowly destroy the data structure through an iterative process of diffusion. The reverse diffusion process can then be learned by reconstructing the previously destroyed data structure - this method creates an extremely flexible and easy-to-manage generative model [24].

The diffusion process (forward process) is a Markov chain [15] and is performed by gradually adding Gaussian noise to the image data based on the noise variance schedule (*variance schedule*) $\beta_1, \beta_2, ..., \beta_T$, where T is the duration of the schedule (how many times noise will be added to the data). Figure 3 diffusion process is marked $q(x_t|x_{t-1})$, where $x_0 \sim q(x_0)$ - initial data; $x_1, x_2, ..., x_{t-1}, x_t, ..., x_T$ - latents (data with applied Gaussian noise).

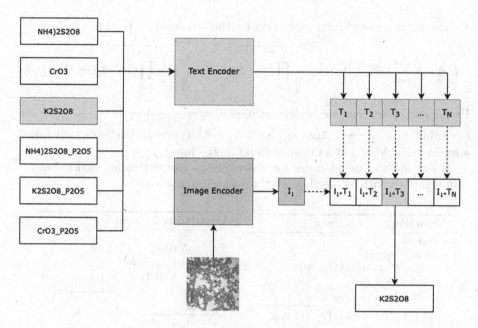

Fig. 2. CLIP inference

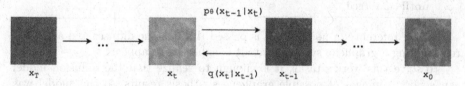

Fig. 3. Diffusion process - Markov chain

Mathematically, the diffusion process $q(x_{1:T}|x_0)$ is written [12]:

$$q(x_{1:T}|x_0) := \prod_{t=1}^{T} q(x_t|x_{t-1}), \qquad q(x_t|x_{t-1}) := \mathcal{N}(x_t|\sqrt{1-\beta_t}x_{t-1}, \beta_t\mathbf{I}) \qquad (1)$$

The reverse process Fig. 3 is denoted by $p_\theta(x_{t-1}|x_t)$. Is run to reconstruct image data destroyed by noise and is described by a Markov chain with learned Gaussian transitions starting from $p(x_T) = \mathcal{N}(x_T|\mathbf{0}, \mathbf{I})$:

$$p_\theta(x_{0:T}) := p(x_T) \prod_{t=1}^{T} p_\theta(x_{t-1}|x_t), \qquad p_\theta(x_{t-1}|x_t) := \mathcal{N}(x_{t-1}|\boldsymbol{\mu}_\theta(x_t,t), \boldsymbol{\Sigma}_\theta(x_t,t))$$

$$(2)$$

We can write a closed-form expression for the loss function [12,13]:

$$L = c + \sum_{t=1}^{T} \kappa_t \mathbb{E}_{x_0, \epsilon} \left\| \epsilon - \epsilon_\theta \left(\sqrt{\prod_{s=1}^{t}(1-\beta_t)} \cdot x_0 + \sqrt{1 - \prod_{s=1}^{t}(1-\beta_t)} \cdot \epsilon, t \right) \right\|_2^2 \quad (3)$$

Here, c and κ_t are constants independent of θ. $\kappa_t = \frac{\beta_t}{2(1-\beta_t)(1-\prod_{s=1}^{t}(1-\beta_{t-1}))}$ when $t > 1$, but $\kappa_1 = \frac{1}{2(1-\beta_1)}$. Also, $\epsilon_\theta : \mathbb{R}^L \times \mathbb{N} \to \mathbb{R}^L$ is a neural network that takes a latent variable x_t and a diffusion step t as the input.

Below are the pseudocodes for the diffusion model training and inference (sampling) algorithms based on all the above information.

Algorithm 1. Training	**Algorithm 2.** Sampling

repeat
 $\quad x_0 \sim q(x_0)$
 $\quad t \sim Uniform(\{1, ..., T\})$
 $\quad \epsilon \sim \mathcal{N}(\mathbf{0}, \mathbf{I})$
 $\quad$ Gradient descent step:
 $\quad \nabla_\theta \left\| (\epsilon - \epsilon_\theta(\sqrt{\prod_{s=1}^{t}(1-\beta_t)} \cdot \right.$
 $x_0 + \sqrt{1 - \prod_{s=1}^{t}(1-\beta_t)} \cdot \epsilon, t) \left. \right\|^2$
 until converged
	$x_T \sim \mathcal{N}(\mathbf{0}, \mathbf{I})$
	 for $t = T, ..., 1$ **do**
	 $\quad z \sim \mathcal{N}(\mathbf{0}, \mathbf{I})$ if $t > 1$, else $z = \mathbf{0}$
	 $\quad x_{t-1} = \frac{1}{\sqrt{(1-\beta_t)}}(x_t - \frac{\beta_t}{\sqrt{1-\prod_{s=1}^{t}(1-\beta_t)}} \epsilon_\theta(x_t, t)) + \sigma_t z$
	 end for
	 return x_0

During inference, a noise latent is passed through the diffuser and an image representing a graphene synthesis result is expected to appear.

To solve this work's tasks, it is enough to realize that the diffusion model can generate images of possible graphene synthesis results, so this model was chosen for this very reason. Using the cross-attention layers in the model architecture can influence the model's generative processes, thus, through the input, conditioning the model's output. And this is relevant to the general goal of the work, which will be worked on in the future. This work is limited to research to determine the capabilities of the diffusion model to unconditionally generate images of possible graphene synthesis results.

3   Computational Experiments

All studies were performed in the Google Colab Premium environment using A100 SXM4 40 GB GPU.

3.1   Data

Before conducting the research, 79 images of graphene synthesis results (optical microscopy photographs of graphite bisulfate (GBS) samples) with specified oxidants and reagents, which were used during the synthesis, were obtained from the Faculty of Chemistry and Geosciences of Vilnius University. In GBS synthesis, three different oxidants were used (Fig. 4):

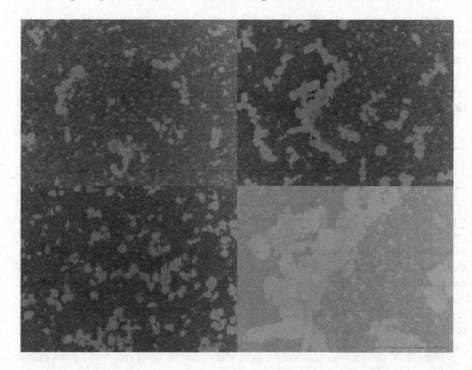

Fig. 4. Optical microscopy samples of graphite bisulfate (GBS)

- Ammonium persulfate - $(NH_4)_2S_2O_8$
- Potassium persulfate - $K_2S_2O_8$
- Chromium trioxide - CrO_3

The GBS synthesis procedure was repeated using all the oxidants and additionally adding a water binding reagent - phosphorus pentoxide P_2O_5 to the oxidizing mixture. Thus, a total of six data classes were obtained, which are named:

- NH4)2S2O8
- K2S2O8
- CrO3
- NH4)2S2O8_P2O5
- K2S2O8_P2O5
- CrO3_P2O5

3.2   Data Processing

The raw data are JPG format photos with a resolution of 2560×1920. The pictures show the result on a large scale, but the number of pictures is not large, so for a better quality of network generalization, a 1000×1000 resolution frame is selected at a random place in the picture and the picture is cropped around the frame. However, the resolution of a 1000×1000 photo is extremely high and its processing would theoretically require a very large neural network, the calculations of which would be extremely expensive, so the resolution of the photo is reduced to 128×128. These augmentations are chosen for optimal model training and subsequent exploitation without damaging essential information in the data.

3.3 CLIP Model Training

The individual components that make up the CLIP model were downloaded using the Hugging Face `transformers` library [26]: the CLIP photo encoder component and the CLIP text encoder component. This decomposition is chosen for convenience in the later implementation of the overall goal of the work, which requires only the CLIP text encoder component. Also downloaded: CLIP feature extractor and CLIP tokenizer in "`openai/clip-vit-base-patch32`" configuration. Settings for the CLIP video encoder and text input encoder components can be found in Appendix A. Projection heads of photo embeddings and text embeddings were also used, which unify the dimensions of the embeddings. These heads are small neural networks consisting of one residual block [9], the diagram is presented in Appendix B.

The CLIP model was trained by providing it with (picture, text) data pairs:

1. The photo is passed through the CLIP image feature extractor
2. The resulting processed photo is passed through the CLIP image encoder
3. The resulting latents are passed through the projection head

1. The text is passed through the CLIP tokenizer.
2. The resulting tokens are passed through the CLIP text encoder
3. The resulting latents are passed through the projection head

Later, the probabilities are calculated, the value of the cross entropy loss function is estimated, the gradient is calculated and the CLIP components' weights are updated. For the training process an Adam optimizer [3] and a cosine learning rate scheduler [10] were used. It was performed using the Hugging Face `accelerate` library for training over 200 epochs [8].

3.4 Diffusion Model Training

The diffusion model was downloaded from the Hugging Face `diffusers` library [16]. Unet architecture for this model was used. A scheme describing the model architecture is given in Appendix D. Settings and hyperparameters are given in Appendix C. Also, from the `diffusers` library a DDPM noise scheduler according to [12] was downloaded.

The diffusion model was trained by loading batches of pictures to it:

1. Noise is added to photos according to the noise schedule
2. Noisy pictures are passed through a diffusion model

3. The obtained noise estimate of the model is evaluated with the noise schedule reference and the value of the mean square error (MSE) loss function is calculated

4. Based on the received value, the gradient is calculated and the model weights are updated

For training, over 200 epochs, the Adam optimizer [3] with decoupled weight loss regularization [11], the cosine learning rate scheduler [10] were used, and the training process was performed using the Hugging Face `accelerate` library [8]. Afterward, the diffuser images were generated using the `DDPMPipeline` class from the `diffusers` library. The implementation code will be shared in the future.

4 Results

4.1 CLIP Results

For this work, CLIP model inference provided disappointing results. Currently, the model is not able to reliably classify the imaging data of graphene synthesis - for each of the classes of oxidants with which graphene was synthesized, the model assigns approximately equal probabilities during classification, trying to guess the real oxidant.

Table 1. CLIP inference 1

Top predictions	
NH4)2S2O8_P2O5	16.76%
NH4)2S2O8	16.71%
K2S2O8_P2O5	16.67%
Actual text prompt	NH4)2S2O8_P2O5

Table 2. CLIP inference 2

Top predictions	
CrO3	16.78%
K2S2O8	16.74%
CrO3_P2O5	16.72%
Actual text prompt	NH4)2S2O8

Table 3. CLIP inference 3

Top predictions	
NH4)2S2O8_P2O5	16.73%
NH4)2S2O8	16.70%
K2S2O8_P2O5	16.67%
Actual text prompt	NH4)2S2O8

The figures above show three attempts to determine the oxidant, used to synthesize graphene, from a photograph. The "*Actual text prompt*" box in the tables shows the actual graphene synthesis result class, and above it, in descending order, the class predictions with probabilities. From the Table 1 example, we can see that the model correctly identified the oxidant class, but the other guesses get slightly lower probabilities. Similar classification probabilities are also seen in the Table 2, 3 examples, where the model failed to correctly identify the graphene synthesis oxidant, and Table 2 the correct oxidizer class is not even among the three most likely guesses. It is believed that in this design, the CLIP model fails to draw strong differences between graphene synthesis samples synthesized with different oxidants, and in the latent space, all latents: both text and photos, are too closely spaced.

4.2 Diffusion Model Results

A mathematical expression for evaluating the quality of the model, except for the loss function, was not designed, therefore the most optimal hyperparameter values were selected by comparing the values of the loss function calculated after the appropriate number of training epochs. Forty (40) training epochs were found to suit the need for training the generative diffuser, as the following number of them did not make significant learning progress. It was found that the most optimal tested learning rate value for this model is equal to $3 * 10^{-4}$. With values of this hyperparameter higher than 10^{-3}, the model diverged. The study results are provided in Fig. 5.

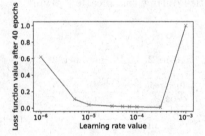 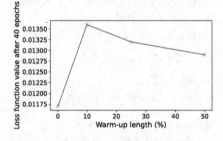

Fig. 5. Learning rate study for the diffuser

Fig. 6. Learning rate scheduler warm up study for the diffuser

The warm-up [10] of the cosine training step planner was also investigated. Warm-up values of different lengths were tested as a percentage (%) of the number of warm-up steps from the total number of model training steps. The optimal value of this hyperparameter was found to be 0%. A warm-up study of the cosine training step scheduler is displayed in Fig. 6.

The results of diffusion model generation were evaluated visually and by Frechet Inception Distance (FID) metric [25]. The images generated unconditionally by the model look similar to the images that were fed to the model during training. However, a closer look at the results reveals a lack of detail clarity - it is difficult to accurately distinguish graphene regions in generated images. This problem may be caused by the small amount of data and the low resolution of the generated photo (128×128). Examples of generated images are provided in Fig. 9.

Frechet Inception Distance was calculated by comparing original photos (training data), disregarding the classes of the sample, with:

1. The same sample of original data
2. The same sample of original data, but within the same classes
3. Random images
4. Generated images by the first iteration model
5. Generated images by the best model

The results are shown in Fig. 7 - original FID score values and Fig. 8 - percentages of the scores compared with the highest score. The highest FID score, indicating the worst result, is reached by the first iteration's model-generated images. The best model's generated images reach 68.51% of the highest score and indicate a better resemblance of the generated photos to the original than comparing original data with random images. Yet the score is far from the original data's comparison as it reaches negligible FID value and is the target of this task.

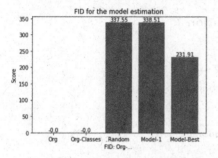

Fig. 7. Frechet Inception Distance scores

Fig. 8. Frechet Inception Distance score percentages

Fig. 9. Generated samples (resolution 128×128)

Conclusion

The novel text2image usage to encode chemical formulas for the generation of chemical images was proposed and successfully implemented. The results demonstrated the image generation part to be working more sufficiently than the graphene synthesis results classification. The encoder model assigns similar probabilities to all guess variants of the class. Furthermore, the diffusion model (generator) generates images similar to what would be expected, but these images are not detailed enough, lacking resolution. However, based on the tasks of text2image, these two models can be adapted to predict the results of graphene synthesis reactions.

On the other hand, for the development of a high-quality image prediction tool, there remain a number of tasks related to the adaptation of the CLIP model and the generative diffuser.

A   The Used CLIP Model Configuration

A.1   CLIP Image Encoder Configuration

See Table 4.

Table 4. CLIP image encoder configuration

Hidden layers dimension (*hidden_size*)	768
Encoder's feed-forward layer dimension (*intermediate_size*)	3072
Number of hidden layers (*num_hidden_layers*)	12
Number of attention heads (*num_attention_heads*)	12
Number of image color channels (*num_channels*)	3
Resolution of images (*image_size*)	128×128
Resolution of an attention patch (*patch_size*)	32
Hidden layers' activation function (*hidden_act*)	"quick_gelu"
Normalization layers' epsilon ϵ value (*layer_norm_eps*)	1e-05
Dropout probability in fully connected layers (*dropout*)	0.0
Dropout probability in attention layers (*attention_dropout*)	0.0
The standard deviation of model weights initiation (*initializer_range*)	0.02
Model weights initiation factor (*initializer_factor*)	1.0

A.2   CLIP Text Encoder Configuration

See Table 5.

Table 5. CLIP text encoder configuration

Vocabulary size (*vocab_size*)	49408
Hidden layers dimension (*hidden_size*)	512
Encoder's feed-forward layer dimension (*intermediate_size*)	2048
Number of hidden layers (*num_hidden_layers*)	12
Number of attention heads (*num_attention_heads*)	8
The maximum tokenized sequence length (*max_position_embeddings*)	77
Hidden layers' activation function (*hidden_act*)	"quick_gelu"
Normalization layers' epsilon ϵ value (*layer_norm_eps*)	1e-05
Dropout probability in fully connected layers (*dropout*)	0.0
Dropout probability in attention layers (*attention_dropout*)	0.0
Standard deviation of model weights initiation (*initializer_range*)	0.02
Model weights initiation factor (*initializer_factor*)	1.0
The ID of the padding token (*pad_token_id*)	1
The ID of the beginning-of-stream token (*bos_token_id*)	0
The ID of the end-of-stream token (*eos_token_id*)	2

A.3   General CLIP Configuration and Hyperparameters

See Table 6.

Table 6. General CLIP configuration and hyperparameters

Training batch size	26
Number of training epochs	200
Gradient accumulation steps	1
Learning rate	$5 \cdot 10^{-2}$
CLIP temperature	0
Learning rate warm-up steps	0
Latents dimension	256

B   Embeddings Projection Head Scheme

See Fig. 10.

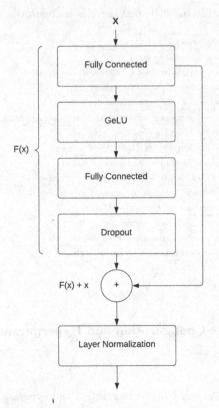

Fig. 10. Embeddings projection head scheme

C   Diffusion Model Configuration and Hyperparameters

See Table 7.

Table 7. Diffusion model configuration and hyperparameters

Resolution of generated photos	128
Number of input image channels	3
Number of output image channels	3
Number of training epochs	200
Gradient accumulation steps	1
Learning rate	$3 \cdot 10^{-4}$
Learning rate warm-up steps	0
Number of diffusion steps	1000

D   Architecture of the Diffusion Model - Unet

See Fig. 11.

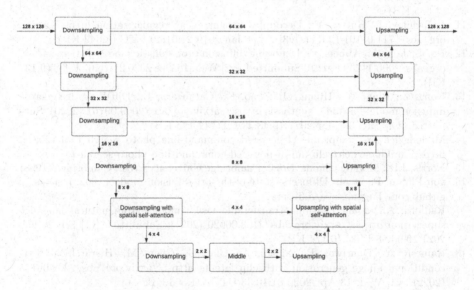

Fig. 11. Architecture of the diffusion model - Unet

References

1. Geim, A.K., Novoselov, K.S.: The rise of graphene. Nat. Mater. **6**, 183–191 (2007)
2. Nukrai, D., Mokady, R., Globerson, A.: Text-only training for image captioning using noise-injected clip https://arxiv.org/abs/2211.00575 (2022). [v1] Tue, 1 Nov 2022 16:36:01 UTC (798 KB)

3. Kingma, D.P., Ba, J.: Adam: a method for stochastic optimization https://arxiv. org/abs/1412.6980 (2014). [v9] Mon, 30 Jan 2017 01:27:54 UTC (490 KB)
4. Kingma, D.P., Welling, M.: Auto-encoding variational bayes https://arxiv.org/ abs/1312.6114 (2013). Submitted: [v11] Sat, 10 Dec 2022 21:04:00 UTC (3,451 KB)
5. Akyon, F.C., Temizel, A.: Deep architectures for content moderation and movie content rating https://arxiv.org/abs/2212.04533 (2022). [v2] Mon, 12 Dec 2022 07:53:17 UTC (801 KB)
6. Geim, A.K.: Graphene: status and prospects. Science **324**, 1530–1534 (2009)
7. Goodfellow, I.J., et al.: Generative adversarial networks https://arxiv.org/abs/ 1406.2661 (2014). Submitted: [v1] Tue, 10 Jun 2014 18:58:17 UTC (1,257 KB)
8. Gugger, S., Debut, L., Wolf, T., Schmid, P., Mueller, Z., Mangrulkar, S.: Accelerate: Training and inference at scale made simple, efficient and adaptable (2022). https://github.com/huggingface/accelerate
9. He, K., Zhang, X., Ren, S., Sun, J.: Deep residual learning for image recognition. In: 2016 IEEE Conference on Computer Vision and Pattern Recognition (CVPR), pp. 770–778 (2016). https://doi.org/10.1109/CVPR.2016.90
10. Loshchilov, I., Hutter, F.: SGDR: stochastic gradient descent with warm restarts https://arxiv.org/abs/1608.03983 (2016). [v5] Wed, 3 May 2017 16:28:09 UTC (1,385 KB)
11. Loshchilov, I., Hutter, F.: Decoupled weight decay regularization https://arxiv. org/abs/1711.05101 (2017). [v3] Fri, 4 Jan 2019 21:01:49 UTC (8,347 KB)
12. Ho, J., Jain, A., Abbeel, P.: Denoising diffusion probabilistic models https://arxiv. org/abs/2006.11239 (2020). Submitted: [v2] Wed, 16 Dec 2020 21:15:05 UTC (9,137 KB)
13. Kong, Z., Ping, W., Huang, J., Zhao, K., Catanzaro, B.: DiffWave: a versatile diffusion model for audio synthesis https://arxiv.org/abs/2009.09761 (2020). Submitted: [v3] Tue, 30 Mar 2021 19:48:38 UTC (1,145 KB)
14. Mikhalevich, Y.: rclip: an AI-powered command-line photo search tool (2021). https://mikhalevi.ch/rclip-an-ai-powered-command-line-photo-search-tool/
15. Norris, J.R.: Markov Chains. No. 2, Cambridge University Press, Cambridge (1998)
16. von Platen, P., et al.: Diffusers: state-of-the-art diffusion models (2022). https:// github.com/huggingface/diffusers
17. Radford, A., et al.: Learning transferable visual models from natural language supervision https://arxiv.org/abs/2103.00020 (2021). Submitted: [v1] Fri, 26 Feb 2021 19:04:58 UTC (6,174 KB)
18. Ramesh, A., Dhariwal, P., Nichol, A., Chu, C., Chen, M.: Hierarchical text-conditional image generation with clip latents https://arxiv.org/abs/2204.06125 (2022). [v1] Wed, 13 Apr 2022 01:10:33 UTC (41,596 KB)
19. Reed, S., Akata, Z., Yan, X., Logeswaran, L., Schiele, B., Lee, H.: Generative adversarial text to image synthesis https://arxiv.org/abs/1605.05396 (2016). [v2] Sun, 5 Jun 2016 13:39:27 UTC (2,147 KB)
20. Rimkutė, G., Gudaitis, M., Barkauskas, J., Zarkov, A., Niaura, G., Gaidukevič, J.: Synthesis and characterization of graphite intercalation compounds with sulfuric acid. Crystals **12**, 421 (2022). https://doi.org/10.3390/cryst12030421
21. Rimkutė, G., Niaura, G., Pauliukaite, R., Gaidukevič, J., Barkauskas, J.: Wet synthesis of graphene-polypyrrole nanocomposites via graphite intercalation compounds. Crystals **12** (2022). https://doi.org/10.3390/cryst12121793
22. Rombach, R., Blattmann, A., Lorenz, D., Esser, P., Ommer, B.: High-resolution image synthesis with latent diffusion models https://arxiv.org/abs/2112.10752 (2022). [v2] Wed, 13 Apr 2022 11:38:44 UTC (38,971 KB)

23. Sankaran, K.J., et al.: Laser-patternable graphene field emitters for plasma displays. Nanomaterials **9** (2019). https://www.mdpi.com/2079-4991/9/10/1493

24. Sohl-Dickstein, J., Weiss, E.A., Maheswaranathan, N., Ganguli, S.: Deep unsupervised learning using nonequilibrium thermodynamics https://arxiv.org/abs/1503.03585 (2015). Submitted: [v8] Wed, 18 Nov 2015 21:50:51 UTC (6,095 KB)

25. Szegedy, C., Vanhoucke, V., Sergey Ioffe, J.S., Wojna, Z.: Rethinking the inception architecture for computer vision https://arxiv.org/abs/1512.00567 (2015). [v3] Fri, 11 Dec 2015 20:27:50 UTC (228 KB)

26. Wolf, T., et al.: Transformers: state-of-the-art natural language processing. In: Proceedings of the 2020 Conference on Empirical Methods in Natural Language Processing: System Demonstrations, pp. 38–45. Association for Computational Linguistics, Online, October 2020. https://www.aclweb.org/anthology/2020.emnlp-demos.6

27. Wu, G.Y., Lue, N.Y.: Graphene-based qubits in quantum communications https://arxiv.org/abs/1204.6365 (2012). Submitted: [v2] Mon, 9 Jul 2012 02:32:19 UTC (1,426 KB)

Bird Species Recognition in Soundscapes with Self-supervised Pre-training

Hicham Bellafkir[✉], Markus Vogelbacher, Daniel Schneider, Valeryia Kizik, Markus Mühling, and Bernd Freisleben

Department of Mathematics and Computer Science, University of Marburg, Marburg, Germany
{bellafkir,vogelbacher,schneider,kizik,muehling, freisleb}@informatik.uni-marburg.de

Abstract. Biodiversity monitoring related to bird species is often performed by identifying bird species in soundscapes recorded by microphones placed in the birds' natural habitats. This typically produces a large amount of unlabeled data. While self-supervised machine learning methods have recently been successfully applied to computer vision and natural language processing tasks, state-of-the-art automatic approaches for bird species recognition in audio recordings mainly rely on transfer learning using pre-trained ImageNet models. In this paper, self-supervised learning is leveraged to improve bird species recognition in soundscapes. Specifically, we use a novel self-supervised approach to pre-train a self-attention neural network architecture on the target domain to take advantage of the vast amount of unlabeled and weakly labeled data. Experiments on data sets from different recording environments show the effectiveness of our approach. In particular, self-supervised pre-training on the target domain improves the cross-domain recognition quality.

Keywords: bird species recognition · self-supervised training · deep neural networks · transformer architectures · self-attention

1 Introduction

Birds play a crucial role in linking habitats, resources, and biological processes in ecosystems. Thus, detecting temporal and spatial changes in bird populations as early as possible is important to obtain meaningful alerts for a potential decline of ecosystem health and a loss of biodiversity. For this purpose, autonomous recording units (ARU) are placed in the natural habitats of birds to record soundscapes that are subsequently analyzed to identify the bird species occurring in the audio recordings. This approach typically produces large amounts of unlabeled data.

Bird species recognition in audio recordings is currently often performed by human experts, but deep neural networks are increasingly used to perform this task [16,17], as shown in Fig. 1. The input to a deep neural network model

© The Author(s), under exclusive license to Springer Nature Switzerland AG 2024
A. Bennour et al. (Eds.): ISPR 2023, CCIS 1941, pp. 60–74, 2024.
https://doi.org/10.1007/978-3-031-46338-9_5

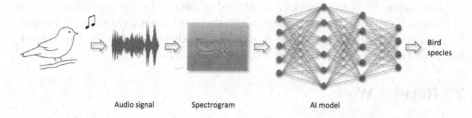

Audio signal   Spectrogram   AI model

Fig. 1. Bird species recognition in audio data based on a deep neural network.

is usually not the raw audio signal, but a spectrogram that can be generated using a Fourier transform. State-of-the-art approaches predominantly rely on convolutional neural networks (CNNs). Yet, deep learning methods based on self-attention mechanisms have outperformed neural network models constructed with convolutional layers on various vision tasks [4, 33]. However, neural networks based on self-attention (i.e., transformers) tend to require more labeled training data compared to CNNs. Furthermore, the manual acquisition of ground truth data for supervised learning of bird recognition models is very time-consuming. As a consequence, existing data sets contain only weak labels. On the one hand, most recordings are only labeled with the predominantly audible bird species, while background species are neglected. On the other hand, there is usually no temporal information about the annotations, i.e., there are no exact time stamps for each annotation. One way to address this problem is to perform a self-supervised pre-training of the models, i.e., encouraging the model to understand the data distribution of the target domain without using any labels.

In this paper, we present a novel approach to automatically recognize bird species in audio recordings using self-attention neural networks and self-supervised pre-training. The bird sound data used to realize our approach comes from more than 11,000 h of audio recordings recorded in a forest ecosystem, of which only a very small fraction is labeled. Furthermore, we crawled training data from publicly available sources like the Xeno-canto data collection [36], which are only weakly labeled. To meet the data requirements of transformers and to boost the recognition performance, we use a self-supervised approach to pre-train a transformer model on the target domain to leverage the vast amount of unlabeled data. In particular, our contributions are as follows:

- To the best of our knowledge, our approach is the first approach that uses self-supervised pre-training to improve bird species recognition quality.
- We perform a cross-domain adaptation via self-supervised pre-training on the target domain.
- We make our self-supervised pre-trained models for bird species recognition publicly available at https://github.com/umr-ds/ssl-bird-species-recognition. In this way, we enable other researchers to build on our work and fine-tune our models with their own data. Furthermore, we release our fine-tuned transformer models that are ready to use for bird species recognition in soundscapes.

The remainder of this paper is structured as follows. In Sect. 2, related work is discussed. Section 3 introduces the concept and implementation of our approach. Experimental evaluations of the recognition quality are presented in Sect. 4. Section 5 concludes the paper and outlines areas of future work.

2   Related Work

This section outlines current trends in deep neural network architectures (Subsect. 2.1) and state-of-the-art approaches for recognizing bird species in soundscapes (Subsect. 2.2).

2.1   Deep Neural Network Architectures

Convolutional neural networks (CNNs) are the key building block of several solutions to different computer vision tasks. In recent years, transformer architectures with self-attention mechanisms were used with great success in natural language processing tasks [35], and several researchers [4,6,18,33] successfully applied transformer networks also in the field of image and audio processing. Dosovitskiy et al. [4] introduced the Vision Transformer (ViT) model showing that relying on convolutions is not necessary and that neural networks based solely on self-attention can be sufficient to perform well in computer vision tasks. The main idea of the ViT model is to divide an image or spectrogram into patches of fixed size to create a sequence of patches on which the self-attention mechanism can operate in the same way as in the context of natural language processing.

Touvron et al. [33] presented a training process and a distillation method that incorporated an extra distillation token. This token has a similar function to the class token introduced by ViT, but instead attempts to replicate the label vector distribution estimated by a teacher network. This fully attentional approach, called DeiT, achieves state-of-the-art performance on ImageNet. Furthermore, the model was used in the Audio Spectrogram Transformer (AST) [6] designed for audio event tagging on the AudioSet[1] dataset [5].

Moreover, Gong et al. [7] showed the effectiveness of self-supervised pre-training with transformer architectures. This is important, since manual annotation of training data is a very time-consuming and hence expensive task. Self-supervised training of transformer networks is performed by masking several regions of an input image that the model attempts to reconstruct using the unmasked regions as context. In this way, the model learns the underlying distribution of the data.

2.2   Bird Species Recognition

In recent years, machine learning methods and especially CNNs brought great advances in the field of automated bird species recognition in soundscapes [31].

---

[1] https://research.google.com/audioset.

Most approaches rely on data that is weakly labeled by human experts or user communities, e.g., Xeno-canto. A widely used approach is BirdNET [17] that is trained on a large annotated dataset. The data is extensively pre-processed and augmented to enrich the training material. Following the advances of vision transformer architectures in computer vision tasks [4] and audio event classification [6], STFT [26] was introduced. This approach performs bird song recognition based on a ViT architecture, while reaching the recognition performance of CNNs.

Nevertheless, CNNs are still quite dominant in bird species recognition, as the annual bird call identification challenge (BirdCLEF [14–16]) confirms. The best approaches in recent years are based on ensembles of several CNNs in combination with extensive hyperparameter tuning.

This trend is also evident in the most recent editions of the annual BirdCLEF challenge. For example, the winning approach in 2021 uses a network architecture based on ResNet-50 [8] in combination with gradient boosting. The utilization of relevant metadata can further improve the final results [23]. Henkel et al. [10] proposed the usage of nine CNNs in an ensemble, operating on Mel spectograms. Furthermore, a binary classifier deciding on whether a snippet contains any bird vocalization contributes to the final result. Heavy data augmentation in the form of a novel mixup scheme for 30 s snippets was conducted to tackle the problem of weakly labeled data.

In the latest edition of BirdCLEF, many competitors used the approach by Henkel et al. as a foundation for their work. They extended it with various new post-processing steps, like class-wise thresholds [2, 19] or a penalization that reduces output probabilities proportional to the amount of training data [2]. Sampathkumar et al. [28] conducted extensive data augmentation experiments on their single model approach. They found that Gaussian noise, loudness normalization, and *tanh* distortion have the biggest impact on training an EfficientNet-B0 [32]. Due to great data imbalances, Miyaguchi et al. [21] developed an unsupervised feature extraction method based on Motif mining [30]. A Neural Architecture Search (NAS) method operating on raw audio data [22] won the BirdCLEF 2020 challenge [22]. Bird@Edge [12] is a solution to enable real-time monitoring of biodiversity in remote locations. This approach is based on a model for bird species recognition in audio data which has been optimized for the usage in embedded edge devices. A custom binary cross-entropy loss is used to train an EfficientNet-B3 [32] that was pre-trained on ImageNet. Autonomous recording units (ARU) are used gather huge amounts of unlabeled audio data. However, such data can not easily be used for training the mentioned machine learning approaches, since labeling bird sounds in audio files is a tedious task. Up to now, only little attention has been paid to unsupervised or weakly-supervised deep learning methods for bird species recognition. Ryan et al. [27] used a model pre-trained on bird songs for an industrial downstream task, and Michaud et al. [20] attempted to improve the quality of bird sound data sets through unsupervised classification with a novel data-centric labelling function. Furthermore, to get around the lack of labeled data, Cohen et al. [1] introduced TweetyNet, an approach to automatically label and segment large amounts of bird sound recordings. However, to the best of our knowledge, there is no competitive approach in current research that works with unlabeled data to pre-train a model for bird species recognition.

3   Method

In this section, we present our novel approach for building a bird species recognition model for audio recordings. For this purpose, we use a self-attention neural network architecture. Instead of performing a supervised pre-training on ImageNet, we leverage a self-supervised approach to pre-train the model on the vast amount of unlabeled and weakly labeled data to eventually improve the bird species recognition performance. Overall, our approach consists of two phases: (1) a self-supervised pre-training phase using unlabeled and weakly labeled data to initialize the model, and (2) a supervised fine-tuning phase to build the final bird species recognition model.

In the following, we describe the preprocessing steps (Sect. 3.1), including the generation of audio spectrograms as well as data augmentation, the self-attention neural network architecture, and the training workflow, divided into the self-supervised pre-training approach (Sect. 3.3) and the supervised fine-tuning step (Sect. 3.4), in more detail.

3.1   Pre-processing

First, we resampled the audio recordings to a common sampling rate of 32 kHz, which captures all possible frequencies of bird sounds, mostly in the range of 50 Hz to about 12 kHz.

During pre-training and fine-tuning, we randomly took a sample of duration x from each training recording, or padded it with zeros if they are short recordings. We used log-Mel filterbank features instead of other methods, e.g., MFCCs (Mel frequency cepstrum coefficients) or PLP (perceptual linear prediction) coefficients, to generate the spectrograms, since they proved to be useful in various (bird-related) audio recognition tasks [10,12,17]. However, there is an ongoing discussion in the literature on how audio signals should be processed by deep neural networks. For example, HEAR [34] is a competition based on a benchmark suite to evaluate audio representations across a variety of domains, including speech, environmental sound, and music. Essentially, there are three basic approaches: (a) feeding raw audio files (i.e., waveforms) directly into a deep neural network without any feature extraction [3], (b) applying hand-crafted feature extraction methods [29] (e.g., the log-scaled Mel spectrograms used by us), and (c) extracting filterbank features automatically by learning them from the raw audio data [25,38]. Furthermore, Stowell [31] discusses several issues of processing animal vocalizations and natural soundscapes with deep neural networks.

In our work, we applied a Hanning window with a size of 25 ms and a hop length of 10 ms to the audio sample resulting in a sequence of 128-dimensional

Table 1. Overview of DeiT models.

Model	Layers	Embedding dimension	Heads	Parameters
Deit-Base	12	768	12	86 M
Deit-Small	12	384	6	22 M
Deit-Tiny	12	192	3	5 M

log-Mel filterbank features. This transformation results in a spectrogram with a resolution of $W \times 128$, where W is the temporal resolution for a given snippet with duration x. Note that during supervised fine-tuning we performed data augmentation operations, such as time and frequency masking and noise augmentation, that were disabled during pre-training, since we assume a sufficient variation in our data sets used for pre-training.

3.2 Self-attention Neural Network Architecture

Our approach is based on a fully attentional neural network that relies on a self-attention mechanism, as proposed by Vaswani et al. [35]. The attention function is calculated simultaneously for a number of queries, which are summarized in a matrix Q. The keys and values are also summarized in the matrices K and V, respectively. The output of the attention function is calculated as follows:

$$Attention(Q, K, V) = Softmax(QK^\intercal \sqrt{d})V \qquad (1)$$

In self-attention, the query, key, and value matrices Q, K, and V, respectively, are computed from the given input sequence X of N vectors by using trainable linear transformation functions W_Q, W_K, and W_V:

$$Q = XW_Q, \quad K = XW_K, \quad V = XW_V \qquad (2)$$

Each layer of the model contains multiple self-attention heads h, where each head is applied to the input separately, generating a sequence of $N \times d$, with d as the embedding dimension. Finally, the h output sequences are re-ranged to $N \times dh$ and linearly projected to a $N \times d$ sequence. ViT architectures, such as DeiT [33], operate by dividing images into patches of a predetermined size. These patches are then linearly embedded and the position embeddings are added. The resulting sequence of vectors is subsequently passed to a self-attention model. Note that the position embeddings are a way to incorporate the order of the patches into the model, since the self-attention model itself does not have any sense of position or order of the patches. For classification, a learnable classification token is added to the sequence. In particular, we use the DeiT transformer architecture [33], evaluating the Base, Small and Tiny versions of the DeiT model in our experiments. The differences between these architectures are shown in Table 1.

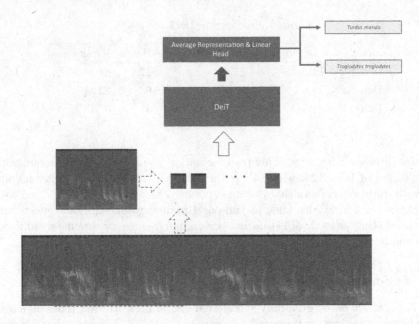

Fig. 2. Bird species recognition workflow. A 5 s snippet is converted to a Mel-spectroram, patchified, and passed to the trained model for prediction.

3.3   Self-supervised Pre-training

Our self-supervised training method is based on the joint discriminative and generative masked spectrogram patch modeling (MSPM) approach introduced by Gong et al. [7]. For a given patchified spectrogram with patch size 16 × 16, a set of patches is masked and replaced by a learnable mask embedding E_{mask}. The position embeddings are then added to the patch embeddings and fed into the DeiT model, as described in Sect. 3.2. For each masked patch x_i, the corresponding model output is O_i. Each O_i is then fed into a classification and a reconstruction head to obtain the outputs c_i and r_i of dimension 256, respectively, which equals the dimension of a patch of size 16 × 16. The model is expected to match the correct (x_i, c_i) pairs, which is enhanced by using the contrastive loss InfoNCE [9] L_d:

$$L_d = \frac{1}{M} \sum_{i=1}^{M} \log \left(\frac{\exp(c_i^T x_i)}{\sum_{j=1}^{M} \exp(c_i^T x_j)} \right), \tag{3}$$

where M is the number of masked patches. For the reconstruction head, the mean squared error (MSE) loss L_g is used:

$$L_g = \frac{1}{M} \sum_{i=1}^{M} (r_i - x_i)^2. \tag{4}$$

We trained our models based on 10 s snippets sampled as described in Sect. 3.1, resulting in a spectrogram with a resolution of 128×1024. Since we used a patch size of 16×16, we ended up with $(128/16) \times (1024/16) = 512$ patches for each 10 s spectrogram. During pre-training, we masked 400 patches for each spectrogram, which is 78.1% of the input spectrogram, yielding the best results [7].

3.4 Supervised Fine-Tuning

After the self-supervised pre-training phase, we trained the model in a supervised manner using the entropy loss with the weakly annotated Xeno-canto data for 82 species occurring in our region. To achieve this, we initialized the model with the best weights according to the scoring metric used during pre-training and added a classification header on the average vector representation generated by the model, as shown in Fig. 2. For fine-grained temporal predictions, we trained the model based on 5 s snippets, as described in Sect. 3.1.

4 Experiments

In this section, we show that self-supervised pre-training improves bird species recognition performance on different data sets compared to models that were pre-trained on ImageNet. First, we describe the data sets used during the pre-training phase and the details regarding the pre-training and fine-tuning phases. Then, we describe the quality metrics used in our experiments, as well as our evaluation results.

All our experiments were performed on a system with an AMD EPYC^{TM} 7702P 64-Core CPU, 256 GB RAM, and four NVIDIA® A100-PCIe-80GB GPUs. We used the PyTorch [24] deep learning framework to implement the proposed methods. For audio and signal processing, we used the Torchaudio library [37]. Our implementation is based on code provided by the SSAST repository released by Gong et al. [7].

4.1 Data Sets

To evaluate our approach, data sets from slightly different domains and recording environments were used, as described below. For self-supervised pre-training, we used three data sets summarized in Table 2. While the fine-tuning relies on audio recordings from the Xeno-canto website, three data sets from different domains were used for testing, as shown in Table 3. In the following, we describe the data sets in more detail.

MOF consists of soundscapes passively recorded with AudioMoth devices [11] during 2020 and 2021 in the Marburg Open Forest (MOF), an open research forest belonging to our university and located near Marburg, Germany. MOF is a densely vegetated mixed forest of roughly 2 km^2, with some clearings. The vast

Table 2. Datasets used in the pre-training phase.

Dataset	Training samples	Duration (h)
MOF	691,624	11,497
Xeno-canto	667,038	10,733
AudioSet	1,743,534	4,823

majority of these soundscapes is unlabeled. We used 691,624 files, with an overall duration of 11,497 h, in the pre-training phase. However, a small fraction of all files was labeled in terms of 5 s snippets by human experts with respect to 82 bird species. We used these 7,020 audio snippets for validation after fine-tuning our model.

Xeno-Canto is used to acquire further data, i.e., we crawled data from the bird sound collection Xeno-canto [36]. The collection includes bird species from all five major world regions (i.e., Africa, America, Asia, Australasia, Europe) and is organized as a citizen science project. Meanwhile, with more than 600,000 recordings of over 10,000 species, Xeno-canto is one of the largest and most used bird song databases. Many of the recordings in Xeno-canto were made using a directional microphone, which puts emphasize on foreground birds. It is also noteworthy that the soundscapes provided by Xeno-canto contain only weak labels. These labels indicate the species audible in a soundscape, but there are neither temporal clues nor complete annotations on species occurring in the background. We used 667,038 files or 10,733 h of audio from Xeno-canto in the self-supervised pre-training phase and 128,115 files in the fine-tuning phase.

AudioSet is used to increase data variation during the self-supervised learning phase. This is an extensive collection of more than 2 million 10 s sound clips from YouTube videos with 527 sound classes, including human sounds, animal sounds, music, natural sounds, and environmental sounds. For our pre-training, we used 1,743,534 sound files from AudioSet with an overall duration of more than 4,823 h.

FVA contains 2,357 audio files with a duration of 30 s each. Every file is completely annotated, i.e., there is a label for any bird that is audible in the corresponding file. All files were recorded in a forest in south-western Germany with AudioMoth devices [11] by a forest research institute, namely the Forstliche Forschungs- und Versuchsanstalt (FVA) Baden-Württemberg, Germany[2], and annotated by human experts. We used the FVA dataset only for evaluation.

---

[2] https://www.fva-bw.de.

Table 3. Datasets used for fine-tuning and testing.

Dataset	Training samples	Test samples
Xeno-canto	128,115	-
MOF	-	7,020
FVA	-	2,357
iNaturalist	-	1,365

iNaturalist is a social network for naturalists and a citizen science community. Therefore, the audio quality is rather diverse, however, many files contain clear recordings with a dominant species. Up to now, iNaturalist [13] offers over 15 million observations of animals, plants, and other organisms around the globe, which are supported by images or audio recordings. We crawled 1,365 audio files from iNaturalist, which we used exclusively for validating the fine-tuned models.

4.2 Quality Metrics

We used the following quality metrics in our evaluation.

Average Precision. We evaluated our recognition model based on average precision (AP). The precision-recall curve is summarized by AP as a weighted average of precision values attained at various thresholds. The weight is determined by the change in recall from the previous threshold. AP is computed as follows:

$$AP = \sum_n (R_n - R_{n-1})P_n,$$

where R_n and P_n are recall and precision at the n-th threshold. AP is calculated for each bird species separately. The mean AP (mAP) is obtained by taking the mean of the AP values from all species.

Weighted-Averaged F1-Score. The second quality metric used in our experiments is the weighted-averaged variant of the F1-score. It is calculated by taking the mean of the F1-scores of all classes while considering the number of each class's test samples. The F1-score is defined as:

$$F_1 = \frac{TP}{TP + \frac{1}{2}(FP + FN)},$$

where TP, FP, and FN are true-positives, false-positives, and false-negatives, respectively.

Table 4. Pre-training results on the test set.

Method	Masked patches	Masked patch ratio	Accuracy	MSE
DeiT-Base	400	0.78	**0.400**	**0.019**
DeiT-Small	400	0.78	0.357	0.020
DeiT-Base	200	0.39	0.399	**0.019**
DeiT-Small	200	0.39	0.356	0.020
DeiT-Tiny	200	0.78	0.178	0.032

4.3   Results

As mentioned above, our pre-training was carried out using the data sets presented in Table 2. We trained our models for 10 epochs. During training, we randomly masked 78.1% (39.05%) of the input patches, i.e., 400 (200) masked patches for an input duration of 10 s. Note that DeiT-Base and DeiT-Small were trained using 10 snippets, while DeiT-tiny is based on 5 s snippets. The results of the self-supervised pre-training are shown in Table 4. The validation split used for the pre-training corresponds to 10% of the Xeno-canto dataset. The results show that the

$$\text{patch accuracy} = \frac{\text{number of correctly assigned patches}}{\text{total number of masked patches}}$$

and the mean squared error (MSE, see Eq. (4)) are in favor of the DeiT-Base model and are therefore better in reconstructing the masked patches.

The bird species recognition model relies on 82 bird species that are relevant for our target domain. We carried out fine-tuning using the Xeno-canto data set with the corresponding audio recordings. The models were trained for 50 epochs using the preprocessing approach described in Sect. 3.1. We trained three models using different weight initializations, namely self-supervised models as a starting point compared to ImageNet and random initialization. The results of our experiments in terms of mean average precision (mAP) and weighted-averaged F1-score are shown in Table 5. Altogether, we evaluated our models using three different test sets from different recording environments and hence from different domains. Our DeiT-Base model based on self-supervised pre-training weights outperforms the DeiT-Base model fine-tuned from scratch by 41.7% in terms of mAP and by 33% in terms of weighted F1 score on the MOF dataset, illustrating the effectiveness of our pre-training approach. Moreover, the comparison with supervised pre-training is in favor of self-supervised training using unlabeled data from the target domain. Specifically, our DeiT-Base model outperforms the model fine-tuned based on ImageNet pre-training by 4.3% in terms of mAP on the MOF dataset, by 3.1% on the FVA dataset, and by 2.3% on the MOF+iNat dataset. Considering the domain differences between the Xeno-canto data set and the MOF/FVA data, these results show that without using labels from the target domain (MOF and FVA), we improved the cross-domain recognition performance via self-supervised pre-training.

Table 5. Fine-tuning results.

Method	Pre-training	Mean Average Precision			Weighted F1-score		
		MOF	FVA	MOF+iNat	MOF	FVA	MOF+iNat
DeiT-Base	Scratch	0.588	0.497	0.574	0.612	0.473	0.592
DeiT-Base	ImageNet	0.799	0.622	0.818	0.776	**0.606**	0.771
DeiT-Tiny	Ours	0.833	0.585	0.807	0.787	0.570	0.771
DeiT-Small	Ours	0.827	0.603	0.827	0.812	0.585	**0.812**
DeiT-Base	Ours	**0.833**	**0.641**	**0.837**	**0.814**	0.602	0.801

5 Conclusion

While self-supervised learning approaches have recently been successfully applied to computer vision and natural language processing tasks, state-of-the-art approaches for bird species recognition in audio recordings mainly rely on transfer learning using pre-trained ImageNet models. In this paper, we used self-supervised pre-training to improve bird species recognition in soundscapes by taking advantage of the vast amount of unlabeled and weakly labeled data. In particular, we used a self-supervised approach to pre-train a self-attention neural network architecture on the target domain. The experiments showed that using self-supervised pre-training improved the bird species recognition performance compared to an ImageNet initialization by up to 4.3% in terms of mean average precision and by up to 4.9% in terms of the weighted F1-score. In particular, the experiments revealed that we could improve the results on the MOF and FVA datasets without using labeled data from the target domain.

There are several areas for future work. First, a further subdivision of the bird species into different call types like song, flight call, or juvenile calls would give added value to the community. Second, since there are very few labeled training samples in this context, it would be interesting to investigate semi- or unsupervised approaches for learning the different call types of bird species in a hierarchical setting. Finally, another interesting field is active and/or continual learning via integration of user feedback and/or citizen science approaches.

Acknowledgments. This work is funded by the Hessian State Ministry for Higher Education, Research and the Arts (HMWK) (LOEWE Natur 4.0, LOEWE emergenCITY, and hessian.AI Connectom AI4Birds, AI4BirdsDemo), and the German Research Foundation (DFG, Project 210487104 - SFB 1053 MAKI).

References

1. Cohen, Y., Nicholson, D.A., Sanchioni, A., Mallaber, E.K., Skidanova, V., Gardner, T.J.: Automated annotation of birdsong with a neural network that segments spectrograms. eLife **11**, e63853 (2022). https://doi.org/10.7554/eLife.63853
2. Conde, M.V., Choi, U.: Few-shot long-tailed bird audio recognition. In: Proceedings of the Working Notes of CLEF 2022 - Conference and Labs of the Evaluation Forum, Bologna, Italy. CEUR Workshop Proceedings, vol. 3180, pp. 2036–2046. CEUR-WS.org (2022). http://ceur-ws.org/Vol-3180/paper-161.pdf
3. Dai, W., Dai, C., Qu, S., Li, J., Das, S.: Very deep convolutional neural networks for raw waveforms. In: 2017 IEEE International Conference on Acoustics, Speech and Signal Processing (ICASSP), pp. 421–425 (2017). https://doi.org/10.1109/ICASSP.2017.7952190
4. Dosovitskiy, A., et al.: An image is worth 16x16 words: transformers for image recognition at scale. In: 9th Int. Conference on Learning Representations (ICLR), Austria (2021). https://openreview.net/forum?id=YicbFdNTTy
5. Gemmeke, J.F., et al.: Audio set: an ontology and human-labeled dataset for audio events. In: IEEE International Conference on Acoustics, Speech and Signal Processing (ICASSP), pp. 776–780 (2017). https://doi.org/10.1109/ICASSP.2017.7952261
6. Gong, Y., Chung, Y., Glass, J.R.: AST: audio spectrogram transformer. In: Interspeech 2021, pp. 571–575 (2021). https://doi.org/10.21437/Interspeech. 2021–698
7. Gong, Y., Lai, C.I., Chung, Y.A., Glass, J.: Ssast: self-supervised audio spectrogram transformer. Proc. AAAI Conf. Artif. Intell. **36**(10), 10699–10709 (2022). https://doi.org/10.1609/aaai.v36i10.21315
8. He, K., Zhang, X., Ren, S., Sun, J.: Deep residual learning for image recognition. In: IEEE Conference on Computer Vision and Pattern Recognition (CVPR) (2016). https://doi.org/10.1109/CVPR.2016.90
9. Hénaff, O.J., et al.: Data-efficient image recognition with contrastive predictive coding. In: III, H.D., Singh, A. (eds.) Proceedings of the 37th International Conference on Machine Learning. Proceedings of Machine Learning Research, vol. 119, pp. 4182–4192. PMLR (13–18 Jul 2020). https://proceedings.mlr.press/v119/henaff20a.html
10. Henkel, C., Pfeiffer, P., Singer, P.: Recognizing bird species in diverse soundscapes under weak supervision. In: Working Notes of CLEF 2021 - Conference and Labs of the Evaluation Forum, Bucharest, Romania. CEUR Workshop Proceedings, vol. 2936, pp. 1579–1586. CEUR-WS.org (2021). http://ceur-ws.org/Vol-2936/paper-134.pdf
11. Hill, A.P., Prince, P., Snaddon, J.L., Doncaster, C.P., Rogers, A.: Audiomoth: a low-cost acoustic device for monitoring biodiversity and the environment. HardwareX **6**, e00073 (2019). https://doi.org/10.1016/j.ohx.2019.e00073
12. Höchst, J., et al.: Bird@Edge: bird species recognition at the edge. In: Koulali, M.-A., Mezini, M. (eds.) Networked Systems: 10th International Conference, NETYS 2022, Virtual Event, May 17–19, 2022, Proceedings, pp. 69–86. Springer, Cham (2022). https://doi.org/10.1007/978-3-031-17436-0_6
13. iNaturalist: A community for naturalists. https://www.inaturalist.org/
14. Kahl, S., et al.: Overview of BirdCLEF 2020: bird sound recognition in complex acoustic environments. In: Working Notes of CLEF 2020 - Conference and Labs of the Evaluation Forum, Thessaloniki, Greece. CEUR Workshop Proceedings, vol. 2696. CEUR-WS.org (2020). http://ceur-ws.org/Vol-2696/paper_262.pdf

15. Kahl, S., et al.: Overview of BirdCLEF 2021: bird call identification in soundscape recordings. In: Working Notes of CLEF - Conference and Labs of the Evaluation Forum, Bucharest, Romania. CEUR Workshop Proceedings, vol. 2936, pp. 1437–1450. CEUR-WS.org (2021). http://ceur-ws.org/Vol-2936/paper-123.pdf

16. Kahl, S., et al.: Overview of BirdCLEF 2022: endangered bird species recognition in soundscape recordings. In: Proceedings of the Working Notes of CLEF 2022 - Conference and Labs of the Evaluation Forum, Bologna, Italy. CEUR Workshop Proceedings, vol. 3180, pp. 1929–1939. CEUR-WS.org (2022). http://ceur-ws.org/Vol-3180/paper-154.pdf

17. Kahl, S., Wood, C.M., Eibl, M., Klinck, H.: Birdnet: a deep learning solution for avian diversity monitoring. Eco. Inform. **61**, 101236 (2021). https://doi.org/10.1016/j.ecoinf.2021.101236

18. Liu, Z., et al.: Swin transformer: hierarchical vision transformer using shifted windows. In: IEEE/CVF International Conference on Computer Vision (ICCV), Montreal, QC, Canada, pp. 9992–10002. IEEE (2021). https://doi.org/10.1109/ICCV48922.2021.00986

19. Martynov, E., Uematsu, Y.: Dealing with class imbalance in bird sound classification. In: Proceedings of the Working Notes of CLEF 2022 - Conference and Labs of the Evaluation Forum, Bologna, Italy. CEUR Workshop Proceedings, vol. 3180, pp. 2151–2158. CEUR-WS.org (2022). http://ceur-ws.org/Vol-3180/paper-170.pdf

20. Michaud, F., Sueur, J., Le Cesne, M., Haupert, S.: Unsupervised classification to improve the quality of a bird song recording dataset. Eco. Inform. **74**, 101952 (2023). https://doi.org/10.1016/j.ecoinf.2022.101952

21. Miyaguchi, A., Yu, J., Cheungvivatpant, B., Dudley, D., Swain, A.: Motif mining and unsupervised representation learning for birdclef 2022. In: Proceedings of the Working Notes of CLEF 2022 - Conference and Labs of the Evaluation Forum, Bologna, Italy. CEUR Workshop Proceedings, vol. 3180, pp. 2159–2167. CEUR-WS.org (2022). http://ceur-ws.org/Vol-3180/paper-171.pdf

22. Mühling, M., Franz, J., Korfhage, N., Freisleben, B.: Bird species recognition via neural architecture search. In: Working Notes of CLEF 2020 - Conference and Labs of the Evaluation Forum, Thessaloniki, Greece, September 22–25, 2020. CEUR Workshop Proceedings, vol. 2696. CEUR-WS.org (2020). http://ceur-ws.org/Vol-2696/paper_188.pdf

23. Murakami, N., Tanaka, H., Nishimori, M.: Birdcall identification using CNN and gradient boosting decision trees with weak and noisy supervision. In: Working Notes of CLEF 2021 - Conference and Labs of the Evaluation Forum, Bucharest, Romania. CEUR Workshop Proceedings, vol. 2936, pp. 1597–1608. CEUR-WS.org (2021). http://ceur-ws.org/Vol-2936/paper-136.pdf

24. Paszke, A., et al.: PyTorch: an imperative style, high-performance deep learning library. In: Advances in Neural Information Processing Systems 32, pp. 8024–8035. Curran Associates, Inc. (2019). http://papers.neurips.cc/paper/9015-pytorch-an-imperative-style-high-performance-deep-learning-library.pdf

25. Prashanth, H., Rao, M., Eledath, D., Ramasubramanian, V.: Trainable windows for sincnet architecture. EURASIP J. Audio Speech Music Process. **2023**(1) (2023). https://doi.org/10.1186/s13636-023-00271-0

26. Puget, J.F.: STFT transformers for bird song recognition. In: Working Notes of CLEF 2021 - Conference and Labs of the Evaluation Forum, Bucharest, Romania. CEUR Workshop Proceedings, vol. 2936. CEUR-WS.org (2021). http://ceur-ws.org/Vol-2936/paper-137.pdf

27. Ryan, P., Takafuji, S., Yang, C., Wilson, N., McBride, C.: Using self-supervised learning of birdsong for downstream industrial audio classification. In: ICML Workshop on Self-supervision in Audio and Speech (2020). https://openreview.net/forum?id=_P9LyJ5pMDb

28. Sampathkumar, A., Kowerko, D.: TUC media computing at BirdCLEF 2022: Strategies in identifying bird sounds in a complex acoustic environments. In: Proceedings of the Working Notes of CLEF 2022 - Conference and Labs of the Evaluation Forum, Bologna, Italy. CEUR Workshop Proceedings, vol. 3180, pp. 2189–2198. CEUR-WS.org (2022). http://ceur-ws.org/Vol-3180/paper-174.pdf

29. Sharma, G., Umapathy, K., Krishnan, S.: Trends in audio signal feature extraction methods. Appl. Acoust. **158**, 107020 (2020). https://doi.org/10.1016/j.apacoust.2019.107020, https://www.sciencedirect.com/science/article/pii/S0003682X19308795

30. Silva, D.F., Yeh, C.M., Zhu, Y., Batista, G.E.A.P.A., Keogh, E.J.: Fast similarity matrix profile for music analysis and exploration. IEEE Trans. Multim. **21**(1), 29–38 (2019). https://doi.org/10.1109/TMM.2018.2849563

31. Stowell, D.: Computational bioacoustics with deep learning: a review and roadmap. PeerJ **10**, e13152 (2022). https://doi.org/10.7717/peerj.13152

32. Tan, M., Le, Q.V.: Efficientnet: rethinking model scaling for convolutional neural networks. In: Proceedings of the 36th International Conference on Machine Learning, (ICML) Long Beach, California, USA. Proceedings of Machine Learning Research, vol. 97, pp. 6105–6114. PMLR (2019). 1905.11946

33. Touvron, H., Cord, M., Douze, M., Massa, F., Sablayrolles, A., Jégou, H.: Training data-efficient image transformers and distillation through attention. In: Proceedings of the 38th International Conference on Machine Learning (ICML). 139, pp. 10347–10357 2021. http://proceedings.mlr.press/v139/touvron21a.html

34. Turian, J., Schuller, B.W., Herremans, D., Kirchoff, K., Perera, P.G., Esling, P. (eds.): HEAR: Holistic Evaluation of Audio Representations (NeurIPS 2021 Competition), Proceedings of Machine Learning Research, vol. 166. PMLR (2022)

35. Vaswani, A., et al.: Attention is all you need. In: Advances in Neural Information Processing Systems, pp. 5998–6008 (2017). https://doi.org/10.5555/3295222.3295349

36. Xeno-canto: Sharing bird sounds from around the world. https://www.xeno-canto.org/

37. Yang, Y., et al.: Torchaudio: building blocks for audio and speech processing. In: IEEE International Conference on Acoustics, Speech and Signal Processing (ICASSP), Virtual and Singapore, pp. 6982–6986. IEEE (2022). https://doi.org/10.1109/ICASSP43922.2022.9747236

38. Zeghidour, N., Teboul, O., de Chaumont Quitry, F., Tagliasacchi, M.: LEAF: a learnable frontend for audio classification. In: International Conference on Learning Representations (2021). https://openreview.net/forum?id=jM76BCb6F9m

On the Different Concepts and Taxonomies of eXplainable Artificial Intelligence

Arwa Kochkach[1]([✉]), Saoussen Belhadj Kacem[2], Sabeur Elkosantini[2],
Seongkwan M. Lee[3], and Wonho Suh[4]

[1] Higher Institute of Management of Tunis, University of Tunis, Tunis, Tunisia
arwa.kochkache@gmail.com
[2] Faculty of Economics and Management of Nabeul, University of Carthage, Tunis,
Tunisia
saoussen.belhadjkacem@fsegn.u-carthage.tn,
Sabeur.elkosantini@fsegn.ucar.tn
[3] Colleage of Engineering, United Arab Emirates University, Al Ain,
United Arab Emirates
MarkLee@uaeu.ac.ae
[4] Hanyang University, ERICA Campus, Seoul, South Korea
wonhosuh@hanyang.ac.kr

Abstract. Presently, Artificial Intelligence (AI) has seen a significant shift in focus towards the design and development of interpretable or explainable intelligent systems. This shift was boosted by the fact that AI and especially the Machine Learning (ML) field models are, currently, more complex to understand due to the large amount of the treated data. However, the interchangeable misuse of XAI concepts mainly "interpretability" and "explainability" was a hindrance to the establishment of common grounds for them. Hence, given the importance of this domain, we present an overview on XAI, in this paper, in which we focus on clarifying its misused concepts. We also present the interpretability levels, some taxonomies of the literature on XAI techniques as well as some recent XAI applications.

Keywords: EXplainable Artificial Intelligence · Interpretability ·
Explainability · Post-hoc explanation techniques

1 Introduction

Recently, the sophistication and advancement of the Artificial Intelligence (AI)-powered systems has increased exponentially. Indeed, it reached a scope that "almost no human intervention is required for their design and deployment" [2]. However, although these models exhibit high performance, many of them are opaque in terms of explainability, i.e. they are not able to provide an explanation of their outputs. In this context, many Machine Learning (ML) models are considered as black boxes such as Artificial Neural Networks (ANN) or gradient boosting machines [22].

A. Bennour et al. (Eds.): ISPR 2023, CCIS 1941, pp. 75–85, 2024.
https://doi.org/10.1007/978-3-031-46338-9_6

It is significant to note that there exist numerous cases and situations where the explanation of the AI application is not necessary or needed at all. This is especially true when the problem under study and treatment is well represented, well-known, complete and its consequences are not critical (e.g. movie recommendation or mail sorting) [8]. However, explanations are crucial for the user to comprehend and trust the decisions yielded from a system for many other critical cases [14]. This is mainly true when these latter have an effect on humans' lives (e.g. healthcare, medicine, defense or law). Moreover, as declared by Zhu et al. [30], "humans are reticent to adopt techniques that are not directly interpretable, tractable and trustworthy". Hence, most of the community members stand, nowadays, in front of the barrier of "explainability". Paradigms underlying this issue fall within the so-called eXplainable Artificial Intelligence (XAI) field. Gunning [13] states that XAI "will create a suite of machine learning techniques that enables human users to understand, appropriately trust, and effectively manage the emerging generation of artificially intelligent partners". Nevertheless, although it may not necessarily occur in many datasets [27], there is a widespread belief that there exists a trade-off between the interpretability of a model and its performance (e.g. predictive accuracy) [9]. For this reason, XAI's goal is the creation of more interpretable ML systems while preserving their high level of learning performance in order to avert the limitation and the sacrifice of the effectiveness of the current AI-powered systems [2].

Within the field of XAI, there is a lot of concepts which need to be well identifed and understood in order to make correct and flawless insights. Indeed, many of them are interchangeably misused in the literature. The most notable ones are "interpretability" and "explainability". This issue, mainly, obstructed the foundation of common grounds for XAI's nomenclatures. For this reason, it is crucial to make a clear distinction between them. Therefore, this will be the goal of this paper. In fact, our aim is to present a basic overview and to summarize the main nomenclature frequently utilized in XAI community. We also clarify the distinctions and similarities among them.

The remaining of this article is structured according to the following steps. Section two is dedicated for the definition of some XAI-related concepts as well as giving a clear line to distinguish between them according to our proposed criteria. Then, the levels of interpretability of a model are presented in Sect. 3. Section 4 will contain our proposed taxonomy to categorize XAI techniques within the ML field based on different ones from the litterature. Lastly, some of the application domains of XAI will be shown in Sect. 5. And then, we end this paper with a conclusion.

2   On the Concepts of eXplainable Artificial Intelligence

The term "explainability", being relevant in the literature, gave rise to the direction of XAI [29]. However, there is some ambiguity in this field with regard to terminology [6]. This was well outlined by Lipton [20] who notes that also "the term interpretability holds no agreed upon meaning, and yet machine learning conferences frequently publish papers which wield the term in a quasi-mathematical way".

Indeed, in the literature, there were several attempts to define the concepts of "explainability" and "interpretability". Despite that, generally speaking, there is no consensus within the ML community on their definitions [8,20,22]. Hence, many researchers consider them as the same [6,7] and they usually use them interchangeably in their works "in the broad general sense of understandability in human terms" [12]. According to our researches, the most popular definition, considered for both of them, adopted for numerous authors [2,6,9,12,19,22] is that of Doshi-Velez and Kim [8] stating that they represent "the ability to explain or to present in understandable terms to a human" . Besides, we note that there exist other terms, in the literature, such as "intelligibility" [5,21], "understandability" [20] and comprehensibility [9,11] that are considered synonymous to interpretability and they are often used interchangeably also [22].

However, many other authors argue that there are important reasons to distinguish between the concepts of "explainability" and "interpretability" [12,22,27]. Hence, they try to identify their differences in their works. Indeed, there is no concrete mathematical definition for both of the concepts i.e. they have not been measured by metrics [19]. Nevertheless, many tries have been made in order to make them clear. For example, authors in [12] assume that "explainable models are interpretable by default but the reverse is not always true". Moreover, in [4], authors discern that "interpretability is usually used in terms of comprehending how the prediction model works as a whole" while explainability "is often used when explanations are given by prediction models that are incomprehensible themselves". We believe that the most relevant distinction is made by Rudin [27] who marks a clear line between explainable and interpretable techniques. He states that the latter ones "focuse on designing models that are inherently interpretable"; whilst the former ones "try to provide post hoc explanations for existing black box models". For the sake of convenience, in our paper, we adhere this dissimilarity. We believe that this is the most pertinent one with regards to the field of XAI. Hence, we present these concepts more thoroughly in this section according to this belief.

2.1   Interpretability

Interpretability is defined by Miller [23] as "the degree to which a human can understand the cause of a decision". The property of the interpretability of a ML model stems from the model itself, i.e. it is intrinsic and built-in. In other words, this concept regards a model that is clear by design and does not need to be explained by another technique. Kim et al. [17] describe interpretability, in a more accurate way, as "the degree to which a human can consistently predict the model's result". Based on the above, interpretability is then chiefly attached to the intuition behind a model's outcome [1]. Thus, it is higher if it is easier for a human user to identify the causes and the effects the inputs have on the outputs of the model, i.e it is easier to trace back why a given prediction was output [6,19].

We note that, although being intuitive, the aforementioned definitions clearly lack mathematical rigour and formality i.e. there is no mathematical definition of interpretability [1,6,20]. In fact, none of them is specific or restrictive enough to enable formalization. Besides, many authors assume that "interpretability is a very subjective concept" hence it is not that easy to state it formally [28].

An interpretable model is a ML model that is interpretable by design. It has inherent/ in-model interpretability in it i.e. the interpretability process takes place during the building of the ML model. It is also called an "interpretable by nature" [4], intrinsically interpretable [6], transparent [2], or white-box model [19]. This can be achieved if the model is simple enough by nature or by imposing some relevant constraints on the complexity of the ML model to be developed. These constraints can be causality, monotonicity, sparsity, additivity, and/or other desirable properties [6]. The model can be structured to reflect some physical constraints coming from the domain knowledge too [27]. This can also be done via the hybridization or the mapping of the black-box system with a more interpretable ML model (a white-box twin) [2]. For example, in [16], a Deep Neural Network and a Case Based Reasnoning model (a k-Nearest Neighbors) are paired with a view to enhance the interpretability of the former while maintaining its high level of accuracy. Neural networks can also be explained by mixing them with fuzzy systems giving rise to the famous interpretable Neuro-Fuzzy Systems [26]. There exists also the process of the "deep formulation" of the classical ML models [2]. A prominent example of the latter model improved by its Deep Learning (DL) counterpart is Deep KNN (DkNN) [25].

2.2 Explainability

The issue of explainability was not present in the last miscellany of AI techniques (namely rule based models and expert systems) and it came to light recently. This problem arises and explainability becomes required when there is some degree of incompleteness in a problem formalization making direct optimisation and validation impossible [4,8,22] in some concrete situations such as Scientific Understanding, Safety and Ethics [22]. Incompleteness denotes that "there is something about the problem that cannot be sufficiently encoded into the model". It is different from uncertainty which hints at "something that can be formalized and handled by mathematical models" [8]. In fact, for some incomplete problems or prediction tasks, the resulting output representing the prediction solely (the "what") it is not enough. Hence, the model should also explain and makes clear how it came to this prediction (the "why") [6].

Explainability establishes an interface of interaction between humans and a decision-making model (the ML model). Hence, it has to be an accurate proxy of this latter while being well comprehensible to humans. However, each users group may vary in their background knowledge and may have a preferred explanation type that is able to communicate information in the most effective way [14]. For this reason, the concept of "audience" is being the cornerstone of XAI [2]. Hence, authors in [2] argue that its definition must be rephrased to reflect this dependence explicitly as follows: "Given a certain audience, explainability refers to the details and reasons a model gives to make its functioning clear or easy to understand" [2]. Put it differently, explainability is "the ability a model has to make its functioning clearer to an audience" [2]. Indeed, authors in [19] point out that it concerns the internal logic and procedures which are executed while training the model and making decisions. Hence, its goal is to describe them in a way that is comprehensible to humans [12]. A more relevant definition is given by Rudin [27] which assumes that "explainable ML focuses on providing explanations for existing black box (opaque) models by means of training another surrogate model post hoc". These post hoc explainability techniques (also called "explanation methods" or "post-modeling explainability") are specially devised to explain/ improve the interpretability of an existing opaque ML model, which do not meet any criterion allowing to deem it interpretable, after its building and training by analyzing it [6,24]. For this reason, they have to be able to summarize the reasons for its behavior, produce insights and communicate understandable information about the causes of their decisions (predictions) and gain the trust of users [12]. Many authors argue that this notion frequently depends on the domain of application [11,27,28], the user [6] (his abilities, expertise, preferences and expectations [9,14]), the context that depends on the task and other contextual variables [14]. Therefore, an all-purpose definition might be unnecessary [22] or infeasible [27].

2.3   Explainable vs Interpretable Machine Learning

As can be seen in the previous subsection, there are notable differences among XAI concepts, mainly Interpretability and Explainability. In table 1, we propose some criteria that help to distinguish between them with the intent to draw a clear line discerning between the two of them.

W ithin the XAI domain, there exist several other concepts which often appear in the literature such as:

- Transparency, which can be considered as the obverse of "black-box-ness", is "a feature that a model can feature by itself" [2]. It can also be defined as "the search for a direct understanding of the mechanism by which a model works" [20]. Authors in [9] state that a transparent model is both interpretable and explainable. Others use it interchangeably with interpretable models [2]. In this review we adopt that transparency refers to the highest level of interpretability i.e. simulatability.
- Understandability which is also ofen used interchangeably with interpretability and intelligibility [5,20–22]. Authors in [2] state that understandability is a two-sided matter: model understandability vs. human understandability. Model understandability is" the characteristic of a model to make a human user understand its functioning without any need for explaining its internal structure". However, "human understandability measures the degree to which a human user can understand a decision made by a ML model" i.e. relies on the capability of the specific audience.

2.4 Explanation

As we noted previously, the concept of explanation is relevant in the field of XAI since explainability is clearly associated with it. Miller [23] presents a simple goal-oriented definition of an explanation stating that "it is the answer to a "Why" question". Within the XAI community, an explanation is defined as "the means by which a ML model's decisions are explained" [6] in a humanly understandable fashion. Taking the abovementioned definitions into consideration, an explanation is deemed to be pertinent if it rejoin to the needs and goals of the user[4] (providing as much information as possible while being as short as possible), allows a tradeoff between explainability and completeness (descriptions with a high level of details) [12] and convincing to the user (he accepts it) [4].

There exists a wide range of post-hoc explanation techniques in the literature. Among them we can mention Feature Relevance Explanation techniques which measure the influence or importance which every managed input feature (variable) has on the output coming out of the opaque ML model. Another example can be the Local Explanations which process first by segmenting the solution space then generating explanations for some less complex relevant subspaces. Explanation by Simplification functions by rebuilding a whole simple (less complex) new system to explain the original one while preserving a similar performance score [2]. Then we can cite Visual Explanation techniques which

Table 1. Interpretable vs. Explainable Machine Learning.

Criterion	Interpretable	Explainable
Characteristic of the model	Passive characteristic: "the level at which a given model makes sense for a human observer" [2].	Active characteristic: an external action taken by a model in order to clarify its internal functioning [2].
Means of explanation	Models are interpretable by design [2].	Models are explained via external XAI techniques [2].
Level of the explanation	Intrinsic/ Inherently interpretable: explanations exist in the model itself [2, 22, 27].	Post hoc: the explanation is given by another model "after" the training of the existing black-box one [27].
Which question it raises?	"How does the model work?" [20]	"What else can the model tell me?" [20]
Extent of the explanation	Is not necessarily a model that human users are able to comprehend its internal logic and processes [19].	Deal with the internal logic and procedures of a black-box system [19].
Focus	Model-centric [10]	Subject-centric [10]
Mode of explanation	Users can mathematically analyze the mappings [7]	Models should output symbols, rules or figures with their prediction to help the user to understand the rational behind the input-output mappings being made [7].
Need for the explanation	Used to comprehend how the model works superficially and as a whole [4].	Used when a model is incomprehensible itself and its formalization is incomplete [4].

goal is the visualizing the ML model's comportment, Explanations by Example which "extractes representative examples that grasp the inner relationships and correlations found by the model" and Text Explanations [3].

3 Levels of Interpretability

An AI model can feature different levels of Interpretability [2]. Hence, within Interpretable models, three levels are contemplated with regards to the scope of interpretability: the portion of the prediction process targeted for explanation [6]. This gives rise to three different categories of AI models: simulatable, decomposable and algorithmically transparent models [20]. Authors in [2] introduce thoroughly these three classes as follows:

1. *Algorithmic transparency* is the lowest level of interpretability. "It deals with the ability of the user to understand the process followed by the model to produce any given output from its input data" [2]. It also refers to the way works the algorithm which generates or learns the ML model from the data [24]. Thus, it only requires acquaintance with the model's algorithm and it does not concern the managed data. Thus, for a model to be algorithmically transparent, it has to be "fully explorable by means of mathematical methods and analysis" [2].
2. *Decomposability* is the second level of interpretability (on a modular level). It refers to the ability to give an explanation about each part of a ML model such as inputs or parameters. Hence, it responds to the question "how do parts of the model affect predictions" [24]. However, this property does not feature for every ML model since it requires every input used to train it to be readily interpretable which does not apply for cumbersome features for example.
3. *Simulatability* (Global Model Interpretability [6], transparency) is the highest level of interpretability. It denotes the capability a model have to be wholly and strictly simulated by they user. Moreover, Lipton [20] defines it as "the ability to comprehend the entire model at once". To do so, the trained model and the knowledge of the algorithm (each of the learned components such as parameters, weights, etc) as well as a holistic view of the data features are needed [6]. Therefore, it is very difficult to attain this level in practice [24]. Honegger [15] acknowledges that, for a model to fulfill this criteria, it must be simple enough.

4 Taxonomies of Interpretable and Explainable Machine Learning

ML interpretability and explainability techniques can be classified according to different criteria. In the following, we review some of them.

4.1   Model-Specific vs. Model-Agnostic Techniques for Post-hoc Explainability

This is an important criterion distinguishing between two different algorithmic approaches as follows.

1. *Model-agnostic methods* are techniques which are designed to be linked seamlessly to any ML model (black box or not) with the intention of extracting knowledge about the decision-making procedure [2]. Hence, they count just on analyzing pairs of input and output [6,9] and they cannot have access to the internal representations or the inner working process of the black-box ML model such as weights or structural information [2,24].
2. *Model-specific explanation methods* are tailored or specifically designed for a specific ML model or one class of them [22] since they are based on some particular model's internals [24] and it uses idiosyncrasies of its representation [9]. Hence, they cannot be directly plugged to any other model [2].

4.2   Post-hoc Explainability Techniques for Shallow vs. Deep Machine Learning

Authors in [2] propose a taxonomy of the post-hoc explainability techniques which divide the literature into two main categories: techniques devised for shallow ML models and others devised for DL models. In the following, both of them are presented.

1. *Post-hoc explainability in shallow ML models:* Shallow models collectively refer to all ML models which structure is not layered following neural processing units i.e. they have a relatively straightforward structure. Within these models, there are the strictly interpretable (transparent) ones (e.g. Decision Tree or K-NN) and others that have more sophisticated or complex learning algorithms. Hence, the latter ones require additional post-hoc explanation (e.g. Support Vector Machines or Tree Ensembles)[2].
2. *Post-hoc explainability in deep learning:* The most common DL models are multi-layer neural networks (MLNN), Recurrent Neural Networks (RNN) and Convolutional Neural Networks (CNN). They are efficient and widely used thanks to their great capability of deducing such complex connections among different and numerous variables. However, their structures are extremely complex and difficult to understand which makes them always considered as black-box models. Hence, they require explainability techniques to make their functioning clear for the user [2].

Figure 1 provides one possible proposed taxonomy of the explainable and interpetable models that we exposed in this section. It is noteworthy that this taxonomy is not a disjoint partition of the suite of all the techniques. In fact, almost all the discussed criteria in this sections are related to each other in some way. In fact, the interpretability of interpretable models lies in the core of the model (intrinsic) and is present while training it (In-model) hence it is always model specific. This interpretability can be global touching the whole model or local in some of its parts. In deep models, the structure is always complex and hard to understand. Hence, they always need post-hoc techniques to explain them which are resorted to after the training of the model (Post-model). These latter techniques can be either model agnostic or model specific. We note that, although there are few model-specific techniques that are deployed post-hoc, most of them are achieved through intrinsically interpretable models. In an analogous way, most of post-hoc methods are separated from the models and, hence, they are model-agnostic [6].

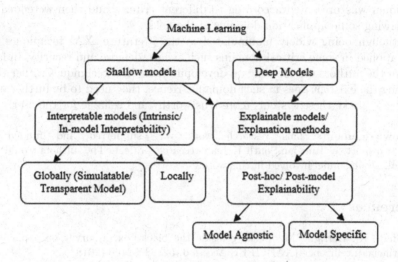

Fig. 1. An overall taxonomy of Interpretable and Explainable ML techniques.

5 Application Domains of eXplainable Artificial Intelligence

At present, XAI is becoming paramount and needed in many application domains. Mainly, when the result (prediction, assessment, decision or classification) of the ML model affects human's life, its explanation to the user becomes crucial to enhance the faithfulness of this model. Indeed, it is always needed in medicine and healthcare. For example, it was applied for the predicting of pneumonia risk [5] and surgical effort in advanced stage epithelial ovarian cancer [18]. Moreover, XAI was also employed in military simulations and computer games. For example, in [29], it was used to provide explanations utilized by a training framework specially designed for the U.S. Army for small-unit tactical behavior. Besides, XAI can be used in some non-critical domains like entertainment. For example, in [30], it was used for games designers to help them better employ AI and ML in their design tasks through co-creation.

6 Conclusion

Our paper has overviewed the field of XAI which is being more and more identified as an essential requirement, in some real-life applications, within the community of AI. Our research study has elaborated on this theme by, first, shedding light on some XAI-related concepts and by clarifying their interchangeable misuse. This is mainly done for interpretable models which are interpretable by design and explainable models which need an external XAI technique to explain them after their training. Secondly, we presented the three levels of interpretability. Thereafter, a proposed overall taxonomy of recent literature dealing with XAI techniques was presented according to different criteria and then we ended by overviewing some application domains of the field.

Although being widely manipulated in the literature, XAI techniques are fewly applied in some critical domains such as law, defense and security. In fact, most of the authors are focusing on developing new XAI techniques rather then applying the existent ones in such domains. Hence, they need to be further used in order to make these systems more trustworthy and reliable for the user.

Acknowledgements. This research work was supported and funded by Data4Transport, a Tunisian-South Korean research project. The authors would like to thank all personnel involved in this work.

References

1. Adadi, A., Berrada, M.: Peeking inside the black-box: a survey on explainable artificial intelligence (XAI). IEEE Access **6**, 52138–52160 (2018)
2. Arrieta, A.B., et al.: Explainable artificial intelligence (XAI): concepts, taxonomies, opportunities and challenges toward responsible AI. Inf. Fusion **58**, 82–115 (2020)
3. Bennetot, A., Laurent, J.L., Chatila, R., Díaz-Rodríguez, N.: Towards explainable neural-symbolic visual reasoning. arXiv preprint arXiv:1909.09065 (2019)
4. Burkart, N., Huber, M.F.: A survey on the explainability of supervised machine learning. J. Artif. Intell. Res. **70**, 245–317 (2021)
5. Caruana, R., Lou, Y., Gehrke, J., Koch, P., Sturm, M., Elhadad, N.: Intelligible models for healthcare: predicting pneumonia risk and hospital 30-day readmission. In: Proceedings of the 21th ACM SIGKDD International Conference on Knowledge Discovery and Data Mining, pp. 1721–1730 (2015)
6. Carvalho, D.V., Pereira, E.M., Cardoso, J.S.: Machine learning interpretability: a survey on methods and metrics. Electronics **8**(8), 832 (2019)
7. Doran, D., Schulz, S., Besold, T.R.: What does explainable AI really mean? A new conceptualization of perspectives. arXiv preprint arXiv:1710.00794 (2017)
8. Doshi-Velez, F., Kim, B.: Towards a rigorous science of interpretable machine learning. arXiv preprint arXiv:1702.08608 (2017)
9. Došilović, F.K., Brčić, M., Hlupić, N.: Explainable artificial intelligence: a survey. In: 2018 41st International Convention on Information and Communication Technology, Electronics and Microelectronics (MIPRO), pp. 0210–0215. IEEE (2018)
10. Edwards, L., Veale, M.: Slave to the algorithm: why a right to an explanation is probably not the remedy you are looking for. Duke L. Tech. Rev. **16**, 18 (2017)

11. Freitas, A.A.: Comprehensible classification models: a position paper. ACM SIGKDD Explor. Newsl. **15**(1), 1–10 (2014)
12. Gilpin, L.H., Bau, D., Yuan, B.Z., Bajwa, A., Specter, M., Kagal, L.: Explaining explanations: an overview of interpretability of machine learning. In: 2018 IEEE 5th International Conference on Data Science and Advanced Analytics (DSAA), pp. 80–89. IEEE (2018)
13. Gunning, D.: Explainable artificial intelligence (XAI). Defense Advanced Research Projects Agency (DARPA), nd Web **2**(2), 1 (2017)
14. Gunning, D., Stefik, M., Choi, J., Miller, T., Stumpf, S., Yang, G.Z.: XAI-Explainable artificial intelligence. Sci. Robotics **4**(37), eaay7120 (2019)
15. Honegger, M.: Shedding light on black box machine learning algorithms: development of an axiomatic framework to assess the quality of methods that explain individual predictions. arXiv preprint arXiv:1808.05054 (2018)
16. Keane, M.T., Kenny, E.M.: The twin-system approach as one generic solution for XAI: an overview of ANN-CBR twins for explaining deep learning. arXiv preprint arXiv:1905.08069 (2019)
17. Kim, B., Khanna, R., Koyejo, O.O.: Examples are not enough, learn to criticize! criticism for interpretability. In: Advances in Neural Information Processing Systems 29 (2016)
18. Laios, A., et al.: Factors predicting surgical effort using explainable artificial intelligence in advanced stage epithelial ovarian cancer. Cancers **14**(14), 3447 (2022)
19. Linardatos, P., Papastefanopoulos, V., Kotsiantis, S.: Explainable AI: a review of machine learning interpretability methods. Entropy **23**(1), 18 (2020)
20. Lipton, Z.C.: The mythos of model interpretability: in machine learning, the concept of interpretability is both important and slippery. Queue **16**(3), 31–57 (2018)
21. Lou, Y., Caruana, R., Gehrke, J.: Intelligible models for classification and regression. In: Proceedings of the 18th ACM SIGKDD International Conference on Knowledge Discovery and Data Mining, pp. 150–158 (2012)
22. Marcinkevičs, R., Vogt, J.E.: Interpretability and explainability: a machine learning zoo mini-tour. arXiv preprint arXiv:2012.01805 (2020)
23. Miller, T.: Explanation in artificial intelligence: insights from the social sciences. Artif. Intell. **267**, 1–38 (2019)
24. Molnar, C.: Interpretable machine learning. Lulu.com (2020)
25. Papernot, N., McDaniel, P.: Deep k-nearest neighbors: towards confident, interpretable and robust deep learning. arXiv preprint arXiv:1803.04765 (2018)
26. Rajurkar, S., Verma, N.K.: Developing deep fuzzy network with takagi sugeno fuzzy inference system. In: 2017 IEEE International Conference on Fuzzy Systems (FUZZ-IEEE), pp. 1–6. IEEE (2017)
27. Rudin, C.: Stop explaining black box machine learning models for high stakes decisions and use interpretable models instead. Nat. Mach. Intell. **1**(5), 206–215 (2019)
28. Rüping, S., et al.: Learning interpretable models (2006)
29. Van Lent, M., Fisher, W., Mancuso, M.: An explainable artificial intelligence system for small-unit tactical behavior. In: Proceedings of the National Conference on Artificial Intelligence, pp. 900–907. Menlo Park, CA; Cambridge, MA; London; AAAI Press; MIT Press; 1999 (2004)
30. Zhu, J., Liapis, A., Risi, S., Bidarra, R., Youngblood, G.M.: Explainable AI for designers: a human-centered perspective on mixed-initiative co-creation. In: 2018 IEEE Conference on Computational Intelligence and Games (CIG), pp. 1–8. IEEE (2018)

Classifying Alzheimer Disease Using Resting State Coefficient of Variance BOLD Signals

Youssef Hosni[1]([⊠]) and Ahmed Elabasy[1,2][iD]

[1] Center for Machine Vision and Signal Analysis, University of Oulu, 90014 Oulu, Finland
Youssef.Honsi95@outlook.com,
https://www.oulu.fi/en/research-groups/oulu-functional-neuroimaging-ofni
[2] Oulu Functional NeuroImaging, Diagnostic Radiology, Medical Research Center, Oulu University Hospital, 90029 Oulu, Finland
https://www.oulu.fi/en/university/faculties-and-units/faculty-information-technology-and-electrical-engineering/center-machine-vision-and-signal-analysis

Abstract. There is a strong relationship between neurodegenerative disease and altering brain clearance. Brain clearance was shown to be driven by three physiological brain signals. In this paper, we show the importance of abnormality detection of physiological pulsations which is effective in detecting neurodegenerative diseases. We use the coefficient of variance of rs-fMRI BOLD signal (CV_{BOLD}) as the physiological signal distribution that reflects brain activity. In this work, we found relations between the CV_{BOLD} values in specific brain regions of Alzheimer's subjects. We provided a comparison of different models that were used to handle this challenge and the brain regions that are most affected.

Keywords: Machine Learning · Deep Learning · Alzheimer's disease · Coefficient of variance · Brain connectivity · Feature Selection

1 Introduction

The human body's intricate functions are all under the control of the brain. Structurally, the brain exhibits diverse regional specializations, each of which caters to the processing of distinct neural signals. Additional specialization occurs in the brain to facilitate the implementation of perceptual and cognitive functions. Neuronal units, which constitute the fundamental building blocks of the brain, do not store internal energy sources like glucose or blood. Instead, when these neurons are activated, they receive an increased energy supply through neighboring capillaries via a physiological mechanism termed the hemodynamic response. This process involves the delivery of cerebral blood and oxygen to the activated neurons, thereby sustaining their heightened metabolic demands.

A. Bennour et al. (Eds.): ISPR 2023, CCIS 1941, pp. 86–100, 2024.
https://doi.org/10.1007/978-3-031-46338-9_7

This changes the levels of oxyhemoglobin and deoxyhemoglobin and this can be detected using functional magnetic resonance imaging, due to the different magnetic properties of both of them which is known as Blood Oxygenated Level Dependent (BOLD) contrast imaging. The changing in the BOLD signal is the main block of fMRI imaging, as it is used to construct a map of the activated brain region by activity or stimulus at low frequency (0.1–0.01 Hz) [11]. The BOLD signal was found also to be a proxy indicator for the neural activity and it reflects the vascular, respiratory, and physiological activity [11].

Resting-state fMRI (rs-fMRI) is acquired without a baseline condition. The principle of the rs-fMRI depends on BOLD signal fluctuations. The rs-fMRI is an imaging technique used to characterize standard and aberrant functional connectivity in various clinical conditions. The data for rs-fMRI can be collected through targeted scans, where participants are directed to engage in a state of rest or the periods of rest are inferred from within a series of tasks. The analysis of rs-fMRI data involves various approaches and techniques, which are continuously expanding. The complementary nature of these methods suggests that applying multiple methods to the same data set could result in better outcomes compared to using a single method. Understanding each processing method is essential to interpret the divergent or common findings reported in the literature on rs-fMRI [7]. The physiological pulsations in the brain have been shown to play an important role in the homeostasis of the brain, as brain clearance was shown to be driven by these physiological signals, and it was found that there is a strong relation between neurodegenerative diseases and altering in the brain clearance. The pathological process may also modify the physiological noise characteristics or variability of the BOLD signal similar to its effects on low-frequency connectivity [11].

Recently, the coefficient of variant of the BOLD signal (CV_{BOLD}) has been employed to examine alterations in noise properties of the BOLD data, as it was found that based on some studies on the characteristics of the abnormal noise of the BOLD signal, the structure of noise of the BOLD signal measured using CV_{BOLD} is altered in AD patients in a distinct approach [11]. Based on these findings these characteristics can be used to train machine learning models to distinguish between healthy and AD subjects, so the machine can do this task after that with high accuracy without the need for experts and to detect the disease at an early stage. In this paper CV_{BOLD} data was used to classify the data set given into both AD and healthy subjects and to know the voxels of the brain which the classifiers rely on to make its decision, which will be the region of the brain that was most affected with the disease.

The paper consists of 3 sections, first, we explain the materials and methods used to first prepare the data second then classify it using different machine learning models and feature engineering methods. In the subsequent segments, we demonstrate the results we got from these different methods, and the third section is the discussion section in which we discuss the results we got and what we have achieved, and the future work that could be done.

2   Materials and Methods

2.1   Datasets

Two different datasets were used in the practical work of this paper; the first one was provided by the local Institute (Oulu Hospital) and this was used to train the models used to classify the healthy and AD subjects. The second one is the Alzheimer's Disease Neuroimaging Initiative (ADNI) database (adni.loni.usc.edu) which is used was used as a test data set, to evaluate the performance of the trained models.

Local Institute Data. In this study, a total of 41 individuals were recruited, comprising 17 patients diagnosed with Alzheimer's disease (AD) and 24 healthy control subjects. The AD patients underwent comprehensive assessments, including neurological and neuropsychological evaluations, routine laboratory tests, and brain magnetic resonance imaging (MRI) at the Memory Outpatient Clinic of the Department of Neurology, Oulu University Hospital. Every individual diagnosed with Alzheimer's disease fulfilled the probable AD criteria established by the National Institute of Neurological and Communicative Disorders and Stroke, as well as the Alzheimer's Disease and Related Disorders Association. In addition, cerebrospinal fluid (CSF) AD biomarkers were analyzed in this study. Table 1 provides a summary of the attributes of these datasets.

Table 1. The study groups' descriptive attributes within the local institute's dataset.

Local institute data	Control subjects	AD patients
Participants	24	17
Age (years)	7	78
Female	12(50%)	11(65%)
Disease(yrs)	n.a	2.6 ± 1.3
MMSE	29 ± 1.1	23 ± 2.6

Where MMSE is the Mini-Mental State Examination (maximum total score is 30) and n.a. means not applicable. The age and disease duration are expressed as mean $\pm$ SD.

ADNI Data Set. A standardized protocol was employed to collect magnetic resonance imaging (MRI) data using Philips 3 T MRI systems (Philips, Amsterdam, The Netherlands) across a total of 13 sites (http://adni.loni.usc.edu). Table 2 summarizes the characteristic of this data set.

The data sets used were the coefficient of the variant of these BOLD signals, where the

$$CV = (\sigma(x)/\mu(x)) \tag{1}$$

Table 2. The study groups' descriptive attributes within the ADNI dataset.

Local institute data	Control subjects	AD patients
Participants	40	30
Age(yrs)	72.7 ± 4.3	71.8 ± 6.5
Female	25(62%)	15(50%)
Disease(yrs)	–	2.6 ± 1.3
MMSE	29 ± 1.1	21.9 ± 3.6

where x is the voxel time series, σ is the standard deviation for one voxel along the time axis and the μ is the mean of each voxel along the time axis. So the data are given in the form of a CV for each voxel along the time dimension, which means that the data has no longer information along the time axis. Although using only the CV of the data makes the data less informative, it makes it much easier to deal with and less complex and at the same time needs less storage and less processing power, and this is possible to try different methods in a short time and without the need to high computational process or large storage facilities.

2.2 Prepossessing

Data Cleaning. In this study, BOLD rs-fMRI data were preprocessed using a standard pipeline from the FMRIB Software Library (FSL) version 5.0.8 (http://www.fmrib.ox.ac.uk/fsl). The pipeline encompassed several stages, including head motion correction (using FSL 5.0.8 MCFLIRT motion estimates to assess motion differences between groups), brain extraction (f=0.5 and g=0), spatial smoothing (using a 5-mm full width at half maximum Gaussian kernel), and high-pass temporal filtering with a cutoff of 100 s. Furthermore, multi-resolution affine co-registration within FSL FLIRT software was utilized to align the mean, non-smoothed fMRI volumes to three-dimensional (3D) FSGR volumes of the corresponding subjects and to co-register anatomical volumes to the Montreal Neurological Institute's (MNI152) standard space template. [12].

This previous method was done on the raw FMRI data and other methods were used after that on the CV_{BOLD} data to make both the train and test data similar in their characteristics. Since both of them came from different devices, and this leads to both of them will be different in their properties and their statistical characteristics and therefore some preprocessing methods were used to overcome this problem as it affects the performance of the classifiers after that.

Standardization. Standardization, also known as Z-score normalization, transforms the features to adhere to the characteristics of a standard normal distribution with a mean (μ) of 0 and a standard deviation (σ) of 1. Here, μ represents the mean (average), and (σ) is the standard deviation from the mean. The calculation of the standard scores (z scores) for each sample is accomplished by subtracting the mean (μ) and dividing the result by the standard deviation (σ):

$$z = (x - \sigma/\mu) \tag{2}$$

Applying feature standardization centers the features around a mean of 0 and scales them to have a standard deviation of 1. This practice proves crucial not only when comparing measurements with diverse units but also serves as a fundamental prerequisite for numerous machine learning algorithms. Given that the two data sets in this problem originate from different sources, standardizing the data becomes essential to ensure both sets exhibit the same statistical distribution.

Min Max Scaling. Unlike standardization, this bounded range incurs the cost of yielding smaller standard deviations, which, in turn, can diminish the impact of outliers. Thus, this specific type of standardization was employed to mitigate the influence of outliers effectively.

A Min-Max scaling is typically done via the following equation:

$$X_{norm} = (X - X_{min})/(X_{max} - X_{min}) \tag{3}$$

Kolmogorov-Smirnov Test (KS Test): The Kolmogorov-Smirnov (KS) test is a non-parametric statistical test used to compare two empirical distributions. This test is based on calculating the maximum difference between the two cumulative distribution functions, which serves as a measure of the degree of disagreement between the two distributions. [6]. Smirnov test is a two-sample test, used to determine if two samples follow similar distribution or not [1]. After that, the voxels which seem to follow a different distribution are removed from both sets, so as to make sure that both of the data sets have similar distributions. So this statistical test was used to overcome the covariate shift problem which is the case of distinct training and test distributions in a learning problem.

In machine learning, it is commonly assumed that the training data follows the same distribution as the data that the model will encounter during deployment. However, in practice, it can be challenging to have complete control over the data generation process. The training data may be obtained under laboratory conditions that are not representative of the real-world scenarios that the system will encounter post-deployment [1]. Since in this case, the training set is different from the test data as they both came from different laboratories and this causes some difficulty for the classifier to distinguish between both of them. So using KS Test will remove the features which do not have a similar distribution from both data sets, and this will make the two data sets more similar and will improve the overall performance.

2.3   Feature Selection and Machine Learning Model Training

As mentioned above the main target of the project is to train machine learning models to distinguish between the healthy and the AD subjects. This will be achieved by training different classifiers and finding the best one. The second main goal of this research is to build a mask of the important brain voxels which was used to distinguish between healthy and AD brains, and this will be done by the feature selections, as the feature used is the brain voxels so choosing some of them and neglecting other will provide the mask required. The following methods were used to accomplish these tasks.

Handcrafted Voxel Selection. Each voxel is considered a feature for the classifiers. There are voxels that are effective in making classifications decision and other voxels are redundancies that should be ignored, in order to provide robust test performance metrics and speed up the processing of the model training.

The voxel selection process involved the use of recursive forward-backward feature selection. This method, also known as recursive feature elimination (RFE), is a backward selection method for identifying features. Initially, the model is built using all the features, and an importance score is computed for each predictor. The least significant feature(s) are removed, and the model is rebuilt. This process is repeated until the desired number of predictor subsets is evaluated, with each subset having a predetermined size specified by the analyst. The subset size serves as a tuning parameter for RFE. It determines the optimal number of features to consider, based on performance criteria, for constructing a model. After evaluating different subset sizes, the one that optimizes the performance criteria is selected, and the corresponding subset of predictors is then used to train the final model [2]. It has one hyperparameter which is the min number of selected voxels. The only drawback of using this method, it was slow in processing because it uses one feature elimination in a single time to provide robust voxels, and the number of brain voxels was 26257. Therefore, we needed a faster method.

For feature selection, the second approach involved regularization, wherein a basic linear SVM model was utilized, augmented with regularization parameters. We applied two types of regularization: L1 regularization, which imposes a penalty term to promote small sums of the absolute values of model features, and L2 regularization, which encourages small sums of the squares of model features. Notably, L1 regularization often leads to many features having a value of zero, effectively producing a sparse parameter vector [8]. The max number of features that we could use with L1 regularization was 200 voxels, which was not enough to make decisions. Therefore, we used the L2 penalty, however, we threshold the importance of the selected features to get a limited number of selected voxels. The hyperparameter of regularization models is the penalty parameter, which determines strictly the model is in converging toward zero values.

To apply the first method, the data was divided into five folds in which one fold will always be left to validate the classifier and the rest will be used to choose the voxels from them. The recursive forward-backward feature selection algorithm takes these 4 folds of the data and then divides it into five folds, one for validation and the rest will be used to train an SVM classifier on which the feature selection algorithm is applied. It is applied k times on the 4 folds from above, in which k=5. After being applied it returned the important features, the minimum number of features was used as a hyper-parameter so it can be tuned to give the best performance.

After applying this recursive model on all the folds, there will be five masks of selected voxels. These masks are summed together, and a threshold is applied to take the most common voxels through all the masks. The chosen voxels will be used to train a Gaussian process classifier (GPC) [9], and the main reason to use this classifier was that the GPC catches the data uncertainty, and this was important in our case as the data set was small and it has a large number of features with uncertainties. The GPC is known to be computationally expensive, so it has taken a long time to be trained with the chosen due to a large number of features, so some methods of feature reduction such as PCA and ICA but they did not improve the results.

Development in Mask Creation. Instead of creating one mask for testing as in the previous method, we created two masks, one mask was created using the training data inliers, and the second using the training data outliers. This improved the performance significantly since each mask was suitable for each type of data. We improved the mask creation process further by creating two nested cross-validations tasks. The inner cross-validation determined the best hyperparameters. Next, the best hyperparameters were used in the outer cross-validation to get the best mask. Therefore the best mask will not be dependent on one hyperparameter only. This will help in fitting the training data better without overfitting, as cross-validation prevents the model from overfitting [7]. The mask threshold used for both cross-validations was the same which was found to be 5000 voxels using hyperparameter optimization.

Development Regarding Models. The first model has one weakness regarding the prediction probability. The prediction probability of some test data was almost equal. Therefore, we used three models, the first model is made to extract the high-probability predictions. The second model is calibrated using the isotonic calibration method. The third model is a calibrated model to be used on outliers. Calibration was used to increase confidence in each decision. The total test performance metrics were calculated as the mean of the test performance metrics of each model.

2.4   Deep Learning Classifiers

After trying the previous methods, we wanted to improve the performance more, so we thought that using deep learning could improve the performance, as it can understand the relation between the voxels and have the capability to learn more insights about the data than the machine learning classifiers. Since the data used in this study were images Convolution Neural Networks (CNN) were used, as this approach is known to get good performance for images more than the fully connected layers when the input data is images. The first problem we got was the small number of data samples which will not allow the models to be learned well, as the deep learning models need a large amount of data to be learned. Therefore we applied data augmentation methods to increase the data and also we applied transfer learning to overcome the lack of training data.

Data Augmentation. In computer vision, because there are generally millions or even billions of parameters in CNN-based models, data augmentation is critical to accumulate enough data to attain satisfactory performance [3]. We used multiple data augmentation methods as the following.

Flips: For each image in the training set, we implemented a horizontal and vertical flip as a form of data augmentation. While horizontal flips are commonly used in natural images, we found that vertical flips capture a distinctive property of medical images, specifically the lack of vertical reflection invariance.

Gaussian Noise: We create an array, denoted as N, in which each element is drawn from a Gaussian distribution with a $\mu = 0$ and a variance σ^2 within the range of $[0.1, 0.9]$. Subsequently, for every image I, we generate a noisy version, $I' = I + N$, by adding the corresponding elements from array N to it.

Rotation: The following affine transformation,

$$A = \begin{bmatrix} cos\theta & -sin\theta \\ sin\theta & cos\theta \end{bmatrix}$$

where θ applied is 30,-30, 45,-45, 60,-60.

Shifting: where the data was shifted in the four main directions (up, down, right, and left). After applying these augmentation methods, the training data size was doubled more than 25 times the training data was 41 samples and after that, it became approximately 1000.

Deep Learning Models.

LeNet-5 The first deep learning model used was the LeNet-5, since it is a shallow CNN model, its computational time will be small, so it will be good as a first model to prove the concept and see whether it is a good idea to use deep learning or not. The architecture is shown in Fig. 1 [5].

AlexNet The second architecture used was AlexNet which is deeper than the previous one and it is computational time was longer than it [4]. But since it is deeper it is supposed to have higher performance. The architecture of AlexNet is shown in Fig. 2.

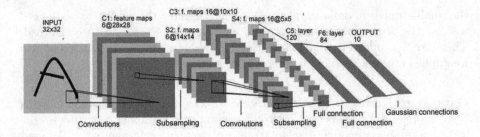

Fig. 1. Illustration of LeNet-5, a Convolutional Neural Network designed for digit recognition. The network comprises feature maps represented as planes, wherein each plane consists of units with identical weight constraints.

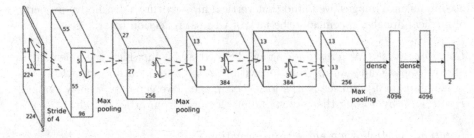

Fig. 2. Diagram depicting AlexNet, a Convolutional Neural Network, showcasing feature maps represented as planes, where each plane contains units constrained to have identical weights.

ZFNet The third architecture used is the ZFNET [13]. It was created by tweaking the hyper-parameters of Alex Net while maintaining the same structure with additional Deep Learning elements. The architecture of ZFNet consists of five convolutional layers, with decreasing filter sizes and increasing number of feature maps, followed by three fully connected layers. The ReLU activation function is used throughout the network, and max pooling is applied after the second, fourth, and fifth convolutional layers.

ZFNet achieved state-of-the-art results on the ImageNet dataset in 2013, and its architecture has been widely adopted and used as a starting point for many subsequent convolutional neural network architectures. The architecture of ZFNet is shown in Fig. 3.

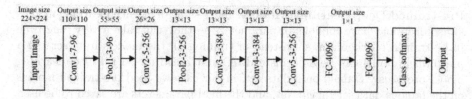

Fig. 3. Visual representation of the ZFNet architecture, featuring 5 convolutional layers that can be shared, along with max-pooling and dropout layers, complemented by 3 fully connected layers.

VGG-16 The fourth architecture used is VGG-16 [10]. The neural network utilized in this study is distinguished by its simplicity, employing a series of 3×3 convolution layers with incremental depth. Reduction in volume size is achieved through max pooling. This is subsequently followed by two fully-connected layers, consisting of 4,096 nodes each, and culminating in a soft-max classifier. A visual representation of the network architecture is depicted in Fig. 4.

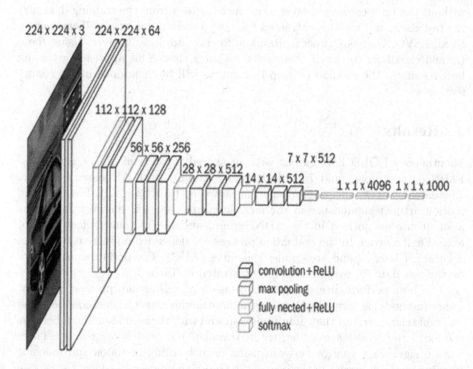

Fig. 4. Visualization of VGG-16, a Convolutional Neural Network (CNN), showcasing feature maps represented as planes, where each plane comprises units with identical weight constraints.

Pre-trained Deep Learning Models. Since some of the deep learning models are so deep and have millions of parameters so they will take a very long time to be trained on the data, so instead of training all the models from scratch a pre-trained model can be used instead. A pre-trained model is a deep learning model previously trained on a data set, in our case we used pre-trained models on the famous data set Image Net, and the final parameters are saved to be used after that on a new data, instead of training the model weights from scratch. The whole pre-trained model can be used or just the upper layers until the fully connected and the fully connected layers will be trained on the new data. The second was used because it will be better to train the fully connected layers on the data we have, as the pre-trained models were trained on the ImageNet data set which consists of natural images and we are working on brain images, so they are of different scopes. The pre-trained models used were VGG16, VGG19, Xception, ResNet50, InceptionV3, InceptionResNetV2, MobileNet, MobileNet V2, DenseNet and NASNet.

Deep Learning as Feature Extractor. Finally, we used the four pre-trained deep learning models as feature extractors. The approach was to train the model on the training data and after that save the model and use the upper layers without the fully connected layer to extract features from the training data and the test data, and use this extracted feature to train different classifiers such as SVM, LSVC, Gaussian process, Random forest, and decision trees. Using these trained classifiers to classify the test data after that. This approach gives the best results in the method of deep learning as will be mentioned in the results section.

3   Results

Spontaneous BOLD fluctuations were measured for a total of 111 control rs-FMRI data samples, and 79 rs-FMRI data samples with Alzheimer's disease. The data set varies in spatial resolution between (1.5 T and 3 T). There was no motion artifact significance in the local institute data set, however, there was a small motion noticed in the ADNI group, and the motion artifact did not exceed half a voxel. In the evaluation process, we depended on three test metrics accuracy, F1-score, and area under the curve (AUC). The performance metrics on the test data for every model are illustrated in Table 3.

The best performance was gotten by using the Gaussian process classifier. Other methods performed well using the first method such as decision trees and ensemble classifier, but they didn't perform well with the second method, because the extracted features were harder in training, but were more robust. There was no significant change between using deep learning methods and machine learning methods except in the case of the Gaussian process classifier. Using deep learning did not improve the performance as expected, due to the small number of data that existed even after applying some data augmentation methods. Using different famous deep learning models such as LeNet-5, VGG-16, AlexNet, and

Table 3. Comparison between different methods to classify patients with Alzheimer's disease from controls.

Method	Accuracy	F1-score	AUC
GP + first method	0.736	0.716	0.71
GP + second methods	0.816	0.809	0.798
SVM + first method	0.588	0.529	0.599
SVM + second methods	0.670	0.662	0.649
Ensemble classifier + first method	0.696	0.678	0.686
Ensemble classifier + second methods	0.606	0.602	0.584
LeNet-5	0.555	0.458	0.43
AlexNet	0.54	0.48	0.52
ZFNet	0.6	0.43	0.556
VGG-16	0.54	0.52	0.577
pre-trained LeNet-5	0.489	0.483	0.527
pre-trained AlexNet	0.534	0.518	0.538
pre-trained ZFNet	0.542	0.473	0.573
pre-trained VGG-16	0.559	0.496	0.516
pre-trained LeNet-5 + GP classifier	0.645	0.66	0.62
pre-trained LeNet-5 + SVM classifier	0.59	0.57	0.56
pre-trained LeNet-5 + Ensemble classifier	0.61	0.568	0.575
pre-trained AlexNet + GP classifier	0.5002	0.4951	0.5809
pre-trained AlexNet + SVM classifier	0.5723	0.4787	0.4638
pre-trained AlexNet + Ensemble classifier	0.5290	0.5696	0.5059
pre-trained ZFNet + GP classifier	0.5495	0.5457	0.5701
pre-trained ZFNet + SVM classifier	0.4864	0.4715	0.4602
pre-trained ZFNet + Ensemble classifier	0.5565	0.5842	0.5453
pre-trained VGG-16 + GP classifier	0.5127	0.5498	0.5043
pre-trained VGG-16 + SVM classifier	0.5081	0.4912	0.5217
pre-trained VGG-16 + Ensemble classifier	0.5149	0.5068	0.5022

ZFNet as classifiers did not improve the performance but the performance was even worse than using machine learning classifiers. Also using some of the famous pre-trained models as mentioned in the Methods section did not improve the results and the results were the same as using different the previous method, and this is probably because of the differences between the data on which these models were trained on and the data we are trying to classify here. As these models were trained on the famous Image-Net dataset, which does not have any classes for medical data. The last method used in deep learning which as mentioned in the Methods section was to use the deep learning models as feature extractors and use machine learning classifiers this improves the overall accuracy and else it can be further increased by using more complex models such as Res-Net 50 and Inception-Net and train them from scratch and use them as feature extractor, but this was not possible to do during working on this project due to the difficulty of accessing the GPUS in the department.

In Figs. 4 and 5, we can find that the extracted ROIs using the second method are Thalamus, Cerebral white matter, lateral Ventricle, Cingulate Gyrus, lateral occipital cortex, Cuneal cortex, Angular gyrus, supramarginal Gyrus, Planum Temporale, parietal operculum cortex, Postcentral Gyrus, Frontal pole, Precuneous Cortex, Supracalcarinre Cortex, Brain stem, Occipital Fusiform Gyrus, Intracalrine Cortex, Lingual Gyrus, Temporale Pole, and Hippo-campus and these ROIs extracted are similar to brain regions affected by Alzheimer disease as mentioned in [11]. The mask used on inliers was quite reasonable regarding symmetry and ROIs, however, the mask used on outliers was not reasonable regarding ROIs. This was due to the inclusion of many noised and filtered versions of training data outliers, which is why we made the L2 penalty stricter to produce about 2700 voxels. In fact, we could not ignore the mask used on outliers, because it didn't provide only better results on outliers, but it gave also close results to the mask used on inliers (Fig. 6).

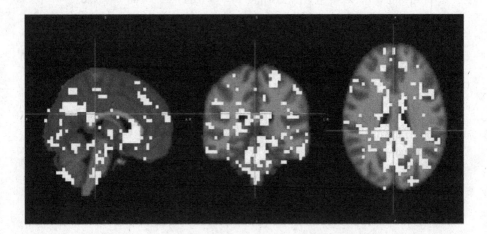

Fig. 5. Mask created using method 2 using inliers, the white voxels represent the most important voxels.

4   Discussion

The aim of this study is to clarify the relationship between Alzheimer's disease and changes in physiological pulsations represented by the CV of rs-fMRI BOLD signal and prove that we can rely on machine learning models to detect brain

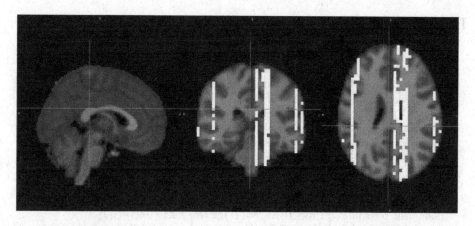

Fig. 6. Mask created using method 2 using noised and filtered outliers.

regions affected by the disease using this data. The best performance we got was by using the Gaussian process classifier because it can fit well training data with just a few samples.

The mask extracted ROIs are Thalamus, Cerebral white matter, lateral Ventricle, Cingulate Gyrus, lateral occipital cortex, Cuneal cortex, Angular gyrus, supramarginal Gyrus, Planum Temporale, parietal operculum cortex, Postcentral Gyrus, Frontal pole, Precuneous Cortex, Supracalcarinre Cortex, Brain stem, Occipital Fusiform Gyrus, Intracalrine Cortex, Lingual Gyrus, Temporale Pole, and Hippocampus. Most are concentrated in the cerebral white matter area. There is a high concentration also around the frontal Pole area. The extension of this work is studying physiological pulsations to all neurodegenerative diseases. That helps in identifying corresponding ROIs for each disease and changes in physiological pulsations. Also, using the same approach we can study the relationship between Alzheimer's disease and caffeine- It is the world's most widely consumed psychoactive drug- consumption, and Study the relation between sugar consumption-which is high across the world- and Alzheimer's disease.

5   Conclusion

Masked CV_{BOLD} can be used to detect abnormal physiological pulsations in Alzheimer's disease (AD) with promising test metrics performance and return the brain regions which the classifiers think are affected by the disease. That can be extended to all neurodegenerative diseases to be detected with simple calculated features (e.g. coefficient of variance) non-invasively.

Acknowledgment. The authors would like to thank Mr. Aleksei Tiulpin for his advice about the methods to be used and their implementations.

References

1. Bickel, S., Brückner, M., Scheffer, T.: Discriminative learning under covariate shift. J. Mach. Learn. Res. **10**(9), 2137–2155 (2009)
2. Butcher, B., Smith, B.J.: Feature Engineering and Selection: A Practical Approach for Predictive Models. In: Max, K., Kjell J. Chapman & Hall/CRC Press (2019), xv+ 297 pp., 79.95 (h), isbn: 978-1-13-807922-9. (2020)
3. Hussain, Z., Gimenez, F., Yi, D., Rubin, D.: Differential data augmentation techniques for medical imaging classification tasks. In: AMIA Annual Symposium Proceedings, vol. 2017, p. 979. American Medical Informatics Association (2017)
4. Krizhevsky, A., Sutskever, I., Hinton, G.E.: ImageNet classification with deep convolutional neural networks. Commun. ACM **60**(6), 84–90 (2017)
5. LeCun, Y., Bottou, L., Bengio, Y., Haffner, P.: Gradient-based learning applied to document recognition. Proc. IEEE **86**(11), 2278–2324 (1998)
6. Lopes, R.H., Reid, I., Hobson, P.R.: The two-dimensional kolmogorov-smirnov test (2007)
7. Lv, H., et al.: Resting-state functional MRI: everything that nonexperts have always wanted to know. Am. J. Neuroradiol. **39**(8), 1390–1399 (2018)
8. Ng, A.Y.: Feature selection, L1 vs. L2 regularization, and rotational invariance. In: Proceedings of the Twenty-first International Conference on Machine Learning, p. 78 (2004)
9. Seeger, M.: Gaussian processes for machine learning. Int. J. Neural Syst. **14**(02), 69–106 (2004)
10. Simonyan, K., Zisserman, A.: Very deep convolutional networks for large-scale image recognition. arXiv preprint arXiv:1409.1556 (2014)
11. Tuovinen, T., et al.: ADNI: altered bold signal variation in Alzheimer's disease and frontotemporal dementia. bioRxiv 455683 (2018)
12. Wolf, D., Bocchetta, M., Preboske, G.M., Boccardi, M., Grothe, M.J., Initiative, A.D.N., et al.: Reference standard space hippocampus labels according to the European Alzheimer's disease consortium-Alzheimer's disease neuroimaging initiative harmonized protocol: utility in automated volumetry. Alzheimer's Dement. **13**(8), 893–902 (2017)
13. Zeiler, M.D., Fergus, R.: Visualizing and Understanding Convolutional Networks. In: Fleet, D., Pajdla, T., Schiele, B., Tuytelaars, T. (eds.) ECCV 2014. LNCS, vol. 8689, pp. 818–833. Springer, Cham (2014). https://doi.org/10.1007/978-3-319-10590-1_53

Proteus Based Automatic Irrigation System

Wafa Difallah[1]([⊠]), Sabira Nour[2,3], Abdeldjalil Yaga[2], and Isaac Elgoul[2]

[1] Faculty of Exact Sciences, Department of Mathematics and Computer Science, Laboratory of Energetic in Arid Zones, Tahri Mohammed University of Bechar, Bechar, Algeria
wafadif@gmail.com
[2] Department of Exact Sciences, Normal Higher School of Bechar, Bechar, Algeria
[3] Laboratory of Semiconductor Devices and Physics, Tahri Mohammed University of Bechar, Bechar, Algeria

Abstract. The Traditional agricultural strategies are not satisfactory to cope with food security. This field must benefit from latest technologies.

Automatic irrigation is one of the most promising solution to maintain food security; recently, there is a growing interest in this system around the world. It has become possible to establish self-contained decision-making systems that monitor various phenomena by relying on wireless sensor networks, with the possibility of connecting them to the Internet.

The object behind this work, is to develop an automatic irrigation system based on DHT11 temperature and humidity sensor, and Arduino microcontroller. The system is tested using Proteus simulation software and the obtained results were satisfactory.

Keywords: Irrigation · Proteus · Sensor · Arduino · Water

1 Introduction

In most countries over the world, agriculture have a significant economic importance. Recently, this sector suffers from numerous challenges such as water scarcity, climatic change and population growth. To facing these challenges, it is obligatory to introduce new technical strategies to the agricultural sector; and especially the irrigation, which is strongly related to food security. Smart irrigation system is one of the most promising solutions to overcome with above mentioned challenges. It is an automated water management strategy, in which irrigation decision is based on the collected environmental data (temperature, soil moisture, humidity, ph,.....etc.). This irrigation system has different forms, and it can be implemented with different tools and technologies like wireless sensor networks, the internet of things, microcontrollers (Arduino, Raspberry Pi, ...) with the object of optimizing irrigation water consumption. In this paper, an automatic irrigation system based on DHT11 humidity and temperature sensor and Arduino microcontroller is presented. The proposed system is simulated in Proteus software. After this introduction, a selection of related works is presented, followed by system description and flow diagram, then results and discussion, and finally the conclusion.

A. Bennour et al. (Eds.): ISPR 2023, CCIS 1941, pp. 101–112, 2024.
https://doi.org/10.1007/978-3-031-46338-9_8

2 Related Works

Automatic irrigation systems is a promising application to optimize water consumption, which justifies the variety of researches in this axis including a portable and wireless sensor network agricultural system was developed in [1]. The system consisting of temperature, humidity, light intensity, and soil moisture sensors with microcontroller and Internet of Things, the sensed information is uploaded wirelessly to the cloud server for future use.

A microcontroller based irrigation system with soil humidity sensor was presented in [2]. The system is controlled using ATmega328 microcontroller based on Arduino platform.

The same idea was used in [3], with GSM standard (Global System for Mobile communication) to transmit information and instruction between the manager and the controller.

To optimize water consumption, an automatic irrigation system was proposed in [4]. The system was tested in Proteus software, in which moisture and temperature sensors are used to collect environmental information. This information is compared with threshold values by the controller to meet the wanted output of the system, and to take an irrigation decision. The system is energized by solar power, battery and has duplex communication using internet. The data in this proposal are analyzed by IoT technique using Blynk software.

In [5], an intelligent irrigation management system was implemented. In this system, PH, soil moisture, nutrients and humidity detectors and a high pixel camera are connected with a solar powered robot to monitor the field. The robot transmit the collected data using the global system for mobile communications module. This system was designed in Proteus with Arduino Mega 2560.

3 Tools and Technologies Used in Automatic Irrigation

This section will not be enough to present all the technologies that have rationalize the irrigation sector, but it summarizes the most important of them.

3.1 Wireless Sensor Networks (WSNs)

As is mention in the second section sensor networks are widely used in the irrigation field. WSN is an infrastructure that offers the user the possibility to observe, instrument, and interact with events and phenomena in a determined environment using inexpensive mobile devices, sensors and microcontrollers [6, 7].

Sensors consist of communication, computing, sensing, actuation, and power units. These units are unified in one or several boards, and packaged in some cubic inches [8] (Fig. 1).

A variety of sensors can be utilized to measuring and gathering different climate and soil data such as:

- Soil humidity: EC-5, 10SH, Watermark 200ss, VH400
- Soil temperature: TDR-3A

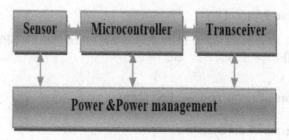

Fig. 1. Sensor node architecture.

- Relative humidity: SHT75, SHT11, DHT22, HS220
- Temperature: DS1822, DS18B20, LM35
- Carbon dioxide: GMW120
- Wind Direction/speed: Met station one (MSO), CM-100 compact
- Illumination intensity: On9558
- Rain flow: Rain Bird RSD-CEx
- Water level: Senix ToughSonic REMOTE 14

Using wireless sensor network with Arduino in the agricultural field is very popular, especially for developing irrigation management system. These two technologies complement each other. The first one feed the system and the second one automatized it.

3.2 Internet of Things (IoT)

IoT is a network of things capable to collect and sense data, as well as communicate with other things and devices [9]. This technology has improved the quality and quantity of the irrigation field. Various platforms support IoT networks like Arduino, Raspberry Pi and Node MCU [10].

3.3 Smart Drones and Robots

Unmanned Aerial Vehicles (UAVs). UAV is an IoT device [11] used in the agricultural field mainly for monitoring crops and spraying pesticides over large areas [12].

Agricultural Robots. It play a significant role in agriculture modernization. Two types of robots are very much useful for agricultural activities:

Smart Tractor Robot. Nowadays, implementation of smart tractor like IoT based tractors is very important in agriculture. It reduces the manual work and working time and lowering the prospect of human error based on real time information [13].

Harvesting Robots. Harvesting robots have a big impact in agricultural production, given that harvesting is the very difficult phase through production procedure. Harvesting crop at right time is much difficult, as doing so either late or early might influence the production expressively [14, 15].

3.4 Transmission Protocol

ZigBee IEEE 802.15.4, Bluetooth IEEE 802.15.1, WIFI IEEE 802.11, GPRS/3G/4G/5G, GSM, WiMAX IEEE 802.16e, …… etc.

4 Description of the Proposed System

4.1 Block Diagram of the System

The proposed architecture is a combination of the following devices:

DHT11 Humidity and Temperature Sensor. This sensor features a humidity and temperature sensor complex and a calibrated digital signal output. The sensor utilizes a dedicated digital modules capture technic with humidity and temperature sensing strategy to guarantee exemplary long-term stability and top reliability. It comprises a resistive-type humidity estimation device with an NTC temperature measurement element, and associates to a high-performance 8-bit microcontroller, giving very good quality, rapid response, cost-effectiveness and anti-interference ability [16, 17].

DHT11 Specifications.

- Operating Voltage: 3.5 V to 5.5 V
- Operating current: 0.3 mA (measuring) 60 uA (standby)
- Output: Serial data
- Temperature limits: 0 °C to 50 °C
- Humidity limits: 20% to 90%
- Resolution: humidity and temperature are 16-bit
- Accuracy: ±1 °C and ±1% [18] (Fig. 2).

Fig. 2. DHT11 temperature and humidity sensor.

Arduino Uno microcontroller. It is an open source board, easily programmed (with just low knowledge of programming), erased and reprogrammed at any time. It was used firstly in 2005 by Massimo Banziand David Cuartielles.

The Arduino platform was destined to bring an economical and simple way for hobbyists, students and professionals to design devices that depend on environmental changes through actuators and sensors [19].There are many kinds of Arduino boards like Arduino Uno, Arduino Lilypad, Arduino Leonardo, Arduino Nano, Arduino Mini and Arduino Mega [20].

The presented system is based on Arduino UNO. It is the latest incarnation of the most popular series of Arduino boards.

"Uno" signifies one in Italian and is called to mark the coming release of Arduino 1.0. This board is founded on the ATmega328. It has fourteen digital input/output pins (of which six can be utilized like PWM outputs), six analog inputs, a sixteen MHz ceramic resonator, an ICSP header, a universal serial bus connection, a power jack, with a reset button. It includes everything needed to support the microcontroller; just link it to a computer using USB connexion or power it through a battery or an AC-to-DC adapter to be activated.

The Arduino Uno board doesn't use the FTDI USB-to-serial driver chip, which make it different from all previous boards [21] (Fig. 3 and Table 1).

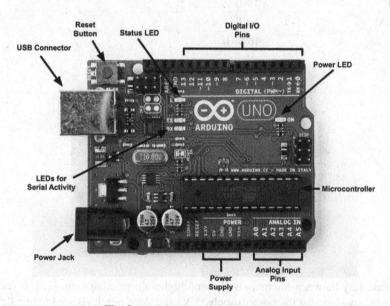

Fig. 3. Arduino Uno components [22].

LCD (Liquid Crystal Display) Screen. A video show uses the light modulating characteristics of liquid crystals to present images or words on a screen. LCD screen was invented in 1964, and its use is continuously and rapidly expanding. This technic is almost included in every electronic devices such as computers, televisions, and dashboards [23]. This screen is widely used with Arduino microcontroller to develop automatic systems.

In this work, LCD is used to display the sensed values of temperature and humidity and the status of the pump (ON/OFF) (Fig. 4).

A Relay. An electrically worked component, it acts as a switch, which can be opened or closed. It is highly used for controlling current signal with a normal voltage of 5 V, which makes it appropriate for use with Arduino based system [24]. It plays the role of

Table 1. Arduino Uno properties [21].

Microcontroller	ATmega328
Normal voltage	5 V
Input voltage (recommended)	7–12 V
Input voltage (limits)	6–20 V
Digital I/O pins	14 (6 of them provide output)
Analog input pins	6
DC current per I/O Pin	40 mA
DC current for 3.3 V Pin	50 mA
Flash Memory	32 KB (ATmega328), 0.5 KB is dedicated for bootloader
SRAM	2 KB
EEPROM	1 KB
Clock speed	16 MHz

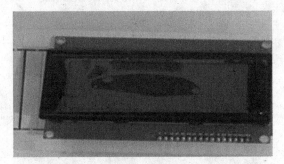

Fig. 4. LCD Screen (20 × 4 characters).

an intermediary between microcontroller and high voltage equipment [25]. It consists of five pins, two pins for the power supply (5 V) and the other three used to handle high voltages (12–130 V).

In the studied system, the relay is connected to the water pump (Fig. 5).

Fig. 5. Relay.

Water Pump. It is necessary to choose a water pump that can work at the same voltage than the microcontroller.

Schematic representation of system components and structure is illustrated in Fig. 6 below.

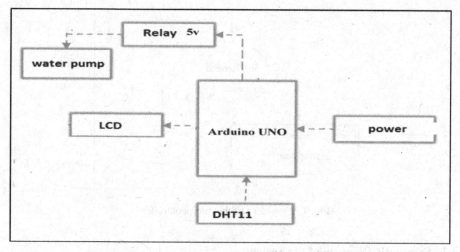

Fig. 6. Block diagram of surveillance and controlling system.

4.2 Flow Diagram of the System

At this stage, the control system (Arduino) checks the collected information. In case these data are lower than the predetermined values of the system, the irrigation process will be started, contrarily; the system stays power off till the comparison shows the different result (see Fig. 7).

There are different considerations that control the selection of threshold values including:

- Meteorological conditions
- Plant kinds
- Soil texture
- Irrigation style.

In the proposed system, threshold values are 30 and 60 for temperature and humidity respectively.

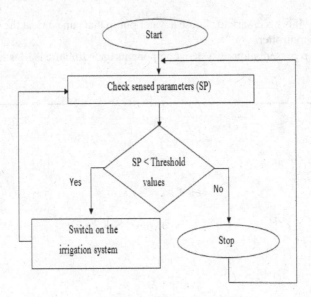

Fig. 7. Flowchart of the microcontroller.

4.3 Schematic Diagram of the System

The system is validated using Proteus. It is a physical simulation and circuit analysis program introduced by British Lab center Company. This software is compatible with the operating system Windows and is based on two packages ARES and ISIS [26, 27]. Proteus professional ISIS software is mainly used for designing circuits diagrams.

Proteus ARES is a high performance PCB design system offering the same user interface as ISIS and full netlist based integration with it [28].

Proteus is a powerful virtual tool with a large number of components, which encourages its use in many fields [27]. It helps to complete successfully the design phase of a project (Fig. 8).

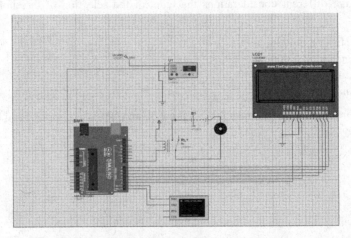

Fig. 8. Schematic diagram of the irrigation system.

5   Results and Discussion

After system activation, DHT11 sensor send humidity and temperature values to Arduino board to compare it with predefined values.

' The system is implemented in four cases as is shown respectively in Fig. 9, Fig. 10, Fig. 11, Fig. 12 and the following results is given:

1. The temperature T = 32, the water pump will be activated and "pump on" appears on the LCD.

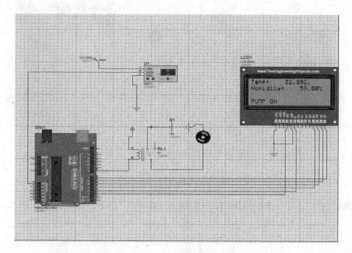

Fig. 9. Simulation output with T > = 30 (case1).

2. The temperature T = 29, the water pump turns off and "pump off" appears on the LCD.

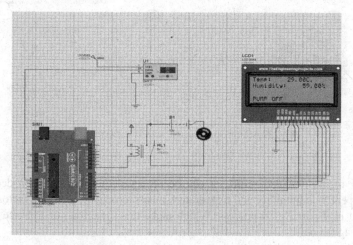

Fig. 10. Simulation output with T < 30 (case 2).

3. The humidity H = 61, the water pump still off or turned off and "pump off" appears on the LCD.

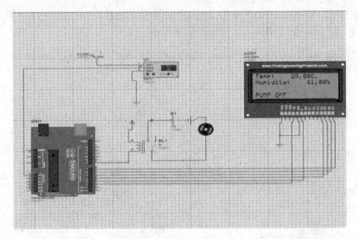

Fig. 11. Simulation output with H > = 60 (case 3).

4. The humidity H = 58 the water pump turns on and "pump on" appears on the LCD.

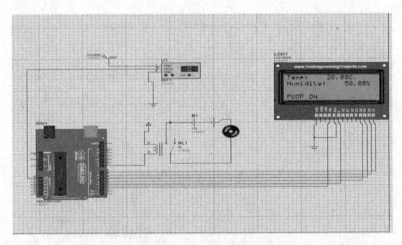

Fig. 12. Simulation output with H < 60 (case 4).

According to the obtained results, it can be observed that the irrigation decision is mainly depends on humidity and temperature measurements. These two factors have a big impact in the irrigation process. Which make this system a good choice for crop water management.

6 Conclusion

In this work, an automatic irrigation process using Arduino UNO microcontroller and DHT11 humidity and temperature sensor is developed. The system is validated using Proteus simulation software with various scenarios, and the obtained results proved its reliability. It is a smart irrigation water strategy based on real time information to reduce water consumption, effort and time required to managing the irrigation operation. In the upcoming works, this system will be developed and enriched with further parameters to give more precision, and hence to be used in experimental study.

References

1. Gsangaya, K.R., Hajjaj, S.S.H., Sultan, M.T.H., Hua, L.S.: Portable, wireless, and effective internet of things-based sensors for precision agriculture. Int. J. Environ. Sci. Technol. **17**(9), 3901–3916 (2020)
2. Al-Rawi, M., Abdulhamid, M., Njoroge, K.: Irrigation system based on Arduino uno microcontroller. Poljoprivredna Tehnika **45**(2), 67–78 (2020)
3. Babaa, S.E., Ahmed, M., Ogunleye, B.S., Al-Jahdhami, S.A., Khan, S.A., Pillai, J.R.: Smart irrigation system using Arduino with solar power. Int. J. Eng. Tech. Res. **9**(5), 91–97 (2020)
4. Shwetha, N., Niranjan, L., Gangadhar, N., Jahagirdar, S., Suhas, A., Sangeetha, N.: Efficient Usage of water for smart irrigation system using Arduino and Proteus design tool. In: 2nd International Conference on Smart Electronics and Communication (ICOSEC) 2021, pp. 54–61. IEEE, New Delhi (2021)
5. Hassan, A., et al.: A wirelessly controlled robot-based smart irrigation system by exploiting Arduino. J. Robot. Control **2**(1), 29–34 (2021)
6. Sohraby, K., Minoli, D., Znati, T.: Wireless Sensor Networks: Technology, Protocols, and Applications. Wiley, Hoboken (2007)
7. Zervopoulos, A., et al.: Wireless sensor network synchronization for precision agriculture applications. Agriculture **10**(3), 1–20 (2020)
8. Difallah, W., Benahmed, K., Bounaama, F., Draoui, B., Maamar, A.: Water optimization using solar powered smart irrigation system. In: 7th International Conference on Software Engineering and New Technologies ICSENT 2018, pp. 1–4. ACM, Hammamet (2018)
9. Abbasi, M., Yaghmaee, M. H., Rahnama, F.: Internet of Things in agriculture: a survey. In: 2019 3rd International Conference on Internet of Things and Applications (IoT), pp. 1–12. IEEE, Isfahan (2019)
10. Kour, V.P., Arora, S.: Recent developments of the IoT in agriculture: a survey. IEEE Access **8**, 129924–129957 (2020)
11. Tsouros, D.C., Bibi, S., Sarigiannidis, P.G.: A review on UAV-based applications for precision agriculture. Information **10**(11), 1–26 (2019)
12. Swamidason, I.T.J., Pandiyarajan, S., Velswamy, K., Jancy, P.L.: Futuristic IoT based smart precision agriculture: brief analysis. J. Mobile Multimedia **18**(3), 935–956 (2022)
13. Tharik Ahamed, S., Vishnu, T., Arun Avinash, M., Gopi, M.: Research review in IOT based smart tractor for field monitoring and ploughing. Int. J. Eng. Res. Appl. **11**(2), 32–38 (2021)
14. Fahmida Islam, S., Uddin, M.S., Bansal, J.C.: Harvesting robots for smart agriculture. In: Uddin, M.S., Bansal, J.C. (eds.) Computer Vision and Machine Learning in Agriculture. Algorithms for Intelligent Systems, vol. 2, pp. 1–13, Springer, Singapore (2022). https://doi.org/10.1007/978-981-16-9991-7_1
15. Bhrugubanda, M., Rao, S.L.A., Shanmukhi, M., Rao, A.S.: Sustainable and intelligent IoT based precision agriculture – smart farming. Solid State Technol. **63**(6), 17824–17833 (2020)

16. DHT11 Humidity & Temperature Sensor. https://www.mouser.com. Accessed 06 Nov 2020
17. Temperature and Humidity Module DHT11 Product Manual. https://components101.com/sites/default/files/component_datasheet/DHT11-Temperature-Sensor.pdf. Accessed 08 Aug 2022
18. DHT11–Temperature and Humidity Sensor. https://components101.com/sensors/dht11-temperature-sensor. Accessed 08 Aug 2022
19. Louis, L.: Working principle of Arduino and using IT as a tool for study and research. Int. J. Control Autom. Commun. Syst. **1**(2), 21–29 (2016)
20. Monk, S.: Programming Arduino™ Getting Started with Sketches. The McGraw-Hill Companies, United States (2012)
21. Arduino Uno. https://www.farnell.com/datasheets/1682209.pdf. Accessed 08 Aug 2022
22. Schmidt, M.: Arduino: A Quick-Start Guide, 2nd edn. The Pragmatic Bookshelf, United States of America (2015)
23. Interfacing to an LCD Screen Using an Arduino. https://www.egr.msu.edu. Accessed 08 Aug 2022
24. Guide for Relay Module with Arduino. https://randomnerdtutorials.com/guide-for-relay-module-with-arduino/. Accessed 08 Aug 2022
25. Arduino-Relay. https://arduinogetstarted.com/tutorials/arduino-relay. Accessed 08 Aug 2022
26. Su, J., Han, H., Liu, Y.: Thinking on the introduction of PROTEUS simulation for single-chip microcomputer teaching. Int. J. Sci. **2**(11), 183–185 (2015)
27. Wu, F., He, T.: Application of proteus in microcontroller comprehensive design projects. In: Zhang, T. (ed.) Instrumentation, Measurement, Circuits and Systems. Advances in Intelligent and Soft Computing, vol. 127, pp. 363–369. Springer, Heidelberg (2012)
28. Intelligent Schematic User Manual. https://www.ele.uva.es/~jesman/BigSeti/ftp/Cajon_Desastre/Software-Manuales/EBook%20-%20Proteus%20Manual.pdf. Accessed 09 Aug 2022

How AI can Advance Model Driven Engineering Method ?

Mohamad Suhairi Md Subhi[1], Willem Nicolas[1], Akina Renard[1,2],
Gabriela Maria Garcia Romero[1], Meriem Ouederni[2(✉)], and Lotfi Chaari[2]

[1] Toulouse INP, Toulouse, France
{mohamadsuhairi.mdsubhi,willem.nicolas,akina.renard,
gabrielamaria.garciaromero}@etu.toulouse-inp.fr
[2] IRIT-Toulouse INP, Toulouse, France
{meriem.ouederni,lotfi.chaari}@irit.fr

Abstract. Artificial intelligence (AI) skills are being increasingly applied in today's field of computer science. This aims at better satisfying customer requirements, reducing errors, improving decision-making, tackling complex problems, system automation, increasing operational efficiencies, etc. To do so, AI implies several sub-fields such as Machine Learning (ML), Deep Learning (DL), Neural Networks (NN), Natural Language Processing (NLP), Robotics, etc. Applications of AI are innumerable, including healthcare and biomedicine, bio-informatics, physics, robotics, geo-sciences and more. Our current paper studies AI applications for modeling IoT systems using Model Driven Engineering (MDE) method. We survey the most significant research work related to our topic and investigate how AI techniques could be used to better resolve software engineering issues. In the context of the current paper, we particularly focus on healthcare systems as an illustrative specific domain.

Keywords: AI · Software Engineering · MDE · IoT · Distributed Systems

1 Introduction

In 1981, Barr & Feigenbaum [21] stated "Artificial Intelligence (AI) is the part of computer science concerned with designing intelligent computer systems, that is, systems that exhibit characteristics we associate with intelligence in human behavior - understanding language, learning, reasoning, solving problems, and so on." In other words, AI is a discipline "concerned with not just understanding but also building intelligent entities-machines that can compute how to act effectively and safely in a wide variety of novel situations". Although this discipline is generally vast, machine learning (ML) is one of its most widely used techniques. ML uses statistical engineering and computational learning technology to emulate human learning processes in order to recognize models from a given dataset.

© The Author(s), under exclusive license to Springer Nature Switzerland AG 2024
A. Bennour et al. (Eds.): ISPR 2023, CCIS 1941, pp. 113–125, 2024.
https://doi.org/10.1007/978-3-031-46338-9_9

On the other hand, Model Driven Engineering (MDE) [24] is a software development methodology where building systems is directed by models of a given domain and their transformations. It was defined by the Object Management Group (OMG) and aims at raising the abstraction level, automation degree in software development, and better complexity handling. In such a context, the Meta Object Facility (MOF) [18] standard from OMG defines pyramidal construction consisting of four levels: the bottom level is referred to as M0 and represents the real-world systems; M1 gives the models of M0 systems; M2 holds metamodels that describe the abstract definition of M1 artifacts; and lastly, M3 is the top level and consists in metametamodel represented with MOF itself.

It is worth noticing that these models, as a main artefact, describe different features of software systems, e.g. functionality, behaviour, structure, and data. Following MDE, models can be later executed and transformed into code, configuration files, or documentation. MDE enables us to build Domain Specific Languages, a.k.a DSLs (contrary to General Purposes Languages a.k.a GPLs), by transforming these models into a language that better expresses the essence of a domain and improves its effectiveness and usability.

Although MDE and AI have been initially two separate fields in computer science, both of them can clearly benefit from cross-pollination and collaboration with each other. There are at least two ways in which such integration can manifest. Firstly, AI concepts can be applied to MDE in order to find the "good" models for an application domain. This can provide advantageous results because the use of AI enables us to create more precise and efficient models, and this eventually implies better data analysis and/or control. On the other hand, MDE can be used to improve the implementation of AI, adding a level of abstraction to AI models. Particularly, using DSLs allows the domain expert to efficiently process their data.

Regarding the aforementioned concepts, i.e. AI and MDE, our paper studies the state-of-the-art on how both of them could be used side-by-side ? [10,14,16,17] to better resolve software engineering issues, such as modelling and properties verification. We then focus on IoT and particularly healthcare systems as an application domain. So far, we illustrate our ideas using two things, namely, Electrocardiograms (ECG) and Electroencephalograms (EEG). ECG and EEG are medical sensors that measure the heart's and brain's electrical signals respectively. While EEG can help diagnose epilepsy and brain tumours, ECG can be used to evaluate the risk of heart attack. As the measure and treatment of these signals occur in real-time, their urgent and correct processing is essential for medical purposes to save human lives. However, the synchronization of EEG and ECG signals is still an open issue such that the users could not fully benefit from the readings during patients' diagnostics. Resolving synchronization would be valuable for the medical community.

The remainder of our paper is organized as follows. Section 2 presents open issues. Then, Sects. 3 studies the application of AI to MDE from the literature. Section 4 shows how AI can benefit from MDE. Later, Sect. 5 presents a survey on signal synchronization using AI. Finally, Sect. 6 sums up our work and sketches some promoting perspectives for future work.

2   Challenges

The challenges in using AI in MDE are due to the lack of datasets that ML models can learn from. Even if the datasets could be available, the heterogeneous nature of the datasets, specifically, for IoT technologies delimiting the application of AI techniques in this domain. The learning granularity of the AI models needs to be fine-grained in order to obtain more accurate and efficient learning. The different notions of models in AI and MDE also bring some difficulties in integrating these two domains together. In the Data Analytics & Machine Learning (DAML) community, ML models, etc. are completely distinct from the models used in MDE: one is a statistical model while another is a structural model. Various works such as ML-Quadrat or MontiAnna [11] are looking forward to creating a synergy between these models, allowing a more seamless model engineering in the domain. We investigate in the following existing work following two directions: we first explore whether the literature applies AI to the MDE methods (Sect. 3); we then study the reverse application (while Sect. 4), namely, first building DSLs by MDE and then running AI techniques on data handled by instantiated systems.

3   AI for MDE

In this Section, we will discuss the question: how to better build models using AI? In other words, how to empower MDE through the application of AI? Regarding the literature, this could imply several approaches given below:

(i) Integration of ML into Domain Modeling. AI is integrated into the meta-meta-model itself, by redefining the modeling language [26].

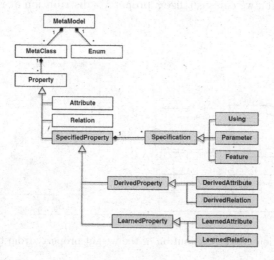

Fig. 1. Meta meta-model [26]

Figure 1 depicts the meta-metamodel proposed in [26] to integrate machine learning in domain modelling. The grey elements in Fig. 1 are the added ML elements, but the rest of the representation resembles the standard MOF/EMOF (variants of EMF). New types of property are defined: learned properties, relations, or attributes learned by a machine learning algorithm, and derived properties, relations, or attributes that are simply computed (not trained or learned). To do so, the authors of this project introduced the concept of microlearning units, which are instances of learning algorithms associated with a learned or derived property. In that respect, "The machine learning algorithm will be "weaved" inside the metamodel instances" ?.

This approach allows for the flexible combination of learned behaviours and domain knowledge and can be more accurate and efficient than learning global behaviours. The approach is demonstrated using a smart grid case study, showing that it can be significantly more accurate than learning a global behaviour, while still being fast enough for live learning.

(ii) Probabilistic Model-Driven Engineering (PSM). As stated in [25], this is a software modelling approach that is intended to complement MDE. The primary goal of PSM is to support software comprehension from a code-centric perspective, which means that it focuses on understanding the software code and its behaviour at runtime. It relies on probabilistic reasoning to analyze the behaviour of software systems. This involves using mathematical models to represent uncertainty and variability in the behaviour of the system.

PSM is used after a system has been implemented, using MDE techniques. The approach provides software engineers with a way to reason about software and its runtime behaviour on the same level as the implementation concepts, including types, properties, and executables. For a given model, a probabilistic model can do predictive and generative tasks. For example, with the following model (on the left), we can visualize a property's distribution at runtime (on the right) (Fig. 2):

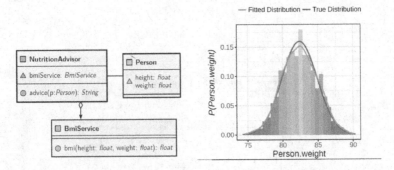

Fig. 2. Visualization of the distribution of the weight property from the class Person

(iii) Creation of Models Dataset. The availability of large and high-quality datasets, such as the ModelSet dataset [10], is crucial for the application of machine learning in model-driven engineering.

Applying ML to software engineering has gained significant attention in recent years, with a focus on three types of ML models, namely: code-generating models, representational models of code, and pattern mining models. These models have various applications, such as code recommendation systems, inferring coding conventions, clone detection, and code-to-text and text-to-code translation. There are several existing datasets related to software models, including the LindholmenDataset and ModelSet, as well as datasets of OCL expressions, BPMN models, and APIs classified using Maven Central tags. However, many of these datasets have limitations, such as a variety of formats and versions, invalid or poor-quality models, or a lack of labels.

ModelSet dataset, which includes over 10,000 labelled models, aims to address the need for a large and high-quality dataset of software models for use in ML research. The dataset is composed of models in the Ecore metamodeling language and includes labels for the model category and domain, as well as structural and textual features. The paper [10] demonstrates the use of the dataset in a classification task, showing that it can be used to improve the accuracy of existing ML models for classifying Ecore models. The developers of ModelSet focused on creating labelled datasets of Ecore and UML models, as both of these languages are widely used. Labelling the models was another challenge because it can be time-consuming to find the usage of a meta-model. They classified the meta-models according to their application domain, their usage, whether the meta-model is part of a tool, and whether a concrete syntax is associated with the meta-model. Hence, an entity of ModelSet will have several labels that describe it.

(iv) Model-Driven Optimization (MDO). The MDO as a field has emerged after the use of model-driven engineering (MDE) in the context of search-based software engineering (SBSE). Design space exploration (DSE), one of the principal branches of SBSE, aims to apply meta-heuristic techniques in order to find a solution to an optimization problem. Here, we consider the optimization problem as finding the optimal representation of models based on a meta-model, with respect to certain objectives. In order to find the solution to this optimization problem, two main approaches have been explored: the model-based approach [2] and the rules-based approach [28]. These two approaches differ in their search for the solution, where the first aim to optimize models directly, and the latter aims to find the optimized model transformation sequence, allowing the user to calculate the cost for the transformation.

(v) Knowledge-Driven Model Augmentation. The implementation of Knowledge Representation (KR) in the domain of AI, or more specifically its resources: ontologies, is taken after the successful debut of knowledge bases in the Web community. The benefits of KR in the domain of AI include the ability to formally structure knowledge, thereby improving its syntax and semantics. KR also facilitates efficient reasoning capabilities, enabling AI systems to better understand, manipulate, and process information. The W3C recommended the Web Ontology Language (OWL) as the default ontologies syntax, where the features of Description Logics [4] i.e. its expressivity and its reasoning complexity are also implemented in the language.

The authors in [13] are looking forward to bringing the KR and MDE together with the aims of bringing forward these benefits: i) correct by design models, ii) sharing and reuse of knowledge, iii) advanced inferences involving complex definitions and high-level abstraction concepts iv) logic rules to further specify the design of a system, v) queries to identify patterns and anti-patterns, and vi) the reduction of heterogeneity in large systems. They managed to introduce a new Drone Components Ontology *ODrone* into the Core Ontology for Robots and Autonomous Systems (CORA - IEE1872) [20].

(vi) Models@runtime. The traditional way of creating models at the development phase has now met its runtime counterpart, where these models are able to adapt to their specific runtime environment. This method of modelization allows the designed models to be reconfigured or redesigned if needed during runtime, without requiring downtime or a complete overhaul of the system. This allows the creation of a more flexible and adaptable software system that can respond to changing requirements or conditions. The AI part of this approach consists of the use of ML to learn the behaviour of a networked system and to apply modifications to the model's design if needed to cater to the need of the system, also called self-adaption. A lot of approaches are introduced in order to apply the technique in the modelling framework, as seen in the state-of-the-art paper [6].

(vii) Generating Models from Natural Language. Another idea when creating a meta-model is to extract UML metamodel from natural language [27]. The conception of a meta-model using UML can be a time-consuming process for developers, thus describing the structure of the meta-model in the natural language can help to speed up the process.

One solution for extracting the UML model from natural language specification was implemented by Song Yang and Houari Sahraoui and can be summarized in Fig. 3. The first step is to create a dataset of UML class diagrams, with an English specification associated with each element. They used the database **AtlanMod Zoo** [5] but any other database could also work, like ModelSet [10] which is larger than the one they used. The English specifications are **processed and fragmented** into sentences. In order to **classify a sentence**, one needs to determine if the sentence describes an object or a relationship. The **UML Model Fragmentation** step creates those fragments from each UML class diagram. They used several well-known classification algorithms and chose to use the Bernoulli Naive Bayes classifier with a tf-idf vectorizer, which is a method specific to natural language processing. After classifying the sentence, a UML

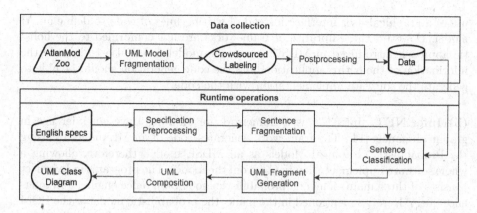

Fig. 3. Overview of the extraction process of UML class diagrams from natural language specification [27]

fragment is **generated**. To do so, they used patterns that are different to identify whether the sentence describes a relationship or a class. For example, a pattern for a relationship would be the verb "have" or the word "with", and a pattern for a class would be the word "there is". The parsing of the classified sentences will match them to one or more patterns, and a UML fragment is generated with rules defined according to the patterns. The last step is to **assemble the generated UML fragments** into a UML class diagram. They created an algorithm that has a complexity of $O((c + r)a^2c)$, where c is the number of classes, r is the number of relationships between classes and a is the number of class attributes [27].

4 · MDE for AI

We will now survey the opposite of our first approach in Sect. 3, namely, MDE for the eventual application of AI techniques on modelled data. The main objective here is to properly integrate the AI component in the system design, allowing a seamless learning process for the AI components thus the need for the abstraction of these AI components themselves. We summarize below recent research results where AI could benefit from integrating concepts and techniques from MDE:

(i) ML Libraries and Frameworks. The starting point of the integration of AI models in MDE starts with providing a level of abstraction for the implementation of machine learning, which is proposed by the libraries or frameworks in the ML domain. For instance, Tensorflow [1] provides a powerful API using an advanced implemented technique. An additional library such as Keras [9] adds another layer of abstraction allowing the use of the APIs of Tensorflow and other API frameworks, e.g. PlaidML [8]. Workflow designers and visualization toolkits such as KNIME [7], RapidMiner [15], and TensorBoard [1] allow the use of a graphical user interface instead of writing lines of code to define an AI model. However, these libraries or frameworks are not conformed to the holistic approach introduced in MDSE, where the models need also to include the information of the entire application, and the code generation should be able to generate the software implementation with the code.

(ii) Infer.NET. Infer.Net is a framework for running Bayesian inference in graphical models [14]. The framework generates code of an ML model by using the Probabilistic Graphical Models as an MDSE model, therefore, allowing a whole software implementation in C# from the model. The programs in Infer.Net consists of three main elements: probabilistic model, inference engine, and inference query. Here, as mentioned in the name, the probabilistic model or the PGM has the closest relation to the model-driven engineering counterpart. The inference engine can then be instantiated and queried using a method called `infer()` to infer the marginal distribution over the variables in the models. However, as Infer.Net only uses PGMs as its source, there may be shortcomings in terms of expressiveness for the whole software system.

(iii) MontiAnna for EmbeddedMontiArc. This is a framework embedded in the component and connector-oriented modelling language family Embedded-MontiArc. It allows developers to design and build Artificial Neural Networks (ANNs) as a component in the same model definition of MDE. As Embedded-MontiArc is a framework where all the software functionality translates into components, the communication between these components is only possible by defining the interface of the component (the input and output ports). The MontiAnna built system will then detect all ANNs in the software architecture and resolve trained model weights for these networks at build time. The automation of the training mechanism of the ANN is triggered when there are no model weights available or the network model has changed.

(iv) ML-Quadrat. This approach [16,17] is based on ThingML, i.e. an IoT modelling language, and it is the most recent technology found to be implementing the holistic nature of the MDSE. This technology is looking forward to creating a synergy between the two different notions of models (the statistical model in DAML and the structural model in MDE), allowing a more seamless model engineering between DAML and MDE. ML-Quadrat is built on top of Ecore Modelling Framework (EMF) and Xtext Framework thus resulting in the possibility of using the features available with these frameworks, i.e. autocompletion, syntax highlighting, and tree-based model editor. The DAML models can be transformed into generated Java or Python code which is seamlessly integrated, and the MDE models can be transformed into a broad range of programming languages such as Java, Python, or C thanks to the code generation features in ThingML.

In order to sum up our study, Table 1 sketches a global view of the aforementioned approaches. We highlight that although Infer.Net and MontiAnna almost provide the same features, both of them are only able to model a specific type of ML model (PGM and ANN respectively). On the other hand, ML-Quadrat allows the modelling of various types of ML models. Thus, to the best of our knowledge, Ml-Quadrat is the most developed technology in the domain.

Table 1. Comparison of Features provided by related Works

Description	Technologies	Full-code generation	Model Type
ML libraries & frameworks	TensorFlow, Keras, etc		DAML models
'Model-based' ML	Infer.Net	✓	SE & ML (PGM) models
MDE4IOT + ML	MontiAnna	✓	SE & ML (ANN) models
MDE4IOT + ML	ML-Quadrat	✓	SE & DAML models

5   Signal Synchronization

We survey here the application of AI to signal synchronization with respect to existing work. Two approaches are presented. The first one relates to the application of AI in signal synchronization. The second one reports on the state-of-the-art approach of synchronizing asynchronous EEG signals and highlights two open-source libraries that implement it. To the best of our knowledge, our study shows that there is no work existing on the synchronisation of signals gathered from different things. Even less, there is a lack of AI applications for synchronising signals coming from distributed things.

(i) AI Application to Signal Synchronization. The work stated in [3] discusses the use of deep learning for real-time anomaly detection in asynchronous multivariate time series data that exhibit regular seasonal variations. The authors propose a new method that involves adding a synchronization layer to a neural network architecture, which is pre-trained to learn the periodic properties of the input data. This method improves learning performance and detection accuracy.

The authors proposed a model for an asynchronous multivariate time series, y in Eq. 1, where the timestamps of the features are inconsistent. The model represents each feature as a combination of S sinusoidal components, indexed by s in Eq. 1, with known angular frequency (ω) and random phase (β) values.

$$y_{i,t_i} = \left[\sum_{s=0}^{S} \alpha_{s,i} sin(\omega_s(t_i + \beta_{s,i})) \right] + \gamma_i + \epsilon_{i,t_i} \qquad (1)$$

In Eq. 1 the amplitude (α), phase (β), and bias (γ) of the sinusoidal components are learned during backpropagation, while the noise term (ϵ) is calculated by subtracting the raw signal from the fitted sinusoidal components. The authors set the ω values based on prior knowledge of the time series seasonality, to reduce the number of learnable parameters. They also express the error term (ϵ) as a function of time as follows :

$$\epsilon_{i,t_i} = \alpha_{\epsilon,i,t_i} sin(\omega_0(t_i + \beta_{\epsilon,i,t_i})$$

where $\beta_{\epsilon,i,t_i} \approx -t_i + \frac{\nu_0}{4}$ is the approximation of the value of β for the error term, which is then used to calculate the noise term (ϵ). These formulas are used to align multivariate time series feature sample times (t_i) and asynchronous phase shifts $(\beta_{s,i}$ and $\beta_{\epsilon,i,t_i})$ to a reference frame.

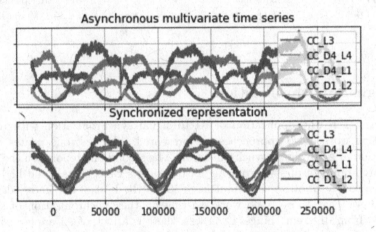

Fig. 4. Synchronization of asynchronous multivariate Series in [3]

Figure 4 shows an example of 4 asynchronous features collected at eBay from users' activity across 4 different time zones and their synchronized representation using $(s \in [0,1])$ as described in [3]. The time-dependent and independent parameters of (1) are in a neural network layer attached to an anomaly detection neural network. The network is pre-trained to learn these parameters by minimizing the noise term ϵ with the 50th quantile loss function. The noise term is then computed with the above formulas to synchronize the time series.

(ii) Application of Middleware for Signal Synchronization. In [12], Lab Streaming Layer (LSL) is used for real-time EEG synchronization. LSL is a cutting-edge solution that is open source and based on clock offset measurements to manage event information and timing [12]. Achieving software synchronization requires transmitting data packets in a dependable and timely manner, which poses a challenge in avoiding additional delays in the application software.

LSL [23] is a system designed to collect time series data in research experiments. It manages networks, time synchronization, real-time access, and centralized data collection. The LSL distribution includes a core library, liblsl, and a set of tools built on top of this library. The library has language interfaces for C, C++, Python, Java, C#, and MATLAB and is cross-platform. The suite of tools includes a recording program, file importers, and applications that make data from various acquisition devices available on the laboratory network.

LSL's built-in time synchronization relies on two pieces of data collected along with the actual sample data: a time stamp for each sample read from a local high-resolution clock on the computer and out-of-tape clock synchronization information transmitted to the receiving computer along with each data stream. This synchronization information consists of measurements of the instantaneous offset between the two clocks involved, taken at regular intervals. Time offsets are measured using a protocol similar to the Network Time Protocol. Synchronization is correct if both the incoming and outgoing routes from the server have a symmetric nominal delay.

We now present two open-source libraries implementing LSL for their signals synchronization mechanism:

(a) Medusa. It is introduced in [22] as a modular and flexible software ecosystem for brain-computer interface and cognitive neuroscience research. It includes a suite of signal processing functions, deep learning architectures, and pre-made experiments, as well as tools for custom experiment development and sharing.

Medusa is designed for developing brain-computer interfaces (BCI) and neuroscience experiments. It consists of two components: Medusa Kernel and Medusa Platform. Medusa Kernel is a Python package that contains ready-to-use methods for analyzing biosignals, including advanced signal processing, machine learning, deep learning, and high-level analyses. Medusa Platform provides high-level functionalities for managing and synchronizing signals in real-time via LSL.

Medusa utilizes packages such as SciPy, Numpy, Scikit-learn, and Tensorflow for implementing state-of-the-art data processing, machine learning, and deep learning. It includes low-level functions that can be used for processing signals in various scenarios, including temporal filters, spatial filters, local activation metrics, and connectivity metrics.

(b) OpenSync. OpenSync [19] is an open-source platform that can synchronize various physiological and behavioural signals, user input responses, and task-related information in neuroscience experiments. It is a software tool that enables users to synchronize and record data from different sources including Python, C, Java, C#, C++, and Matlab. The tool has two main modules, Synchronizer and Recorder, and four submodules - I/O devices, Sensors, Controller, and Marker - that allow for data streaming and synchronization. OpenSync uses Lab Streaming Layer (LSL) as its core protocol for data streaming and synchronization. It can process user response data from I/O devices, read extraneous information, and stream data from various biological sensors. The tool records all the streams from different submodules in a single file with Extensible Data Format (.xdf extension). OpenSync also provides built-in functions for the initialization, configuration, and streaming of EEG devices.

6   Conclusion and Perspectives

Our current paper investigates how both AI and MDE can benefit from each other. Being able to benefit from AI will definitely help the MDE community to establish a better future, allowing a more efficient and self-fixing model. To achieve this objective, we studied the integration in two directions: the implementation of AI in MDE and the use of MDE for AI.

Our main perspective is to provide IoT domain experts with an advanced DSL empowered with AI concepts, and such that signal synchronization is dealt with. To do so, we are working on the development of a generic framework where several domain parameters are considered.

Acknowledgement. This work has been partially supported by the project IBCO-CIMI funded by Toulouse-INP (France).

References

1. Abadi, M., et al.: TensorFlow: Large-Scale Machine Learning on Heterogeneous Distributed Systems (2015)
2. Abdeen, H., et al.: Multi-objective optimization in rule-based design space exploration. In: Proceedings of ACM/IEEE/ASE 2014, pp. 289–300. ASE 2014, Association for Computing Machinery (2014)
3. Abdulaal, A., Lancewicki, T.: Real-time synchronization in neural networks for multivariate time series anomaly detection. In: Proceedings of the ICASSP 2021, pp. 3570–3574. IEEE (2021)
4. Baader, F., Calvanese, D., McGuinness, D., Patel-Schneider, P., Nardi, D.: The Description Logic Handbook: Theory, Implementation and Applications. Cambridge University Press (2007)
5. Barais, O., et al.: AtlanMod: collaborative engineering for complex systems. J. Object Technol. **8**(3), 1–22 (2009)
6. Bencomo, N., Götz, S., Song, H.: Models@run.time: a guided tour of the state of the art and research challenges. Softw. Syst. Model. **18**(5), 3049–3082 (2019). https://doi.org/10.1007/s10270-018-00712-x
7. Berthold, M.R., et al.: KNIME - the Konstanz information miner: version 2.0 and beyond. SIGKDD Explor. Newsl. **11**(1), 26–31 (2009)
8. Chen, H., Cammarota, R., Valencia, F., Regazzoni, F.: PlaidML-HE: acceleration of deep learning kernels to compute on encrypted data. In: Proceedings of the ICCD 2019, pp. 333–336 (2019)
9. Chollet, F., et al.: Keras. GitHub (2015). https://github.com/keras-team/keras
10. Hernández López, J.A., Cánovas Izquierdo, J.L., Sánchez Cuadrado, J.: ModelSet: a dataset for machine learning in model-driven engineering. In: Proceedings of the MODELS 2018, pp. 207–213. ACM (2018)
11. Kirchhof, J.C., et al.: MDE for machine learning-enabled software systems: a case study and comparison of MontiAnna & ML-quadrat. In: Proceedings of the MODELS 2022, MODELS 2022, pp. 380–387. Association for Computing Machinery (2022)

12. Kwong, T., Wong, S., Tsoi, M., Chan, C., Choy, Y., Mung, S.: An EEG device with synchronization of auditory stimuli. In: Proceedings of the 2020 IEEE International Conference on Consumer Electronics, ICCE-Asia 2020, Institute of Electrical and Electronics Engineers Inc., 01 November 2020–03 November 2020. IEEE (2020). Funding Information: This work was supported by The Innovation and Technology Fund (ITF), UIM381
13. Medinacelli, L.P., Noyrit, F., Mraidha, C.: Augmenting model-based systems engineering with knowledge. In: Proceedings of the MODELS 2022, MODELS 2022, pp. 351–358. Association for Computing Machinery, New York (2022)
14. Microsoft Research: Infer.NET (2021). https://dotnet.github.io/infer/
15. Mierswa, I., Wurst, M., Klinkenberg, R., Scholz, M., Euler, T.: RapidMine. SIGKDD Exp. 6(1), 15–19 (2004)
16. Moin, A., Challenger, M., Badii, A., Günnemann, S.: A model-driven approach to machine learning and software modeling for the IoT. In: Proceedings of the CCNC 2018, pp. 1–6. IEEE (2018)
17. Moin, A., Rössler, S., Günnemann, S.: ThingML+: a language and runtime for IoT systems. In: Proceedings of the INDIN 2018, pp. 51–56. IEEE (2018)
18. Object Management Group: Object Management Group Meta Object Facility (MOF) Core Specification. Technical report, Object Management Group (2014). https://www.omg.org/spec/MOF/2.5/PDF/
19. Razavi, M., Janfaza, V., Yamauchi, T., Leontyev, A., Longmire-Monford, S., Orr, J.: OpenSync: an open-source platform for synchronizing multiple measures in neuroscience experiments. J. Neurosci. Meth. 369, 109458–109458 (2022)
20. Rey, V.F., Fiorini, S.R.: IEEE1872-owl (2015). oWL specification of the Core Ontology for Robotics and Automation (CORA) and other IEEE 1872–2015 ontologies. https://github.com/srfiorini/IEEE1872-owl
21. Salveter, S.: The Handbook of Artificial Intelligence. Volume 1 and Volume 2. Avron Barr, Edward A. Feingenbaum The Handbook of Artificial Intelligence. Volume 3. Paul R. Cohen, Edward A. Feingenbaum. Q. Rev. Biol. 58(3), 483–483 (1983)
22. Santamaría-Vázquez, E.: MEDUSA: a novel Python-based software ecosystem to accelerate brain-computer interface and cognitive neuroscience research. Comput. Meth. Program. Biomed. 230, 107357–107357 (2023)
23. Schalk, G., McFarland, D., Hinterberger, T., Birbaumer, N., Wolpaw, J.: BCI2000: a general-purpose brain-computer interface (BCI) system. In: Proceedings of the IEEE Transactions on Biomedical Engineering 2004, vol. 51, pp. 1034–1043. IEEE (2004)
24. Schmidt, D.C., et al.: Model-driven engineering. Computer 39(2), 25 (2006). IEEE Computer Society
25. Thaller, H.: Probabilistic software modeling. arXiv preprint arXiv:1806.08942 (2018)
26. Hartmann, T., Moawad, A.N., Fouquet, F., Traon, Y.L.: The next Evolution of MDE: a seamless integration of machine learning into domain modeling. Soft. Syst. Model. 18, 1–20 (2017)
27. Yang, S., Sahraoui, H.: Towards automatically extracting UML class diagrams from natural language specifications. In: Proceedings of the MODELS 2022, October 2022. ACM (2022)
28. Zschaler, S., Mandow, L.: Towards model-based optimisation: using domain knowledge explicitly. In: Milazzo, P., Varró, D., Wimmer, M. (eds.) STAF 2016. LNCS, vol. 9946, pp. 317–329. Springer, Cham (2016). https://doi.org/10.1007/978-3-319-50230-4_24

Machine and Deep Learning

Detection of DoS Attacks in MQTT Environment

Hayette Zeghida[✉], Mehdi Boulaiche, and Ramdane Chikh

Department of Computer Science, LICUS Lab, Université 20 août 1955-Skikda, Skikda, Algeria
{h.zeghida,r.chikh}@univ-skikda.dz

Abstract. The Message Queuing Telemetry Transport protocol (MQTT) is one of the most widely used application protocols to facilitate machine-to-machine communication in an IoT environment. MQTT was built upon TCP/IP protocol and requires minimal resources since it is lightweight and efficient which makes it suitable for both domestic and industrial applications. However, the popularity and openness of this protocol make it vulnerable and exposed to a variety of assaults, including Denial of Service (DoS) attacks that can severely affect healthcare or manufacturing services. Thus, securing MQTT-based systems needs to develop a novel, effective, and adaptive intrusion detection approach. In this paper, we focus on MQTT's flaws that allow hackers to take control of MQTT devices. We investigate a deep learning-based intrusion detection system to identify malicious behavior during communication between IoT devices, using an open source dataset namely 'MQTT dataset'. The findings demonstrate that the suggested approach is more accurate than traditional approaches based on machine learning; with an accuracy rate greater than 99% and an F1-score greater than 98% .

Keywords: IoT · MQTT · DoS Attack · Deep learning

1 Introduction

The Internet of Things (IoT) system was created in order to increase work efficiency and enhance people's life. Consequently, various devices such as sensors, equipment, and apps may be found in a variety of domains including agriculture, industry, medical, commerce, smart cities, and so on. Most IoT devices have limited memory and a tiny processor, as well as a low power supply. MQTT is one of the protocols that meet these requirements and it is extensively used. Due to the sensitivity of IoT services, the security of devices is a necessary issue today.

MQTT protocol consists of three major entities, which are: publisher, subscriber, and broker. The publisher gathers information from a variety of resources, including sensors installed in equipment or wearables, built-in mobile devices, and so on. Subscribers sign up for the topic on which the publication is producing data by a mobile application for example. The broker is the heart of the MQTT protocol. It acts as a server between publishers and subscribers. The broker saves messages on the cloud and may receive and send several messages at the same time [1].

Furthermore, the MQTT message comprises three (03) levels of quality of service (QoS) to ensure that the message is delivered from the publisher to the subscriber.

A. Bennour et al. (Eds.): ISPR 2023, CCIS 1941, pp. 129–140, 2024.
https://doi.org/10.1007/978-3-031-46338-9_10

(1) Level 0:(default option) at most one delivery where the message is sent only once and with no guarantees, (2) level 1: at least one delivery where, if necessary, the message can be resent after the first transmission, and (3) level 2: exactly one delivery where the message is maintained in this level until the subscriber gets it. To increase MQTT security, the QoS supports reliable transfers utilizing the Secure Sockets Layer and Transport Layer Security (SSL/TLS) with certificates and session key management [2].

Despite the benefits of the MQTT protocol, it is vulnerable to multiple forms of threats, such as data disclosing, data tempering, or forwarding. Particularly, DoS attacks that target the MQTT broker to block access to network resources. As illustrated in Fig. 1, an attacker using different client identifications can access to the broker several times, resulting in server flooding. In other circumstances, the attacker will use high QoS levels with huge payload messages to flood the broker's resources. For instance, level 2 causes the broker to keep messages until they are delivered, and if they are not sent, the message is held, resulting in server resource depletion [3]. One of the DoS attacks is Mirai, flooding, and SlowITe.

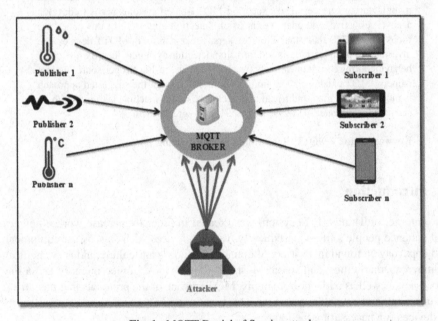

Fig. 1. MQTT Denial of Service attack

Therefore, it is critical to think about finding effective and efficient solutions to maintain a proper functioning of such protocol. In this paper, we proposed an intrusion detection system based on a Deep Learning (DL) approach. Firstly, we use single algorithms such as LSTM, CNN, and GRU, then, we pair the DL algorithms so that we can benefit from the characteristics of the two algorithms together, thus, we used hybrid DL algorithms such as CNN-RNN, CNN-LSTM, and CNN-GRU models. Our experiments were conducted on the database MQTT dataset [6], and the results were very positive, with an accuracy rate exceeding 99% and a very small loss rate with a value of 0.072.

The rest of paper is structured as follows. We presente the MQTT protocol's vulnerabilities in Sect. 2. Then; Sect. 3 presents the related works. While Sect. 4 describes the Dataset used and discusses the experiments and results. Lastly, Sect. 5 provides the paper's conclusion and outlines potential avenues for future research.

2 MQTT Vulnerabilities

To communicate with brokers, MQTT clients utilize TCP/IP port 1883 for unencrypted exchanges; port 8883 is used for encrypted communications relying on SSL/TLS channels. However, the MQTT protocol's security issue remains critical, since the protocol's developers concentrated on its low power and bandwidth settings while overlooking security concerns.

A multitude of threats can compromise data flow. The attacks may be physical on IoT devices when the attacker disables all services on mobile phones, cameras, sensors, and routers. The attacks may be cyber when the attackers hack into a wireless network to change and threaten countless IoT devices.

Andy et al. [15] consider that the weakness of the protocol is related to the following reasons:

- Resource-Constrained Device: Some constrained devices with limited resources, such as memory (RAM and ROM), may be unable to use TLS encrypted protocol since it is a relatively heavy protocol, and these memories are insufficient to perform most security mechanisms, particularly those requiring extensive calculations.
- The number of devices: The sheer number of IoT devices makes device administration extremely challenging. This may adversely affect the security components such as confidentiality, integrity, and availability.
- Lack of security awareness: The developer could put functionality ahead of the device's security by preying on the user's ignorance of security issues. The user may also choose to ignore changing the factory default password and the username, as well as neglecting to update the software. This might damage the device's overall security.

As for Burange et al. [4] vulnerabilities of MQTT protocol are due to:

- Sending information in plain text: MQTT messages are sent in clear text and might include login credentials. Consequently, anybody without authorization can access the data, read the payload, and alter it. This violates the confidentiality of the information.
- Exposed ports: MQTT uses port 8883 by default for encrypted interactions and port 1883 by default for non-encrypted communications. These open ports can be used to launch attacks across networks.

Raikar and Meena [5] add further vulnerabilities:

- Free and open Broker: Many attackers use the Shodan search engine shows information about linked devices all around the world. Attackers look for MQTT public brokers and get information or send false data to their customers, as well as the possibility of launching a DoS attack against that broker.

- The use of wildcards: The potential to exploit wildcards in topic names increases the likelihood of data theft. As an example, to subscribe to all subjects, use the wildcard "#.

3 Related Work

The MQTT protocol may be vulnerable for a variety of reasons, including the use of exposed ports, the ability to subscribe to and publish any topic, sending data in plain text, a lack of user awareness by maintaining the factory's default username and password, and limited resources that prevent the use of encryption STL for MQTT protocol. For these reasons, recent publications provided ideas for fixing security concerns and managing MQTT vulnerabilities. In this part, we present the most recent works, which are mostly based on machine learning (ML) and deep learning (DL) methodologies to enhance MQTT security (Table 1).

Ghannadrad, A. [6] provided an MQTT-based data collection that includes both valid and malicious communication by modeling a smart home scenario. He simulated three distinct types of attackers that use denial-of-service attacks to disseminate malicious material. Afterwards, he presented solutions by using ML techniques to identify and categorize DoS assaults, and the strategy employed yielded positive results.

Alaiz-Moreton et al. [8] created a new MQTT dataset that includes three forms of attacks: DoS, man in the middle, and intrusion. The authors proposed models for detecting assaults in IoT settings, which might be used to create IDS for IoT. They classified the frames that IDS can designate as attack or normal using ML and DL techniques. Owing to the significance of timing and sequencing in network assaults, they picked the LSTM, GRU, and XGBoost models for their classification task. The best results were obtained using this strategy.

Syed et al. [9] created an MQTT data collection and proposed a DoS attack detection technique for MQTT environments. The utility of the recommended feature set was validated using three ML algorithms: AODE based on Naive Bayes, C4.5 based on Decision Tree, and MLP based on ANN count-based flow characteristics and field length features were used to assess the classifiers' effectiveness in detecting normal and attack classes. The suggested MQTT features provided significant detection capabilities when the control packet field size-length based characteristics were used.

Vaccari et al. [10] established an MQTTset, which is almost a real dataset related to the MQTT protocol. They assessed this dataset using numerous balancing methodologies before implementing and comparing different ML algorithms (NN, RF, NB, DT, GB, and MLP). They noted the importance of a well-balanced dataset in producing more realistic results. In the confusion matrices, the balanced dataset gave low metrics but an accurate data distribution. It also has a good accuracy and F1-score.

Kim et al. [11] used the N-BaIoT dataset to create a botnet training model based on five ML models: naive Bayes, K-neighbors Nearest Neighbors, logistic regression, decision tree, and random forest. DL models like CNN, RNN, and LSTM were also employed. ML models like decision trees and random forests performed better in identifying Mirai and Bashlite botnets in N-BaIoT. According to the F1-score measure, CNN performed the best out of all of the DL models.

Hindy et al. [12] comprised both benign and malicious situations within "MQTT-IoT-IDS2020"dataset. They suggested a model that includes six ML algorithms (LR, k-NN, DT, RF, SVM, and NB). Packet-based, unidirectional, and bidirectional characteristics are investigated at three abstraction levels. The studies demonstrated that the suggested ML models were sufficient for MQTT-based networks' Intrusion Detection Systems (IDS) needs. The findings also highlight the need of employing flow-based features to distinguish MQTT-based attacks from benign traffic, whereas packet-based properties are adequate for ordinary networking attacks.

Mosaiyebzadeh et al. [13] proposed a DL-based intrusion detection system for IoT networks to protect the MQTT protocol from attacks. They used the MQTT-IoT-IDS2020

Table 1. Studies on MQTT attacks detection.

Ref	Dataset	Method employed	Performance measures	IoT	MQTT protocol
Ghannadrad, A.[6]	MQTTdataset [7]	RF, SVM, KNN	F1-Score, learningcurve, Confusion matrix		✓
Alaiz-Moreton et al. [8]	MQTT Dataset [16]	SVM, RF, GRUNN, XGBoost, LSTM	Accuracy F-Score		✓
Syed et al. [9]	Dataset Generated [9]	AODE, DT, MLP	Detection Rate (DR), Accuracy, False-Positive Rate (FPR)		✓
Vaccari et al. [10]	MQTTset [17]	NN, RF,NB,DT, GB,and MLP	Accuracy, F1-score		✓
Kim et al. [11]	N-BaIoT [11]	ML(KNN, LR, DT, RF) DL (CNN, RNN, LSTM)	F1-score	✓	
Hindy et al. [12]	MQTT-IoT-IDS2020 [18]	LR, k-NN, DT, RF, SVM, NB	Accuracy Precision, Recall F1-score		✓
Mosaiyebzadeh et al. [13]	MQTT-IoT-IDS2020 [18]	DNN CNN-RNN-LSTM LSTM	Accuracy, Precission, Recall, F1-score Weighted-avg		✓

(continued)

Table 1. (*continued*)

Ref	Dataset	Method employed	Performance measures	IoT	MQTT protocol
Siddharthan et al. [14]	SENMQTT-SET [14]	LR, KNN, SVM, NB DT, RF, GB	Accuracy, ADR, APV, NPV, F1-score, MCC, FAR, ROC, time		✓
Our work	MQTTdataset [7]	LSTM, CNN, GRU, CNN-RNN, CNN-LSTM, CNN-GRU	Accuracy, F1-score, Model loss		✓

dataset to test their proposed DNN, CNN-RNN-LSTM, and LSTM models. Their DL-based Network IDS obtained a performance assessment of very high accuracy and an F1-score in the CNNRNN-LSTM model.

Siddharthan et al. [14] developed and tested a novel dataset called "SENMQTT-SET" for detecting MQTT threats in IoT scenarios. Using (EML) Elite Machine Learning methodologies, they built an intelligent intrusion detection system that can detect or forecast a cyber-attack. The EML is proposed to choose the best model for intrusion detection among ML algorithms. The results showed that DT, RF, and GB classifiers provide better performance when compared to the other LR, KNN, SVM, and NB classifiers.

The use of conventional learning techniques, such as ML, and also the use of an imbalanced data set that leads to questioning the credibility of the results. In this paper, we decided to build IDS utilizing novel learning techniques such as deep learning and its hybrid algorithms to benefit from many types of algorithms. In addition, we used a balanced dataset, which strengthens the veracity of our findings.

4 Proposed Approach

4.1 Dataset Description

A dataset is required to build a good classifier. Ghannadrad, A. [6] provided the MQTT dataset used in this paper. To collect legitimate and malicious data, Ghannadrad, A. [6] simulated the smart home scenario using Wireshark, the world's most popular and widely used packet analyzer and sniffer. At first, there were more than 40 features. Ghannadrad, A. [6] reduced the number of features to 10 by using Analysis of Variance (ANOVA), which was chosen as a filter-based selection method, to reduce training time and improve classifier accuracy. Finally, two types of dataset were created; binary data with two types of traffic Malicious and Legitimate traffic coded respectively by 0 and 1, and multi-value data with four types of traffic; Heavy-Flood, Fast-Flood, Connect-Flood, and Legitimate coded 0, 1, 2, and 3 respectively.

Table 2 and Table 3 describe the top ten features of the two datasets.

Table 2. Binary Classification: Top-10 Best Features dataset [6]

Feature	Description	ANOVA Score
State-CON	Reported transaction is active	1091.797346
State-FIN	Reported transaction is closed	773.240791
DstRate	Destination pkts per second	268.743461
SrcRate	Source pkts per second	252.760566
Rate	Pkts per second	172.975403
TcpRtt	TCP connection setup round-trip time	133.061976
AckDat	The time between the SYN-ACK and the ACK packets	130.633730
SynAck	The time between the SYN and the SYN-ACK packets	120.424206
Load	Bits per second	32.709531
Dur	Record total duration	31.573044

Table 3. Multi-value classification: Top-10 Best Features dataset [6]

Feature	Description	ANOVA Score
Flgs-e *	Both Src and Dst loss/retransmission	1032.465791
Flgs-e	Ethernet encapsulated flow	448.155430
DstLoss	Destination pkts retransmitted or dropped	400.635048
State-FIN	Reported transaction is closed	252.980545
Flgs-e s	Src loss/retransmissions	248.745935
pLoss	percent pkts retransmitted or dropped	207.513853
Load	bits per second	159.877985
State-CON	Reported transaction is active	159.767769
DstPkts	src - > dst packet count	158.117338
DstBytes	dst - > src transaction bytes	146.474182

4.2 Experiments and Results

Although we used the same data as in Ghannadrad, A. [6], our algorithms and evaluation criteria are different. Ghannadrad, A. [6] employed three ML algorithms with only one evaluation criteria (F1-score), whereas we utilized six DL algorithms and three distinct criteria (Accuracy, F1-score, Model loss). In order to prove the efficiency of our DL techniques in the classification of attacks, we tested DL algorithms for binary and multi-value data. In this section, we present and discuss the results obtained from the binary and multi-value classification. we proposed simple DL such as Long Short Term Memory (LSTM), Convolution Neural Network (CNN), and Gated Recurrent Unit (GRU), we proposed also hybrid deep learning such as CNN-RNN (a hybrid DL that combines the Convolution Neural Network with a Recurrent Neural Network), CNN-LSTM (a hybrid DL model that combines the Convolution Neural Network with Long Short Term Memory), and CNN-GRU (a hybrid DL model that combines the convolution neural network with Gated Recurrent Unit). The two datasets were split into two parts, with 30% used for testing and 70% used for training. The proposed models are created with Python and the Keras library. The Adam optimizer has been used, and the Loss function is has ben also set to 'binary crossentropy' for binary classification and 'categorical crossentropy' for multi-value classification, with the number of epochs ranging between 100 and 200 and the batch size ranging from 16 to 64 for each model. Table 4 displays the obtained results.

Table 4. Results of binary classification

Models	Ghannadrad, A. [6]	Our work		
	F1-score	Accuracy	F1-score	Model loss
RF	0.99	/	/	/
SVM	0.84	/	/	/
KNN	0.99	/	/	/
LSTM	/	**0.993**	**0.993**	**0.044**
CNN	/	0.753	0.780	0.493
GRU	/	**0.993**	**0.983**	**0.072**
CNN-RNN	/	0.746	0.775	1.022
CNN-LSTM	/	**0.986**	**0.989**	**0.027**
CNN-GRU	/	**0.980**	**0.980**	**0.011**

As demonstrated in Table 4, in our experiments, the best results for binary classification are the results of LSTM, GRU, CNN- LSTM, and CNN-GRU, with accuracy and an F1-score of around 0.99 for identifying malicious assaults. The model CNN-GRU likewise had the least loss, with a value of 0.011. The CNN and CNN_RNN models, with accuracy ratings of 0.75 and 0.74, respectively, produce the worst outcomes.

Table 5 shows the results of the multi-classification, with the GRU model produced the best results, with an accuracy of 0.97, an F1-score of 0.91, and a loss of 0.14.

While the CNN and CNN-RNN models also produce the lowest results in this case, with accuracy scores of 0.66 and 0.56, respectively.

Comparing the results, we notice that in both types of data, the LSTM and GRU models outperform the CNN model. This difference can be justified by the fact that the CNN model gives the best results when it is applied to picture data and not to tabular data.

The confusion matrices of LSTM and GRU data demonstrate the two models' efficiency in identifying and classifying attacks (Fig. 2). We notice also that the use of the model loss as a criterion for evaluation can help in selecting the most suitable technique

Table 5. Results of multi classification

Models	Ghannadrad, A. [6]	Our work		
	F1-score	Accuracy	F1-score	Model loss
RF	0.91	/	/	/
SVM	0.86	/	/	/
KNN	0.95	/	/	/
LSTM	/	0.906	0.912	0.305
CNN	/	0.663	0.618	0.723
GRU	/	**0.975**	**0.919**	**0.149**
CNN-RNN	/	0.566	0.317	1.062
CNN-LSTM	/	0.913	0.913	0.242
CNN-GRU	/	0.854	0.855	0.388

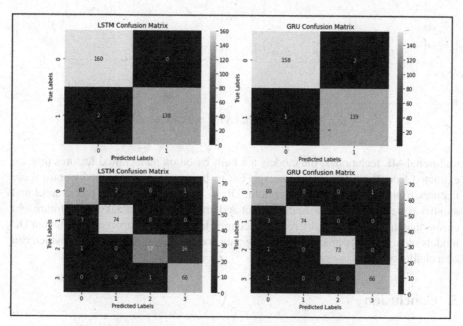

Fig. 2. Confusion matrices of LSTM and GRU models

when the other measurements criteria yield similar results. Figure 3 shows the model loss curve for LSTM and GRU models for binary and Multi classifications.

Overall, we can say that the use of DL techniques compared to ML approaches for MQTT attack detection has the potential to significantly improve detection accuracy and reduce false positives. Indeed, DL techniques can enhance the detection of MQTT attacks by enabling the creation of more sophisticated and accurate detection models. In

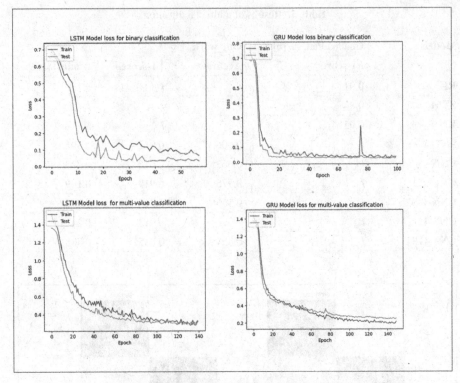

Fig. 3. Model Loss of LSTM and GRU models

traditional ML techniques, the models are built based on handcrafted features that are extracted from the input data. However, DL techniques can automatically learn these features from the raw data. Furthermore, DL models can learn from past attacks and identify patterns that can be used to detect and prevent similar attacks in the future and ensure that the MQTT network remains secure. However, it is important to note that DL models require a large amount of training data and computational resources, which can be a challenge.

5   Conclusion

IoT enables increased connection for services, devices, and systems. Therefore, it encompasses a diverse set of applications and protocols. MQTT's ease of use and open-source code makes it ideal for low-power, low-computing-power, limited-memory, and limited-bandwidth situations such as the IoT. One of the current issues is securing the MQTT protocol. In this paper, we presented an overview of the architecture of the MQTT protocol. We also discussed the major vulnerabilities it faces, particularly, we focused on Dos attacks. Afterwards, we gathered MQTT security enhancement measures.

The major objective of this paper is to identify and classify DoS attacks against MQTT sensor networks. For that, we used simple and hybrid deep learning algorithms such as LSTM, CNN, GRU, CNN-RNN, CNN-LSTM, and CNN-GRU to enhance the detection of DoS attacks in MQTT environments. The results obtained indicate that the models employed were effective, and produced more accurate results.

As part of our future work, we intend to keep working on MQTT security and another possible development of this paper may be the study of a variety of attacks such as 'Brute force', 'Flood', 'Phishing', ' Man in the middle', and so on. Moreover, the incorporation of modern techniques such as blockchain to enhance the security of the MQTT protocol can be considered an extension of this work.

References:

1. Kouicem, D.E., Bouabdallah, A., Lakhlef, H.: Internet of things security: a top-down survey. Comput. Netw. **141**, 199–221 (2018)
2. Dinculeană, D., Cheng, X.: Vulnerabilities and limitations of MQTT protocolused between IoT devices. Appl. Sci. **9**(5), 848 (2019)
3. Firdous, S.N., Baig, Z., Valli, C., Ibrahim, A.: Modelling and evaluation of malicious attacks against the iotmqtt protocol. In: 2017 IEEE International Conference on Internet of Things (iThings) and IEEE Green Computing and Communications (GreenCom) and IEEE Cyber, Physical and Social Computing (CPSCom) and IEEE Smart Data (SmartData), pp. 748–755. IEEE (2017)
4. Burange, A., Misalkar, H., Nikam, U.: Security in MQTT and CoAP Protocols of IOT's Application Layer. In: Verma, S., Tomar, R.S., Chaurasia, B.K., Singh, V., Abawajy, J. (eds.) Communication, Networks and Computing. Communications in Computer and Information Science, vol. 839, pp. 273–285. Springer, Singapore (2019). https://doi.org/10.1007/978-981-13-2372-0_24
5. Raikar, M.M., Meena, S.M.: Vulnerability assessment of MQTT protocol in Internet of Things (IoT). In: 2021 2nd International Conference on Secure Cyber Computing and Communications (ICSCCC), pp. 535–540. IEEE (2021)
6. Ghannadrad, A.: Machine learning-based DoS attacks detection for MQTT sensor networks (2021)
7. https://www.kaggle.com/datasets/alighannadrad/mqttdata
8. Alaiz-Moreton, H., Aveleira-Mata, J., Ondicol-Garcia, J., Muñoz-Castañeda, A.L., García, I., Benavides, C.: Multiclass classification procedure for detecting attacks on MQTT-IoT protocol.Complexity, 2019 (2019)
9. Syed, N.F., Baig, Z., Ibrahim, A., Valli, C.: Denial of service attack detection through machine learning for the IoT. J. Inf. Telecommun. **4**(4), 482–503 (2020)
10. Vaccari, I., Chiola, G., Aiello, M., Mongelli, M., Cambiaso, E.: MQTTset, a new dataset for machine learning techniques on MQTT. Sensors **20**(22), 6578 (2020)
11. Kim, J., Shim, M., Hong, S., Shin, Y., Choi, E.: Intelligent detection of IoT botnets using machine learning and deep learning. Appl. Sci. **10**(19), 7009 (2020)
12. Hindy, H., Bayne, E., Bures, M., Atkinson, R., Tachtatzis, C., Bellekens, X.: Machine learning based IoT intrusion detection system: an MQTT case study (MQTT-IoT-IDS2020 Dataset). In: Ghita, B., Shiaeles, S. (eds.) Selected Papers from the 12th International Networking Conference. INC 2020. Lecture Notes in Networks and Systems, vol. 180, pp. 73–84. Springer, Cham (2021). https://doi.org/10.1007/978-3-030-64758-2_6
13. Mosaiyebzadeh, F., Rodriguez, L.G.A., Batista, D.M., Hirata, R.: A network intrusion detection system using deep learning against MQTT attacks in IoT. In: 2021 IEEE Latin-American Conference on Communications (LATINCOM), pp. 1–6.IEEE (2021)

14. Siddharthan, H., Deepa, T., Chandhar, P.: SENMQTT-SET: an intelligent intrusion detection in IoT-MQTT networks using ensemble multi cascade features. IEEE Access **10**, 33095–33110 (2022)

15. Andy, S., Rahardjo, B., Hanindhito, B.: Attack scenarios and securityanalysis of MQTT communication protocol in IoT system. In: 2017 4th International Conference on Electrical Engineering, Computer Science and Informatics (EECSI), pp. 1–6.IEEE (2017)

16. https://joseaveleira.es/dataset

17. https://www.kaggle.com/cnrieiit/mqttset

18. https://ieee-dataport.org/open-access/mqtt-iot-ids2020-mqtt-internet-things-intrusion-detection-dataset

Staged Reinforcement Learning for Complex Tasks Through Decomposed Environments

Rafael Pina(✉) , Corentin Artaud , Xiaolan Liu , and Varuna De Silva

Institute for Digital Technologies, Loughborough University London, 3 Lesney Avenue E20 3BS, London, UK
{r.m.pina,c.artaud2,xiaolan.liu,v.d.de-silva}@lboro.ac.uk

Abstract. Reinforcement Learning (RL) is an area of growing interest in the field of artificial intelligence due to its many notable applications in diverse fields. Particularly within the context of intelligent vehicle control, RL has made impressive progress. However, currently it is still in simulated controlled environments where RL can achieve its full superhuman potential. Although how to apply simulation experience in real scenarios has been studied, how to approximate simulated problems to the real dynamic problems is still a challenge. In this paper, we discuss two methods that approximate RL problems to real problems. In the context of traffic junction simulations, we demonstrate that, if we can decompose a complex task into multiple sub-tasks, solving these tasks first can be advantageous to help minimising possible occurrences of catastrophic events in the complex task. From a multi-agent perspective, we introduce a training structuring mechanism that exploits the use of experience learned under the popular paradigm called Centralised Training Decentralised Execution (CTDE). This experience can then be leveraged in fully decentralised settings that are conceptually closer to real settings, where agents often do not have access to a central oracle and must be treated as isolated independent units. The results show that the proposed approaches improve agents performance in complex tasks related to traffic junctions, minimizing potential safety-critical problems that might happen in these scenarios. Although still in simulation, the investigated situations are conceptually closer to real scenarios and thus, with these results, we intend to motivate further research in the subject.

Keywords: Reinforcement Learning · Task Decomposition · Multi-Agent Learning

1 Introduction

Reinforcement Learning (RL) is a popular subject in the field of Machine Learning whose many notable applications have raised interest within the scientific community. In fields ranging from robotics, gaming or healthcare, to autonomous vehicles or finance, RL has been proved to have relevant applications [6, 7, 18].

© The Author(s), under exclusive license to Springer Nature Switzerland AG 2024
A. Bennour et al. (Eds.): ISPR 2023, CCIS 1941, pp. 141–154, 2024.
https://doi.org/10.1007/978-3-031-46338-9_11

Due to the constraints that can be lifted in the world of computer games, the success of RL in this area is particularly notorious. For example, works such as [21] describe how RL agents have defeated the human being playing games like Go or achieved super-human performance in multiple Atari Games [14]. However, when applying the same concepts to real-world scenarios, studies have demonstrated how challenging it can be to reproduce the super-human behaviours learnt in simulation [4]. Naturally, the real-world can be very complex and it is often unfeasible to encounter in simulation all the possibilities that may appear in real life. Recent works showed how this can be achieved in controlled scenarios, where the reality is very close to the simulation, making the tasks easier [1,8,26]. However, this procedure is not always feasible outside controlled environments, since it is impossible to predict with maximum accuracy what will happen in our lives as human beings living in an extremely dynamic and stochastic world.

In the context of autonomous vehicle control, there is a wide research that focuses on approximating the use of RL in the real-world [1,9,11,17]. One of the solutions to facilitate learning complex situations is to break complex tasks into simpler tasks. If a given agent starts learning from simpler steps, then it is possible to mix this intermediate knowledge to address a more complex task that can be built from the simpler parts learned. Additionally, when learning simpler parts of a complex task, agents are less likely to cause safety problems when it comes to real applications, since they do not need to attempt and probably fail many times a very complex task that would require a long learning time.

While these problems have been deeply studied in single-agent settings, it is important to note how multi-agent systems are gaining attention, and RL has been proved to be a suitable solution to address challenges that require multiple agents to interact [15]. Multi-Agent Reinforcement Learning (MARL) is the sub-field of RL that studies its applications in muti-agent systems. Learning how to interact as a group brings endless challenges worth investigating. Multiple recent works have shown the potential of using MARL, mostly in controlled simulated environments. Famous games such as StarCraft II [20] or the Google Football environment [12] have served as the workstations for training and studying MARL agents. Other works such as [19,27] demonstrated how MARL can learn complex policies in these kind of computer simulated games. A large part of approaches proposed to tackle these problems follow a general paradigm named Centralised Training Decentralised Execution (CTDE) [10,16]. With this configuration, the agents have access to extra information of the environment during training, but are constrained to their local observations when executing their actions. Yet, when looking at real-world applications, it is often impossible to provide a centralised oracle to the agents that allows them to see the full state of the environment at any moment during their training phases. Another popular convention in these methods that tackle problems in simulated MARL is to share the network parameters of the learning agents (Fig. 1 illustrates the differences between sharing and not sharing parameters in MARL). Logically, sharing a network through a big team of agents in reality can bring

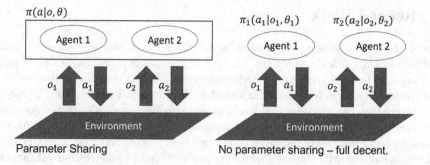

Fig. 1. Illustration of the main differences between sharing parameters (left) and not sharing parameters (right) of the agent networks in MARL.

endless problems. Communication-based approaches can be seen as a solution for MARL in real applications, but in safety critical situations, even very small latency periods can cause tragic events that must be prevented [23].

In this paper, we investigate different applications of RL and MARL in simulated environments in the context of vehicle control, while discussing how we can create methods that are closer to real-world applications. We study the problems of task decomposition in the context of obstacle avoidance and goal reaching, and also demonstrate a method for mixing simpler tasks into the respective complex task. If agents can learn efficiently in a staged manner, learning in reality can also be easier by following simpler steps in order to achieve a more difficult goal, and avoiding potential collisions that could happen if they spend too long learning a difficult task straight away. In addition, we step into the world of multi-agent systems and discuss a training framework in MARL that consists of training agents in simulation following the advantageous CTDE paradigm, and then reuse the learnt policies in a more complex environment that is fully decentralised. By structuring the learning process following this procedure, agents can take advantage of being trained under the CTDE paradigm, and then can be used in more complex settings, and retrained in a fully decentralised basis. The results suggest that this training framework accelerates learning in more complex simulated environments, reducing potential safety problems that would incur from collisions between agents. In this sense, the contributions in this paper are as follows:

- We show that certain simpler tasks can be mixed into more complex tasks to improve learning in the context of obstacle avoidance and goal reaching.
- We investigate a training framework that firstly consists of training agents under the advantageous CTDE paradigm with parameter sharing. Then, the learnt policies can be transferred to more complex environments and continue to be trained but in a fully decentralised setting (no parameter sharing) that is more suitable for future real applications.

2   Related Work

There are several practical problems in adopting RL in real applications, especially in safety critical situations. RL typically requires many iterations of trial and error to learn a given task. Although such trial and error is performed in a simulated virtual environment, transferring the intelligent capabilities learnt in simulation into the physical real world poses a significant challenge [8]. Another problem is that mistakes by robots in physical settings could lead to property damages and even costly damages to human lives [1,4]. Therefore, safety critical guarantees of such RL based controllers are required to ensure that RL agents can perform safely.

It is clear how training agents in a controlled simulated environment is a key factor to create safe entities. For instance, in the concept of Multi-Fidelity Reinforcement Learning (MFRL) [25], an RL agent is trained in multiple simulators of the real environment at varying fidelity levels. It is demonstrated that by increasing simulation fidelity, the number of samples used in successively higher simulators can be reduced [2].

In [8], the authors demonstrated that a vision-based lane following and obstacle avoidance RL agent can learn a suitable steering and obstacle avoidance policy in simulation that then can be transferred to a simple physical setting. However, it is important to note how the used scenarios are controlled and forced to minimise the reality gap (difference between simulation and reality). In [8], the authors concluded that some trained RL agents can be sensitive to problem formulation aspects such as the representation of real world actions. It was shown that, by using domain randomization, a moderately detailed and accurate simulation is sufficient for training such an agent that can operate in a real environment. Domain randomization is another method that minimises the reality gap, allowing to make RL policies learned in simulation closer to reality [26].

With the proven advantages and success of using simulated environments to analyse safety critical problems, the literature regarding approaches that tackle RL problems in more complex simulated environments encompasses a broad range [12,14,24,27]. Works such as [13] analyse how RL agents learn how to act in traffic environments in a safe way. In particular, the authors show how agents learn conventions that are compatible with human rules of driving. In the more complex multi-agent setting, MARL has been well studied. In the work of [3], the authors show how different agents learn to navigate in traffic scenarios and learn driving policies in traffic networks. Under multi-agent settings, the CTDE paradigm is important to scale the approaches [22]. Several works have demonstrated how this setting brings benefits to the performances of trained agents in multi-agent settings that must cooperate [22,24,27]. In addition, sharing the parameters of multiple agents in multi-agent scenarios is also a convention widely used in the literature [5]. Although following such procedures can be very profitable in simulation, it is very challenging to establish a reliable CTDE configuration in real applications, or to have a central entity that allows to share the same network across a number of agents. Instead, the fully decen-

Algorithm 1. Algorithm 1

Let T be a certain decomposable task
Decompose T in n sub-tasks, $\{T_1, \ldots, T_n\}$
for each sub-task in $\{T_1, \ldots, T_n\}$

    Train Q-table Qt_i
    Save Q-table Qt_i

end for
Initialise empty new table Qt_{jt}
for each Q-table in $\{Qt_1, \ldots, Qt_n\}$

    if Qt_{jt} is empty **then**
        Add all Qt_i entries
    else if Qt_i entries in Qt_{jt} **then**
        Combine entries
    else
        Pass

end for
Output joint Q-table Qt_{jt}

---

tralised setting sounds much more inviting, since each agent can be treated as an isolated independent entity.

3   Methods

3.1   Task Factorization

In this section, we introduce a method for solving single-agent tasks using RL, based on the assumption that a certain complex task can be factorized into multiple simple sub-tasks. These sub-tasks are seen as intermediate objectives that must be learned to solve a global task. In this sense, we train multiple Q-tables in separate to solve sub-tasks that together form a main complex task. By learning these tables, it is possible to create a joint Q-table that contains a mix of the Q-tables corresponding to the learned sub-tasks. The agents use Q-learning [28] to learn the Q-tables. This algorithm is part of the foundations of RL, where the updates to optimise a Q-function to solve a certain problem are made by following the update rule

$$Q(s, a) = (1 - \alpha)Q(s, a) + \alpha \left[r + \gamma \max_{a'} Q(s', a') \right] \tag{1}$$

for a certain state and action pair (s, a), and the corresponding next pair (s', a'), where α is a learning rate, γ a discount factor, and r the reward received.

In summary, this task decomposition procedure can be formalized as the following: given a certain task T, it is possible to decompose it into n sub-tasks. Thus, n Q-tables can be learned separately and then exported to form a joint

Q-table, Qt_{jt}, which is a combination of all the sub Q-tables (as described in Algorithm 1).

To demonstrate our assumption, we show ahead in the results section how a complex task can be solved using the joint Q-table that arises from this procedure. This table results from the combination of the separate Q-tables trained in the sub-tasks in which the complex task was decomposed. Hence, the joint Q-table will perform in an environment where it was not trained in advance.

3.2   From CTDE to Full Decentralisation

The second approach that we discuss in this paper is related to MARL problems. This approach consists of a structured training procedure that is composed by two main stages that will be described in detail in this section. In the first stage, we train a team of agents in a simpler environment under the popular CTDE paradigm with parameter sharing. To do so, we use Value Decomposition Networks (VDN) [24], a popular algorithm that is part of a big family of MARL value function factorization algorithms. The key idea of this family of methods is to learn a factorization of a joint Q-function Q_{tot} into a set of individual Q-functions corresponding to each one of the agents in the multi-agent team.

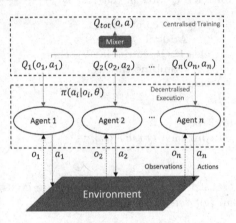

Fig. 2. A simplified overview of how value function factorization methods operate under the CTDE paradigm, considering pairs of observations and actions.

In the case of VDN, this factorization is achieved through addition and can be formally represented by the equation

$$Q_{tot}\left(\tau, a\right) = \sum_{i=1}^{N} Q_i(\tau_i, a_i; \theta_i) \tag{2}$$

for a set of action-observation histories and action pairs (τ_i, a_i), and the θ values are the parameters of the agent neural networks.

To learn this factorization, these methods use this mixer that mixes individual Q-values into the joint Q-value. This mixer can be very different from method to method and it is the key factor to the type of Q-functions that each method can factorise (representational complexity). Formally, a value function factorization method is said to be effective if it satisfies the Individual-Global-Max (IGM) [22] condition

$$\operatorname*{argmax}_{a} Q_{tot}\left(\tau, a\right) = (\operatorname*{argmax}_{a_1} Q_1(\tau_1, a_1), \ldots, \operatorname*{argmax}_{a_N} Q_N(\tau_N, a_N)) \tag{3}$$

In simple terms, this condition means that, for an efficient factorisation, the set of local optimal actions should maximise the joint Q_{tot}. Figure 2 depicts an overview of how value function factorisation methods operate with parameter sharing and under the CTDE paradigm.

The second stage starts after training the agents in the simpler environment following the procedure described above. In this stage, we transfer the trained agents to a more complex environment that is now fully decentralised. This means that, on top of the environment being more complex, the agents only have access to their local observations and do not share parameters anymore (as summarised in Fig. 1). In the considered full decentralised setting, each agent is treated as an independent entity. Thus, we use the famous DQN method [14] to control each one of the agents as an independent agent. Since we do not share parameters, beyond not using any CTDE method in this stage, each agent has its own DQN controller. We refer to this method ahead as IDQL (Independent Deep Q-learning). Following the authors in [14], agents using this method learn by minimising the loss

$$\mathcal{L}(\theta) = \mathbb{E}_{b \sim B} \left[\left(r + \gamma \max_{a'} Q(\tau', a'; \theta^-) - Q(\tau, a; \theta) \right)^2 \right] \tag{4}$$

where b represents a set of experience samples from a replay buffer B and θ^- the parameters of a target network that stabilizes learning.

Since we aim to discuss the application of this procedure towards more real-world friendly applications, we show in particular how this method helps in traffic junction environments to prevent collisions between vehicles. If agents can minimise the collisions on a more complex environment by leveraging the experience gained in a simpler one, then the agents trained are capable of making more secure decisions in these traffic scenarios.

4   Experiments and Results

In this section we conduct a set of experiments to support the hypotheses introduced in this paper. In simulated traffic junction environments we show that, in single agent settings, certain tasks can be decomposed in simpler sub-tasks that are then combined to solve a more complex task. In multi-agent settings, we investigate the discussed training structure that involves leveraging experience from CTDE and using it to help in more complex fully decentralised settings. Note that, while in both subsections of experiments we use a Traffic Junction environment, this environment has different configurations in the different experiments that we clarify ahead.

4.1   Task Factorization

The experiments in this subsection aim to show that a complex task based on a traffic junction (Fig. 3a) can be learned by breaking it in two simpler sub-tasks. In the used environment in this subsection, the task involves one agent (blue

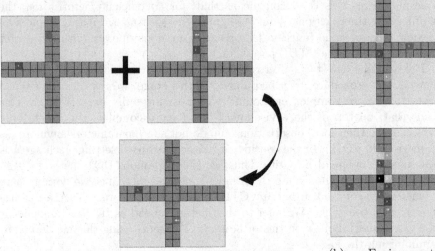

(a) Environments used in section 4.1: sub-tasks of goal reaching and obstacle avoidance (top, from the left to the right) and the joint task (bottom).

(b)   Environments used in section 4.2: Traffic Junction with 4 agents (top) and 10 agents (bottom).

Fig. 3. Environments used in the experiments. Although they all are based on a traffic junction, their settings are different according to the experiments in Sects. 4.1 and 4.2 (see the sections for details). (Color Figure Online)

square in Fig. 3a) that enters the road through one out of four possible roads and has to go to a goal place (red box in Fig. 3a), while avoiding a second vehicle (brown square in Fig. 3a) that is harcoded to try to collide with the agent and thus it has to avoid the collisions. The agent receives a reward of +5 when it reaches the goal and a punishment of -0.2 when it collides with the other vehicle. In addition, note that the environment is not fully observable, meaning that the agent only sees its own location and has a 3×3 observation mask to observe the surroundings. In line with our needed assumption for this experiment, this task can be broken down in two sub-tasks: 1) going from the starting location to the goal (without any obstacles or other vehicles to avoid) (top left in Fig. 3a) and 2) roam around the road while avoiding collisions with the hardcoded vehicle that tries to collide and must be avoided (without any goal place to find) (top right in Fig. 3a). Figure 4a and 4b illustrate the performance of the learned Q-tables on the 2 described sub-tasks for this scenario and Fig. 4c shows the combination of the tables (joint Q-table) compared against a Q-table trained directly in the complex environment (we refer to this as simple Q-table). The illustration of the complex task that results from the sub-tasks can be found in the bottom of 3a.

By looking at the figures, we can see that the agent solves easily the simpler tasks, achieving optimal rewards after some time. However, when looking at the more complex task that is a combination of the two simpler ones, we can see

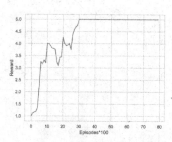

(a) Sub-task 1: Goal reaching.

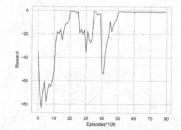

(b) Sub-task 2: Obstacle avoidance.

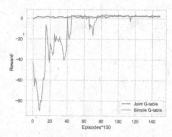

(c) Main task: goal reaching with obstacle avoidance.

Fig. 4. Rewards achieved in the sub-tasks (upper plots) and in the complex task (bottom) as described in Sect. 4.1. The rewards are smoothed with a moving average with a 5-step sliding window.

that an agent trained with the simple Q-table takes longer to learn. Although it eventually learns the task, it takes some time to get there and, if this would happen in a real scenario, the agent would cause a lot of physical damage before learning the task successfully. On the other hand, when we use the joint Q-table that mixes the two simpler learned tasks as per Algorithm 1, we can see that the agent can reuse what was learned in the simpler tasks and use that experience to learn much faster the complex one, avoiding the collisions that incur when using the simple Q-table directly. The results presented enhance that simple tasks can be learned, and the experience can be transferred to complex tasks. This can facilitate solving harder problems by breaking them in easier sub-tasks that can be more easily reproduced rather than stepping directly to a complex task that requires more effort and may cause problems. Also by doing so, the agent avoids many crashes that could happen by training in the complex task blindly.

4.2   CTDE to Decentralisation

As discussed in the previous sections, this training scheme aims to train a team of agents to learn how to negotiate their passages in a traffic junction environ-

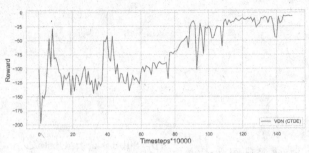

(a) VDN in Traffic Junction with 4 agents.

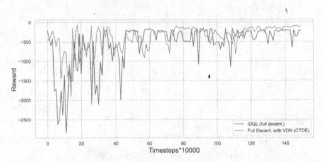

(b) Traffic Junction with 10 agents for IDQL vs IDQL boosted with pre-trained VDN (in the 4 agents environment).

Fig. 5. On the top, the rewards on the simple Traffic Junction with 4 agents (Fig. 3, top), using VDN. On the bottom, the rewards for independent deep Q-learning (without parameter sharing) when starting with the policies learned in the simpler task with 4 agents using VDN (top), but now in a 10-agent scenario (Fig. 3, bottom). The latter is compared against fully independent deep Q-learners (IDQL, also without parameter sharing) in the same 10-agent environment, but without the VDN starting boost.

ment (Fig. 3b) while avoiding collisions with other agents. As mentioned before, note that, in this set of experiments, the configurations of the environment are different from the previous. In this set of experiments, we consider multi-agent settings where the goal is for each agent to reach a pre-defined destination at one of the other ends of the junction that is assigned to them. Each agent receives a punishment of -10 if it collides with another agent and also a penalty of -0.01 every step to incentivize them to reach the goal as fast as possible. In addition, the environment is also partially observable, and thus each agent only sees its own location, together with a 3×3 observation mask to observe the surroundings, and a step counter. Following the training structure described in Sect. 3.2, we start by training VDN in the traffic junction environment with 4 agents. When using 4 agents, the environment is relatively simple to solve, and thus we can see that convergence is reached and remains until the end of training (Fig. 5a). By observing Fig. 5a, we can say that the method has reached an optimal reward at the end of training and the agents have learned how to solve the task. Because the method also uses the parameter sharing convention to speed up training, at the end of training we end up with only 2 different trained networks: a mixing network and a policy network that is shared by all the agents (as summarised previously in Figs. 1 and 2).

At the end of training, the trained policy can be transferred to a more complex environment to be used without a mixing network, i.e., in a fully decentralised fashion without using CTDE. Importantly, since we intend to increase the number of agents in the more complex environment to demonstrate how this training scheme aids in scaling to more agents, in the simpler environment we must pad the policy network to be able to accommodate the additional agents in the harder environment. Then, this network is replicated across the independent agents to boost the training. As Fig. 5a shows, we can see that using padding does not affect the performances in the 4-agent environment since they still show an increasing performance that converges during training. To make the scenario as close to reality as possible, we do not use parameter sharing in the fully decentralised version. This allows us to treat each agent as an isolated unit that does not have any access to any extra information besides their own individual observations of the environment. In Fig. 5b we can see that, when we use the transferred policy networks in the harder environment with 10 agents, the team improves the performances when compared to when they would have not taken advantage of CTDE before. In fact, we can see that, in the latter, agents take much longer to achieve a higher reward, meaning that they would cause much more collisions in the traffic environment before learning. Besides taking longer to learn, with simple fully independent learning the agents fail to achieve as good performance at the end of training as the ones with the prior boost of using policies that were trained before in the advantageous CTDE. These results show that, although CTDE is mostly only feasible in simulation, we can still take advantage of its potential to be used to improve learning in less advantageous scenarios that relate more to reality-compatible applications, helping to prepare agents for more complex situations than what they are used to, and minimising potential safety-critical problems.

5   Conclusion and Further Work

While RL has shown remarkable advances in simulated environments, approximating it to solve real problems is still challenging. In this paper, we have investigated two concepts that, if successfully applied to simulated environments, can ease the reuse of simulated RL in reality. Executing complex tasks in reality can be facilitated if these tasks are broken down to simpler sub-tasks that have been solved before. Logically, mapping simpler tasks in simulation is easier than start solving a complex task straight away. From a multi-agent perspective, with the increasing complexity of multi-agent problems, there is a need to take all the possible advantage from simulated environments. In this sense, it is important to understand how state-of-the-art methods that solve complex simulated environments can be efficiently leveraged to aid in real scenarios. With our preliminary experiments, we introduce how these can be used to tackle problems that are conceptually closer to real scenarios, although still in simulation.

In the future, we aim to study other ways of combining sub-tasks, and how tasks can also be decomposed for multi-agent settings in an automated manner. Furthermore, we intend to extend the described training structuring mechanism for MARL directly from simulation to real-world scenarios. We also intend to study how we can better preserve the learned performances from one stage to the other in this mechanism.

With this work, we aim to incentivise further research on how both single-agent and multi-agent reinforcement learning methods can be trained in simulation and then mapped to reality with reduced losses of performance. While the presented experiments are still in simulation, we believe that these preliminary results can inspire further research in this direction.

References

1. Almási, P., Moni, R., Gyires-Tóth, B.: Robust reinforcement learning-based autonomous driving agent for simulation and real world. In: 2020 International Joint Conference on Neural Networks (IJCNN), pp. 1–8 (2020). https://doi.org/10.1109/IJCNN48605.2020.9207497
2. Chebotar, Y., et al.: Closing the sim-to-real loop: adapting simulation randomization with real world experience. In: 2019 International Conference on Robotics and Automation (ICRA), pp. 8973–8979 (2019). https://doi.org/10.1109/ICRA.2019.8793789
3. Chu, T., Chinchali, S., Katti, S.: Multi-agent reinforcement learning for networked system control. In: International Conference on Learning Representations (2020). https://openreview.net/forum?id=Syx7A3NFvH
4. Dulac-Arnold, G., Mankowitz, D., Hester, T.: Challenges of real-world reinforcement learning (2019). https://arxiv.org/pdf/1904.12901.pdf
5. Gupta, J.K., Egorov, M., Kochenderfer, M.: Cooperative multi-agent control using deep reinforcement learning. In: Sukthankar, G., Rodriguez-Aguilar, J.A. (eds.) Autonomous Agents and Multiagent Systems, pp. 66–83. Springer International Publishing, Cham (2017). https://doi.org/10.1007/978-3-319-71682-4_5
6. Hester, T., Quinlan, M., Stone, P.: RTMBA: a real-time model-based reinforcement learning architecture for robot control. In: 2012 IEEE International Conference on Robotics and Automation, pp. 85–90 (2012). https://doi.org/10.1109/ICRA.2012.6225072
7. Hu, Y.J., Lin, S.J.: Deep reinforcement learning for optimizing finance portfolio management. In: 2019 Amity International Conference on Artificial Intelligence (AICAI), pp. 14–20 (2019).https://doi.org/10.1109/AICAI.2019.8701368
8. Kalapos, A., Gór, C., Moni, R., Harmati, I.: Sim-to-real reinforcement learning applied to end-to-end vehicle control. In: 2020 23rd International Symposium on Measurement and Control in Robotics (ISMCR), pp. 1–6 (2020). https://doi.org/10.1109/ISMCR51255.2020.9263751
9. Kober, J., Bagnell, J.A., Peters, J.: Reinforcement learning in robotics: a survey. Int. J. Robot. Res. **32**(11), 1238–1274 (2013)
10. Kraemer, L., Banerjee, B.: Multi-agent reinforcement learning as a rehearsal for decentralized planning. Neurocomputing **190**, 82–94 (2016)

11. Krishna Lakshmanan, A., et al.: Complete coverage path planning using reinforcement learning for tetromino based cleaning and maintenance robot. Autom. Constr. **112**, 103078 (2020)

12. Kurach, K., et al.: Google research football: a novel reinforcement learning environment. arXiv:1907.11180 (2020)

13. Lerer, A., Peysakhovich, A.: Learning existing social conventions via observationally augmented self-play. In: Proceedings of the 2019 AAAI/ACM Conference on AI, Ethics, and Society, pp. 107–114. AIES 2019, Association for Computing Machinery, New York, NY, USA (2019). https://doi.org/10.1145/3306618.3314268

14. Mnih, V., et al.: Human-level control through deep reinforcement learning. Nature **518**, 529–533 (2015)

15. Nguyen, T.T., Nguyen, N.D., Nahavandi, S.: Deep reinforcement learning for multiagent systems: a review of challenges, solutions, and applications. IEEE Trans. Cybern. **50**(9), 3826–3839 (2020). https://doi.org/10.1109/TCYB.2020.2977374

16. Oliehoek, F.A., Spaan, M.T.J., Vlassis, N.: Optimal and approximate Q-value functions for decentralized POMDPs. J. Artif. Int. Res. **32**(1), 289–353 (2008)

17. Pina, R., Tibebu, H., Hook, J., De Silva, V., Kondoz, A.: Overcoming challenges of applying reinforcement learning for intelligent vehicle control. Sensors **21**(23), 7829 (2021). https://doi.org/10.3390/s21237829

18. Pineau, J., Guez, A., Vincent, R., Panuccio, G., Avoli, M.: Treating epilepsy via adaptive neurostimulation: a reinforcement learning approach. Int. J. Neural Syst. **19**(04), 227–240 (2009)

19. Ruan, J., et al.: GCS: graph-based coordination strategy for multi-agent reinforcement learning. In: Proceedings of the 21st International Conference on Autonomous Agents and Multiagent Systems, pp. 1128–1136. AAMAS 2022, International Foundation for Autonomous Agents and Multiagent Systems, Richland, SC (2022)

20. Samvelyan, M., et al.: The starcraft multi-agent challenge. In: Proceedings of the 18th International Conference on Autonomous Agents and MultiAgent Systems, pp. 2186–2188. AAMAS 2019, International Foundation for Autonomous Agents and Multiagent Systems, Richland, SC (2019)

21. Silver, D., et al.: Mastering the game of go with deep neural networks and tree search. Nature **529**, 484–489 (2016). https://doi.org/10.1038/nature16961

22. Son, K., Kim, D., Kang, W.J., Hostallero, D.E., Yi, Y.: QTRAN: learning to factorize with transformation for cooperative multi-agent reinforcement learning. In: Chaudhuri, K., Salakhutdinov, R. (eds.) Proceedings of the 36th International Conference on Machine Learning. Proceedings of Machine Learning Research, vol. 97, pp. 5887–5896. PMLR (2019). https://proceedings.mlr.press/v97/son19a.html

23. Sun, Y., Kuai, R., Li, X., Tang, W.: Latency performance analysis for safety-related information broadcasting in VeMAC. Trans. Emerg. Telecommun. Technol. **31**(5), e3751 (2020)

24. Sunehag, P., et al.: Value-decomposition networks for cooperative multi-agent learning. In: Proceedings of the 17th International Conference on Autonomous Agents and MultiAgent Systems, pp. 2085–2087. Stockholm, Sweden, (2018)

25. Suryan, V., Gondhalekar, N., Tokekar, P.: Multifidelity reinforcement learning with gaussian processes: model-based and model-free algorithms. IEEE Robot. Autom. Mag. **27**(2), 117–128 (2020). https://doi.org/10.1109/MRA.2020.2977971

26. Tobin, J., Fong, R., Ray, A., Schneider, J., Zaremba, W., Abbeel, P.: Domain randomization for transferring deep neural networks from simulation to the real world. In: 2017 IEEE/RSJ International Conference on Intelligent Robots and Systems (IROS), pp. 23–30 (2017). https://doi.org/10.1109/IROS.2017.8202133

27. Wang, J., Ren, Z., Liu, T., Yu, Y., Zhang, C.: QPLEX: duplex dueling multi-agent Q-learning. In: International Conference on Learning Representations (2021). arXiv: 2008.01062
28. Watkins, C.J., Dayan, P.: Technical note: Q-learning. Mach. Learn. **8**, 279–292 (1992). https://doi.org/10.1023/A:1022676722315

Policy Generation from Latent Embeddings for Reinforcement Learning

Corentin Artaud(✉) ⬤, Rafael Pina⬤, Xiyu Shi⬤, and Varuna De-Silva⬤

Institute for Digital Technologies, Loughborough University London,
The Broadcast Centre Here East, 3 Lesney Ave, E20 3BS, London, UK
{c.artaud2,r.m.pina,x.shi,v.d.de-silva}@lboro.ac.uk

Abstract. The human brain endows us with extraordinary capabilities that enable us to create, imagine, and generate anything we desire. Specifically, we have fascinating imaginative skills allowing us to generate fundamental knowledge from abstract concepts. Motivated by these traits, numerous areas of machine learning, notably unsupervised learning and reinforcement learning, have started using such ideas at their core. Nevertheless, these methods do not come without fault. A fundamental issue with reinforcement learning especially now when used with neural networks as function approximators is their limited achievable optimality compared to its uses from tabula rasa. Due to the nature of learning with neural networks, the behaviours achievable for each task are inconsistent and providing a unified approach that enables such optimal policies to exist within a parameter space would facilitate both the learning procedure and the behaviour outcomes. Consequently, we are interested in discovering whether reinforcement learning can be facilitated with unsupervised learning methods in a manner to alleviate this downfall. This work aims to provide an analysis of the feasibility of using generative models to extract learnt reinforcement learning policies (i.e. model parameters) with the intention of conditionally sampling the learnt policy-latent space to generate new policies. We demonstrate that under the current proposed architecture, these models are able to recreate policies on simple tasks whereas fail on more complex ones. We therefore provide a critical analysis of these failures and discuss further improvements which would aid the proliferation of this work.

Keywords: Reinforcement Learning · Policy Modeling · Deep Generative Models

1 Introduction

As humans, our cognitive capabilities allow us to form concepts, apply logic, recognise patterns, and generate ideas whilst learning from these abilities. Aspiring from these unique processes and their abundant sources of inspiration, areas of machine learning, most notably reinforcement learning and unsupervised learning, have quickly become an important field of research aiming to mimic

A. Bennour et al. (Eds.): ISPR 2023, CCIS 1941, pp. 155–168, 2024.
https://doi.org/10.1007/978-3-031-46338-9_12

such abilities. The reinforcement learning paradigm heavily influences itself on our biological behaviours [16], consisting of an agent that interacts with an environment to receive rewards or regrets. Many real-world scenarios like robotics [8], energy systems [23], and automation [18] apply this particular setting. Unsupervised learning, on the other hand, is an algorithmic technique [19] which discovers underlying patterns, or groupings, in unclassified or unlabelled data. Specifically generative models involve discovering and internalising the essence of data to generate it.

Since its discovery, reinforcement learning has shown breakthroughs in domains that were previously thought unachievable. Lately, with the advancements of computational resources, complex problems that were once unsolvable using reinforcement learning have demonstrated state-of-the-art results by exploiting these advancements [3]. Specifically relevant to this research, the conjunct uses of reinforcement learning and generative models have also become prevalent because of these advancements. Robotics is one such domain where artificially generating data reduces the bottleneck of acquiring enormous amounts of data to train reinforcement learning algorithms, a task especially prevalent for generative models. Studies like [14,20] have shown significant results by using both generative models and reinforcement learning in this specific use case.

Although multiple applications of reinforcement learning and deep generative models conjunct uses have arisen, the approach undertaken remains broadly similar. They primarily extract data from an environment, train a generative model to generate synthetic data, and finally solve a reinforcement learning algorithm using the synthetically generated data. Albeit proven to provide sample-efficient results, this methodology, optimised against the data it was trained on, fails to transfer through domains and restricts the generative model's generalisation abilities. Recent studies presented have shown to alleviate some of these limitations.

A significant part of studies continues to use this underlying methodology due to the abundant requirement of data to solve reinforcement learning problems. However, currently, none have shown a means to use these generative models to generate a solved reinforcement learning problem. Therefore, this research aims to address the question: *can a synthetically generated reinforcement learning algorithm produce better than or equal performance versus its modelled counterpart*. This work intends to discover whether primarily a generative model can find the underlying patterns as to how reinforcement learning algorithms are solved and, secondly, generate close to identical results compared to those modelled.

2   Background and Preliminaries

In this work, we consider *markov decision processes* (MDPs) [17] which fully describes the agent-environment interaction process when dealing with reinforcement learning (RL) and *deep generative models* (DGMs), specifically variational autoencoders when we move to the unsupervised learning section of our proof-of-concept.

MDPs are described as a tuple $\langle S, A, p, r, \gamma \rangle$, where S defines a finite set of states, $s \in S$, where s represents the current state of the environment; A defines the finite set of actions, $a \in A$, where a represents one action the agent can take. r represents the reward function $r(s, a) = \mathbb{E}[r_{t+1}|s_t, a_t]$ for a given state s and action a at time-step t. The transitions to the next state s_{t+1} by following an action a from state s are represented using the state transition probability p where $p(s, a, s') = P(S_{t+1} = s'|S_t = s, A_t = a)$ for time-step $t + 1$ at time-step t. Finally, let γ represent the discount factor where $\gamma \in [0, 1]$.

Q-**Learning.** We aim to learn policies that maximise the expected discounted reward (returns). Q-learning is concerned with learning an optimal Q-function, $Q^* : S \times A \rightarrow \mathbb{R}$, to select actions that maximise the expected return. The optimal Q-function is defined as:

$$Q^*(s, a) := \mathbb{E}\Big[\sum_{t=0}^{\infty} \gamma^t r(s_t, a_t) \mid s_0 = s, a_0 = a, s_{t+1} \sim p(\cdot \mid s_t, a_t) \Big]$$

$$= r(s, a) + \gamma \mathbb{E}[\max Q(s', \cdot) \mid s' \sim p(\cdot \mid s, a)]$$

$$(1)$$

To learn the Q-function tractably, we can represent it using deep reinforcement learning methods that use neural networks as their function approximators that are trained to maximise the loss function:

$$\mathcal{L}_Q(\theta) = \mathbb{E}\Big[(y_t - Q_\theta(s, a))^2 \mid (s, a, r, s') \sim U(D) \Big]$$

$$y_t = r + \gamma \max_{a'} Q_{\bar\theta}(s', \arg\max Q_\theta(s, \cdot))$$

$$(2)$$

where $\bar\theta$ represents the parameters of the target network copied periodically from θ introduced in [12] to improve stability. D is the replay buffer [11] which collects the transitions from an ϵ-greedy policy (or another exploratory policy) with $U(D)$ representing a uniform sampling of these transitions from the buffer.

Conditional Variational Autoencoder. We focus on the generative aspect of our method and to do so turn to unsupervised learning methods namely, the conditional variational autoencoder. A generative model is a probabilistic model that describes the essence of data observed such that it can be sampled to generate new data. Given a dataset $\mathbf{X} = \{\mathbf{x}_i\}_{i=1}^n$ of i.i.d. samples, the goal is to train a network by maximising the marginal likelihood of observing data $\mathbf{x}_i$ under a model. Therefore, a generative model can be viewed as a mapping from low-dimensional space K to high dimensional space D denoted as $g_\theta : \mathbb{R}^K \rightarrow \mathbb{R}^D$ where $K \ll D$ and where the low-dimensional latent variable $\mathbf{z} \mapsto \mathbf{x}$. Since K is a low-dimensional manifold, any sampled x outside this manifold would lead to a likelihood of $p(\mathbf{x}) = 0$ almost everywhere [1].

Although the architectural framework of variational autoencoders has shown tremendous successes, a fundamental issue exists in their generative process whereby they are unable of generating conditioned data. To alleviate the issue, [22] propose conditional variational autoencoder (CVAE). Here the generated output $\mathbf{x}$ from the conditional distribution $p_\theta(\mathbf{x}|\mathbf{z}, \mathbf{y})$, is conditioned on some given observation $\mathbf{y}$ and latent variable $\mathbf{z}$ sampled from the prior $p_\theta(\mathbf{z}|\mathbf{y})$. The learning objective, that aims to maximise $\log p_\theta(\mathbf{y}|\mathbf{x})$ is as follows:

$$\mathcal{L}_{\text{CVAE}}(\theta, \phi; \mathbf{x}, \mathbf{y}) = \mathbb{E}_{q_\phi}[\log p_\theta(\mathbf{x}|\mathbf{z}, \mathbf{y})] - D_{KL}(q_\phi(\mathbf{z}||\mathbf{x}, \mathbf{y})|p_\theta(\mathbf{z}|\mathbf{y})) \qquad (3)$$

3   Policy Generation from Latent Embeddings (PGLE)

We now propose **P**olicy **G**eneration from **L**atent **E**mbeddings (**PGLE**). We aim to discover whether a set of learnt policy models (i.e. network parameters) for the same task present common structures that are exploitable by generative models with the goal of extracting latent embeddings that enable a generative process on a policy-space. However, we assume that these policy-spaces do not only contain optimal solutions. By doing so we hypothesise that providing such differentiable samples and conditining each of these samples on their expected returns would enable our generative model to separate *good* from *bad* policies.

3.1   Method

Given a set of trained policy networks each consisting of l layers with different dimensions, we first require a procedure that will enable us to transform each parameterised layer within each network into a representative *sample* creating a dataset of network parameters for our conditional variational autoencoder. We first retrieve for each layer within a network its set of weights $\mathbf{W} = \{\mathbf{w}^{(i)} \in \mathbb{R}^{m \times n} : i = 1, \ldots, l\}$, where m and n denotes the row and column sizes, respectively, of each layer i, and biases $\mathbf{B} = \{\mathbf{b}^{(i)} \in \mathbb{R}^k : i = 1, \ldots, l\}$ where k denotes the associated row size of each layer i associated to weight i (i.e. $k = n, \forall i \in l$). Our aim is to generate a dataset $\mathbf{X} = \{[\mathbf{W}\ \mathbf{B}]_i : i = 1, \ldots, k\}$ where k denotes the number of samples.

We firstly convert the bias vectors from row to column vectors denoted by $\mathbf{B} = \{\mathbf{b}^{(i)} \in \mathbb{R}^{n \times 1} : i = 1, \ldots, l\}$ and let $\alpha = \max_{i \in [l]} |\mathbf{w}_{n,:}^{(i)}|$ denote the column size of the largest weight matrix. For example, given two matrices $m \times n$ and $m \times k$ with $k > n$, then $\alpha = k$. Further we transpose all weight matrices where the number of column entries for one layer are equivalent to the row entries of its previous layer. More formally, if the number of column entries n in $\mathbf{w}^{(i)}$ is equal to the number of row entries m in $\mathbf{w}^{(i-1)}, \forall i \in l$, then let $\mathbf{w}^{(i)} = (\mathbf{w}^{(i)})^{\mathrm{T}}$. Consequently, expanding the set of weights to α such that $\mathbf{W} = \{\mathbf{w}^{(i)} \in \mathbb{R}^{\alpha \times n} : i = 1, \ldots, l\}$ and the set of biases such that $\mathbf{B} = \{\mathbf{b}^{(i)} \in \mathbb{R}^{\alpha \times 1} : i = 1, \ldots, l\}$ we can then retrieve a dataset $\mathbf{X} = \{[\mathbf{W}\ \mathbf{B}]_i \in \mathbb{R}^{\alpha \times \beta} : i = 1, \ldots, k\}$ concatenated column-wise with β representing the number of column entries after concatenating all layers. We repeat this procedure on all networks individually to retrieve k samples.

3.2 Implementation Details

The model architecture of our policy network consists of a three-layered, fully-connected feed-forward network with a rectifier (ReLU) transform at its middle layer, more details including the hyperparameters used can be found in the Appendix. We use a *conditional variational autoencoder* (CVAE) as our generative model of choice in two variants, a classic fully-connected multi-layered perceptron CVAE (MLP-CVAE) and a convolutional CVAE (Conv-CVAE). Our reasoning behind the use choice of a convolutional CVAE is as follows, after examining a visual representation of the stitched network parameters it clearly showcased image like features such as patched regions of high and low weight values matching the objective of a convolution.

4 Experimental Results

In this section, we present the designed experiments to evaluate the following questions regarding our analysis: 1) Is the underlying learnt policies (i.e. network parameters) structure learnable? and 2) Does generated parameters from the latent representation provide behaviours that are as good as the underlying data? We evaluate our proposed concept using various classic control environments, namely CARTPOLE, MOUNTAINCAR, and LUNARLANDER which are all accessible within the OpenAI Gym [2] framework. All environments are fully-observable and composed each with their respective discrete action-spaces (more details in Appendix A).

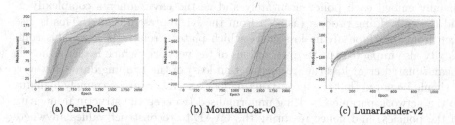

(a) CartPole-v0   (b) MountainCar-v0   (c) LunarLander-v2

Fig. 1. Plots for the results of the experiments in the multiple environments described. The x axis corresponds to the number of epochs. The y axis represents the median of the running mean (bold) reward bounded by their 25th and 75th percentiles (shadows). Each line represents an independent run

Data Collection: Each policy is trained for 2000 epochs for CARTPOLE and MOUNTAINCAR and 1000 epochs for LUNARLANDER. Each environment is evaluated every 50 epoch to record the network parameters and mean accumulated evaluation reward over 100 episodes. Consequently, for each policy trained, 40 individual networks are recorded for CARTPOLE and MOUNTAINCAR, and 20 for LUNARLANDER. The algorithms hyperparameter values are set identically for each environment except for the exploration rate (see Appendix B Table 2 for details). It is the only exception as the exploration-exploitation trade-off depends on the environment's number of time-steps relative to the number of epochs. We showcase five hand-picked runs within each environment in Fig. 1 showcasing the various policies learnt and collected during this stage.

Policy Generation and Evaluation: After training the CVAE variants against a specific environment, the model is conditionally sampled against some desired reward to generate a policy network. The CVAE generates policy networks with dimensions respective to those trained. Therefore, the data transformed described is reversed to assess them (i.e. padded values are stripped, and inverses of the transformations are applied). Once reverted, the generated networks are evaluated using exploitation only for 100 episodes. If the median returned reward meets its conditioned expected reward, the generative process is considered to have generated an optimal imagined policy (see Table 3 for details).

4.1   A Manifold for Network Parameter Sets

The major desiderata of our method relies on the expectation that our trained policies (i.e. the weights and biases of each layer within the network architecture) contain enough information to embed them onto a lower-dimensional manifold enabling a sampling processes to further generate unique, unseen, overperforming policies that would require the equivalent training time as our trained policies used for learning our generative model. At first glance, the results achieved in Fig. 2 contradict our hypothesis laid forth. It suggests that the CVAE was architecturally not robust enough to properly separate and conditionally embed each policy accurately and as the environments complexify so did the embedding process (as seen from the KL representations). This is also seen from the slope of the loss curve which plateau relatively fast. This complexity also impacted the reconstructions of the networks which is seen from the mean-squared error loss (i.e. reconstruction loss) of the training data. However, it should be noted that the reconstruction losses seen here are extremely high due to the network parameters being unnormalised (to keep the structural integrity of the policies) and hence requiring the CVAE to reconstruct values anywhere within the reals.

4.2   Sampled Performance Vs. Optimality

When comparing the results seen in Fig. 2 and the sampled outcomes seen in Table 1 we can see that even though the overall loss is strongly effected by the reconstruction error for all environments, the policies generated for CART-POLE were equivalent to a policy trained from scratch for a minimum of 1000 episodes. This outcome, however, was only possible with the fully-connected model, demonstrating that our initial analysis of patched weighted values, if left unaltered, do not contain the representational structure required for a convolution operation.

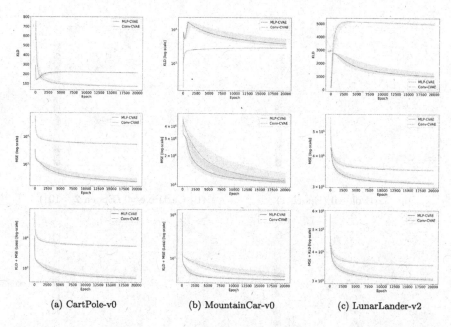

(a) CartPole-v0   (b) MountainCar-v0   (c) LunarLander-v2

Fig. 2. Plots for the results of training MLP-CVAE (blue) and Conv-CVAE (orange). The x axis corresponds to the number of epochs. The y axis represents the running median error for the Kullback-Leiber Divergence, MSE, and overall loss, in log-scale, respective of Eq. 3 bounded by their 25th and 75th percentiles (shadows). (Color figure online)

4.3 Visualising the Embeddings

Since the model for MOUNTAINCAR showcased no generative capabilities whatsoever we now solely concentrate on the other two environments used and further try to understand the latent embeddings performed at two different training stages under the MLP-CVAE. We showcase these embeddings using tSNE plots in Fig. 3. For tSNE the parameters perplexity, learning curve, and the

Table 1. Results of median accumulated rewards rounded to 2 decimal places over 100 consecutive episodes with their respective standard deviations of the sampled Q-networks conditioned on rewards for each environment after training MLP-CVAE and Conv-CVAE.

Environment	Architecture		Conditioned Reward
	MLP-CVAE	Conv-CVAE	
CartPole-v0	150.53 ± 9.28	19.0 ± 3.0	200
MountainCar-v0	-200.0 ± 0.0	-200.0 ± 0.0	-130
LunarLander-v0	-294.72 ± 15.73	-399.25 ± 20.21	200

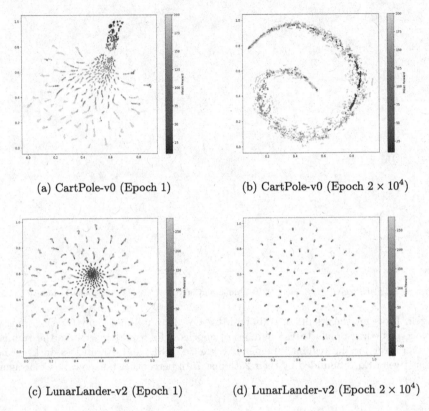

(a) CartPole-v0 (Epoch 1)   (b) CartPole-v0 (Epoch 2×10^4)

(c) LunarLander-v2 (Epoch 1)   (d) LunarLander-v2 (Epoch 2×10^4)

Fig. 3. tSNE plots for CartPole-v0 and LunarLander-v2 at Epoch 1 and 2×10^4 of their respective means (μ) of the encoder networks used for MLP-CVAE

number of iterations were fine-tuned to 100, 10, and 5000, respectively. From these plots, it becomes clearer why the MLP-CVAE was able to generate policies that were above expectation for CARTPOLE while unable to generate them for LUNARLANDER.

The results demonstrate that the manifold created for CARTPOLE separates the learnt policy networks while still keeping a close correlation of the evolution of each trained network that seem to all be learnt in a relatively similar manner. In contrast, the embeddings seen for LUNARLANDER suggest that each policy learnt produces dissimilar policies. These plots reflect the results found by sampling the CVAE and generating untrained resulting networks. It also suggests that the conditioning of the mean evaluation reward did not seem to have an effect. The results attained from these plots indicate that further analysis of the network weight structures should be performed such that deeper conclusions may be attained for these poor generative results.

Together, these results suggest that an MLP-CVAE can generate some policies, but these are sub-optimal, and under the configurations used for these experiments, the generative process only works for simple environment. It is believed that these results are due primarily within the variational objective of CVAEs itself as Fig. 2 indicates a weighting preference either towards the reconstruction loss or the Kullback-Leibler Divergence depending on the complexity of the environment. A further analysis into using a weighted factor such as with the β-VAE [6] would help alleviate this issue.

The analysis demonstrates that DGMs could reconstruct reinforcement learning policies that use neural networks as their underlying structure under specific conditions. However, it has also shown that this framework in its current state is not generalisable. Although not optimal, some policies generated, especially in simple environments produced impressive results that usually take hundreds of steps when trained from scratch. Others, however, demonstrated insufficient results and would require further work. The primary conclusion of the analysis performed indicates that the proper structure of the dataset (i.e. the data transformation applied) is of utmost importance in this context. Secondly, further analysing into the network parameters and their raw structures should be performed, leaving them unmodified may be too strongly hindering the capabilities of the generative model.

5   Related Work

The conjunct use of RL and DGMs is not a new concept and has showcased impressive results when intertwined in some imaginary objective (i.e., generating goals). However, these methods primarily rely on extracting data from an environment, training a DGM to generate synthetic data, and finally solve a reinforcement learning algorithm using the synthetically generated data.

One of the first major applications of using DGMs in non conforming manners involved World Models. In [5] the authors use a lower-dimensional manifold to approximate an environment in an unsupervised manner enabling the generation of artificial experience from previously learned temporal or spatial dimensions, facilitating sample efficiency and proving easier for prediction and control. The authors proposed framework trains an agent in a simulated representation of an environment to then be transferred back into the original one.

In robotics, individually designing reward functions for each task can be expensive and unreliable. Hence, [15], proposed to teach an agent to learn to set its own goals and evaluate its behaviour towards achieving it by interacting with an environment. Another approach proposed in [7] the authors aim to learn disentangled representations of environments being nether task or domain-specific in the aim of discovering whether optimal policies learnt exclusively from a reward structure in a source domain are transferable to a target domain without retraining. There objective is concerned with learning a DGM that can depict disentangled representations of an environment in the aim to zero-shot into unseen/unknown domains.

Another practical RL problem lies in fact that it requires a lot of interactive data. To alleviate this issue another learning method is proposed whereby a policy is learnt entirely from a static dataset, entitled offline reinforcement learning [9]. Current state-of-the-art off-policy algorithms such as DDPG [10] or TD3 [4] fail to learn in this context due to out-of-distribution actions. These out-of-distribution actions lead to extrapolation errors of the Q-values accumulated by the Bellman operator and exacerbated by the policy updates [24]. Therefore, [24], proposed Policy in the Latent Action Space (PLAS), a framework to avoid out-of-distribution actions. To achieve this, they proposed to constrain the policy $\pi(\cdot|s)$ to select actions from a behaviour policy π_B modelled from a static dataset using a Conditional Variational Autoencoder (CVAE). The encoder trains to map an action-space to a latent-space conditioned on states, and the decoder learns the inverse mapping from latent-space to action-space. Intuitively, the goal of the CVAE is to generate actions conditioned on states.

6   Conclusion and Further Work

Reinforcement learning has shown some fascinating results but its challenges are still very present. Most notably, its trail-and-error requirement and unique policy discovery using neural networks as functional approximators hinders its capability of finding the optimal policy under a specific domains. It involved repeatedly training a reinforcement learning algorithm on an environment to assemble a static dataset of unique policies so that it may then be used on a conditional variational autoencoder to regenerate such policies. This process was tested against various environments. The purpose was to determine whether these policies have an underlying structure that can be inferred. We believe that this preliminary work could inspire further research within this direction to develop a viable framework that would solve the challenged posed.

In further works, we aim to study alternative generative architectures, new data transformation techniques and multi-domain generalisation. For architectures we aim to analyse whether the temporal aspect of the reinforcement learning training (i.e. the evolving network parameters over time) would be more suited for recurrent neural networks and as such enable us to use variational recurrent neural network architectures instead. Additionally, with the successes of attention lately, we would also aim to study their applicability to our approach.

A   Environments

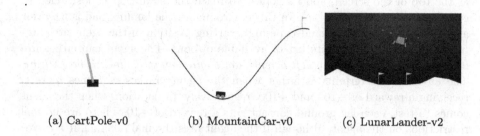

(a) CartPole-v0   (b) MountainCar-v0   (c) LunarLander-v2

Fig. 4. Representations of the environments used in these experiments. From left to right: CartPole-v0, MountainCar-v0, LunarLander-v0

CartPole-v0: This environment corresponds to the cart-pole problem introduced by [21]. An agent (i.e. the cart) has a pole attached by an un-actuated joint and moves along a frictionless track (see Fig. 4a). The agent starts at position $(0, 0)$ with the pole upright and can apply a force of $+1$ or -1 to reduce the agent's velocity to prevent the pole from falling over. The agent receives a reward of $+1$ for every timestep that the pole remains upright, including when it reaches the terminal state. However, the agent velocity is affected by the pole angle and centre of gravity which dictates the force needed to be applied to keep the pole upright. An episode terminates either if the pole surpasses 15 degrees vertically or the agent moves more than 2.4 units from the centre (i.e. the agent reaches the edge of the display). The episode terminates if the agent reaches 200 timesteps and is solved when the agent receives a reward of 195 over 100 consecutive episodes. The observation states of this environment are fully observable. The agent receives a four-dimensional observation space containing its *position, velocity, pole angle,* and *pole angular velocity.*

MountainCar-v0: This second environment is a bit more complex and was first introduced by [13]. It describes an agent (i.e. a car) on a one-dimensional track positioned between two hills (see Fig. 4b). The agent's engine cannot climb the mountain in a single pass. The only way for the agent to succeed is to learn to build momentum by driving back and forth. Its goal is to reach the flag located at the top of the hill on its right. The agent can either *accelerate left, accelerate right,* or *do nothing.* The agent starts at position $(0, 0)$ and receives a reward of 0 if it reaches the flag (i.e. position $(x, 0.5)$) and -1 if the agent's position is less than $(x, 0.5)$. The episode terminates if the agent reaches its goal or the episode length exceeds 200 timesteps. The environment is solved when the agent receives a reward of -110 over 100 consecutive episodes. In addition, the velocity affected by the gravitational pull acting on the agent is not affected by the action taken. For this environment, the observation states are also fully observable. The agent receives a two-dimensional observation space containing both its *position* and *velocity.*

LunarLander-v2: this last environment describes a rocket trajectory optimisation problem and is the hardest out of all. An agent (i.e. the rocket) starts at the top of the screen, and its goal is to reach the landing pad positioned at coordinates $(0, 0)$ (see Fig. 4c). In this environment, the landing pad is fixed for all episodes, whereas the landscape and starting position of the agent are sampled randomly from a finite list of available options. The agent can either fire its *right orientation engine*, *left orientation engine*, *main engine*, or *do nothing*. The environment terminates either when the agent crashes or comes to rest, receiving a reward of -100 and $+100$, respectively. In addition, when the agent comes to rest, each leg ground contact gets a reward of $+10$. There are no fuel restrictions on the agent, although if the agent fires the main engine, it receives a reward of -0.3 for each timestep. The environment is solved when the agent receives a reward of 200 over 100 consecutive episodes. As with CartPole and MountainCar, LunarLander has fully observable observation states. The agent receives an eight-dimensional observation space containing *horizontal* and *vertical* coordinates, *horizontal* and *vertical* speed, *angle* and *angular speed*, and whether the first and second legs have *ground contact*.

B   Experiment Hyperparameters

Table 2. List of hyperparameters and their values for DQN

Hyperparameter	Value	Description
batch size	32	Number of training cases over which each stochastic gradient descent (SGD) update is computed
buffer size	50000	SGD updates are sampled from this number of most recent observations
target network update	1000	The frequency with which the target network is updated
discount factor	0.99	Discount factor γ used in the Q-learning update
learning rate	0.001	The learning rate α used by Adam
initial exploration	0.5	Initial value ϵ in ϵ-greedy exploration
initial exploration	0.05	Final value ϵ in ϵ-greedy exploration
exploration decay (1)	100000	Number of steps to linearly anneal ϵ in *CartPole-V0*
exploration decay (2)	150000	Number of steps to linearly anneal ϵ in *MountainCar-V0*
exploration decay (3)	200000	Number of steps to linearly anneal ϵ in *LunarLander-V0*
evaluation frequency	50	The number of epochs before the learnt Q-network is evaluated with ϵ-greedy exploration set to 0.05
evaluation episodes	50	The number of epsiodes over which the learnt Q-network is evaluated with ϵ-greedy exploration set to 0.05

Table 3. List of hyperparameters and their values for CVAE

Hyperparameter	Value	Description
batch size	40	Number of training cases over which each stochastic gradient descent (SGD) update is computed
learning rate	0.0001	The learning rate α used by Adam
evaluation frequency	100	The number of epoches before the generated Q-network is evaluated with ϵ-greey exploration set to 0.05
evaluation episodes	100	The number of episodes over which the generated Q-network is evaluated with ϵ-greey exploration set to 0.05

References

1. Altosaar, J.: Tutorial - What is a Variational Autoencoder? August 2016. https://doi.org/10.5281/zenodo.4462916
2. Brockman, G., et al.: OpenAI gym (2016)
3. François-Lavet, V., Henderson, P., Islam, R., Bellemare, M.G., Pineau, J.: An introduction to deep reinforcement learning. Found. Trends Mach. Learn. **11**(3–4), 219–354 (2018). https://doi.org/10.1561/2200000071
4. Fujimoto, S., van Hoof, H., Meger, D.: Addressing function approximation error in actor-critic methods (2018)
5. Ha, D., Schmidhuber, J.: Recurrent world models facilitate policy evolution. In: Advances in Neural Information Processing Systems, vol. 31, pp. 2451–2463. Curran Associates, Inc. (2018). https://papers.nips.cc/paper/7512-recurrent-world-models-facilitate-policy-evolution, https://worldmodels.github.io
6. Higgins, I., et al.: beta-VAE: learning basic visual concepts with a constrained variational framework. In: International Conference on Learning Representations (2017). https://openreview.net/forum?id=Sy2fzU9gl
7. Higgins, I., et al.: DARLA: improving zero-shot transfer in reinforcement learning (2018)
8. Kober, J., Bagnell, J.A., Peters, J.: Reinforcement learning in robotics: a survey. Int. J. Robot. Res. **32**(11), 1238–1274 (2013). https://doi.org/10.1177/0278364913495721
9. Levine, S., Kumar, A., Tucker, G., Fu, J.: Offline reinforcement learning: Tutorial, review, and perspectives on open problems (2020)
10. Lillicrap, T.P., et al.: Continuous control with deep reinforcement learning (2019)
11. Lin, L.J.: Self-improving reactive agents based on reinforcement learning, planning and teaching. Mach. Learn. **8**(3), 293–321 (1992)
12. Mnih, V., et al.: Playing Atari with deep reinforcement learning (2013)
13. Moore, A.W.: Efficient memory-based learning for robot control (1990)
14. Nair, A., Bahl, S., Khazatsky, A., Pong, V., Berseth, G., Levine, S.: Contextual imagined goals for self-supervised robotic learning (2019)
15. Nair, A., Pong, V., Dalal, M., Bahl, S., Lin, S., Levine, S.: Visual reinforcement learning with imagined goals (2018)
16. Neftci, E.O., Averbeck, B.B.: Reinforcement learning in artificial and biological systems. Nat. Mach. Intell. **1**(3), 133–143 (2019)
17. Puterman, M.L.: Markov decision processes. Handb. Oper. Res. Manage. Sci. **2**, 331–434 (1990)
18. Raziei, Z., Moghaddam, M.: Adaptable automation with modular deep reinforcement learning and policy transfer (2020)
19. Ruthotto, L., Haber, E.: An introduction to deep generative modeling (2021)
20. Rybkin, O., Zhu, C., Nagabandi, A., Daniilidis, K., Mordatch, I., Levine, S.: Model-based reinforcement learning via latent-space collocation (2021)
21. Sutton, R.S., Barto, A.G.: Reinforcement Learning: An Introduction. MIT Press, Cambridge (2018)
22. Zhang, B., Xiong, D., Su, J., Duan, H., Zhang, M.: Variational neural machine translation (2016)

23. Zhang, X., Jiang, H.: Chapter nine - automated optimal control in energy systems: the reinforcement learning approach. In: Jiang, H., Zhang, Y., Muljadi, E. (eds.) New Technologies for Power System Operation and Analysis, pp. 275–318. Academic Press (2021). https://doi.org/10.1016/B978-0-12-820168-8.00015-8, https://www.sciencedirect.com/science/article/pii/B9780128201688000158
24. Zhou, W., Bajracharya, S., Held, D.: PLAS: latent action space for offline reinforcement learning (2020)

Deep Learning Models for Aspect-Based Sentiment Analysis Task: A Survey Paper

Sarsabene Hammi(✉), Souha Mezghani Hammami(✉),
and Lamia Hadrich Belguith(✉)

ANLP Research Group, MIRACL Lab, FSEGS, University of Sfax, Sfax, Tunisia
hsarsabene@gmail.com, souha.hammami@ihecs.usf.tn,
lamia.belguith@fsegs.usf.tn

Abstract. Due to the significant increase in the volume of data shared on the web, Aspect-Based Sentiment Analysis (ABSA) has become essential. This task ensures a detailed sentiment analysis. It identifies firstly the aspect terms (e.g., price, food, etc.) and then classifies their sentiment polarity as positive, negative, or neutral. Many approaches have been used to treat this task including the machine learning-based approach, the rule-based approach, etc. However, with the important increase in the content of the internet, these approaches became relatively unable to analyze this volume of information, resulting in the emergence of the deep learning-based approach which is the subfield of the machine learning-based approach.

Recently many researchers used the deep learning-based approach to address the ABSA. This paper provides a summary of the deep learning models that have been developed for ABSA, as well as a survey of studies that have employed these models to address different subtasks of the ABSA task. Finally, we discuss the implications of our work and potential avenues for future research.

Keywords: Aspect-based Sentiment Analysis · Aspect Extraction · Aspect Sentiment Classification · Deep Learning

1 Introduction

The Sentiment Analysis (SA) task is a subfield of Natural Language Processing (NLP) that involves using artificial intelligence and information retrieval techniques to identify and extract opinions, emotions, and other subjective information from text. The principal objective of SA is to gain insights into the general sentiment of a group of people toward a specific topic. This research direction has gained significant attention in academia and industry due to its ability to assist in marketing decision-making and track shifts in customer opinions on various subjects, including the medical domain (such as the COVID pandemic). Previous studies divided the SA task into three main levels. The first one is the document-level sentiment analysis, which is focused on identifying the general opinion of a document (such as a tweet, review, or article), and determining whether it is positive, negative, or neutral. The sentence-level sentiment analysis is the second type. It

© The Author(s), under exclusive license to Springer Nature Switzerland AG 2024
A. Bennour et al. (Eds.): ISPR 2023, CCIS 1941, pp. 169–183, 2024.
https://doi.org/10.1007/978-3-031-46338-9_13

concentrates on identifying the opinion of individual sentences within documents. The third one is Aspect-Based Sentiment Analysis (ABSA), which offers a more detailed and precise analysis. This analysis involves two tasks. The first one is the Aspect Extraction (AE) task, which identifies the aspect terms of a certain entity. The second one is the Aspect Sentiment Classification (ASC) task and it tends to determine the opinion related to aspects identified in the AE task. Taking as an example the following comment, the ABSA determines first the aspects "*camera*" and "*fingerprint reader*". Then classifies sentiment polarity related to these aspects. In this case, positive sentiment is given to the aspects "*camera*" and "*fingerprint reader*" of the entity "*smartphone*".

*The **camera** is very good. The **fingerprint reader** works well.*

Early studies categorized the approaches of ABSA task into four main approaches. The rule-based approach, machine learning-based approach, deep learning-based, and hybrid approach. In this survey paper, we focused only on the research papers that employed deep learning models to tackle the ABSA task.

The organization of the rest of this paper is as follows: In Sect. 2, we provide a broad summary of the deep learning models that are utilized in ABSA. Section 3, and 4 summarizes the different studies proposed respectively for AE and ASC tasks. Section 5 describes research papers that treat simultaneously the AE and ASC. Section 6, discusses different models used and gives statistics about the most performant model in the ABSA task. Section 7 concludes this study.

2 Deep Learning Models

Deep learning is a category of machine learning that draws inspiration from the organization and operation of the mind, particularly its neural networks. Deep learning models use multiple layers of artificial neural networks to learn and make decisions. Each layer receives information from the preceding layer and exploits it to produce another information beneficial to the classification.

The deep learning models have many advantages in comparison to the other machine learning models. Firstly, these models are able to improve their performance over time through the process of training. During this process, the network's biases and weights are adjusted based on the accuracy of predictions. One of the significant advantages of these models is their ability to enhance their performance gradually by means of training. During this process, the network's weights and biases are adjusted based on the precision of its forecasts, resulting in improved accuracy over time. Also, the deep learning models are self-adaptive. They are able to adapt to the data and find features on their own, without the need for the functional or distributional form of the model to be defined beforehand. So, the ability to learn, adapt to the data, and improve the performance without explicit programming makes deep learning models particularly well-suited for sentiment analysis tasks, where the features and relationships in the data may be complex and difficult to specify in advance.

Several deep-learning models have been employed in the ABSA. The upcoming subsections will offer a detailed account of each deep learning model.

2.1 Classical Recurrent Neural Network Model

The Recurrent Neural Network (RNN) model is a popular deep learning model. Recently, it has been greatly utilized in several NLP tasks, including the ABSA task [1]. This big use can be explained by the good results achieved by this model in the treatment of sequential data. The principal idea of the RNN model lies in the treatment of tokens that compose the input in a sequential manner. The classical RNN model follows the mechanism of forward propagation. At instant t, the RNN model feeds tokens of input sequence (X) into a neural network architecture which is composed of nodes interconnected between them. Then, assuming the connection between the different tokens of the sequence, the RNN model uses the outputs of previous nodes (h_{t-1}) to estimate the value of current node (h_t). The final node contains all information about the tokens appearing with the aspect, treated from left to right. This information is included, finally in an output layer to predict the label. The RNN model's architecture is presented in Fig. 1.

In addition, RNN model can follow also the mechanism of backward propagation. A backward RNN works in the same manner as a forward RNN but in the opposite direction: from right to left (the prediction is made from the end towards the beginning of the sequence). It aims in recovering the information from the next node to calculate the value of the current node at instant t. An RNN model that used the mechanism of forward propagation, in addition to backward propagation is named the bidirectional RNN model (Bi-RNN). This model is very performant and surpasses the classical RNN model thanks to its performant architecture.

The RNN model is a performant model that uses the contextual information between words, nevertheless, it still suffers from problems such as the vanishing gradients problem and the long-term dependencies learning.

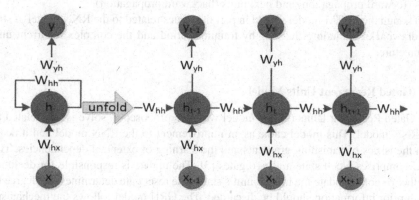

Fig. 1. RNN model's architecture.

2.2 Long Short-Term Memory Model

To address some of the problems faced by the RNN model, a variant called the Long Short-Term Memory (LSTM) was introduced [2]. As we mentioned above these problems are related mainly to the vanishing gradients and long-term dependencies learning.

To address these problems, researchers replaced the classic recurrent hidden unit with a memory unit. The LSTM unit is composed of a central node, containing the internal state (or memory) of the unit, and three gates. The input gate decides whether the cell's state must be updated or not. The output gate detects the next hidden state's value. The forget gate chooses which information should be ignored. Like RNNs, LSTMs follow the mechanism of forward propagation and treat the input sequence in a unidirectional manner. The LSTM model's architecture is presented in Fig. 2.

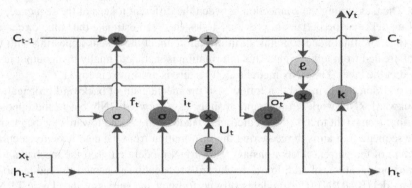

Fig. 2. LSTM unit's architecture.

Besides the LSTM model, we can find also a bidirectional LSTM (Bi-LSTM) model. It determines a word's label by leveraging the information coming from the previous units (forward propagation) and next units (backward propagation).

Though the LSTM model solved in part the issues related to the RNN model, it still has drawbacks, including the lengthy training period and the complex recurrent unit architecture.

2.3 Gated Recurrent Units Model

The Gated Recurrent Units (GRU) model was also proposed to solve issues related to the RNN model. This model came as an improvement to the RNN model and it deals with the issues of vanishing gradients and the learning of extended dependencies. The GRU comprises a cell state and two gates [3]. The update is responsible for deciding whether or not to update the hidden unit's state. The reset gate determines the degree to which prior information should be discarded. The GRU model follows the mechanism of forward propagation. Also, it can be improved by a bidirectional GRU (Bi-GRU) model that treats the input sequence according to forward propagation and backward propagation. Figure 3 presents the GRU model's architecture.

This model is a very performant model which is characterized by a simple architecture compared to the LSTM model, however, it has some drawbacks such as the small learning ability.

2.4 Convolution Neural Networks Model

The Convolution Neural Networks (CNN) model specializes in multi-layered networks. It is probably used when the input is structured according to a grid (e.g. an image). These networks were inspired by the work of [4] on the visual cortex of animals.

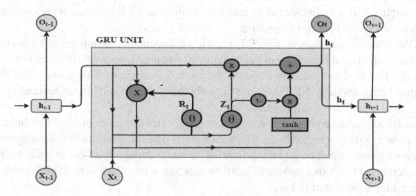

Fig. 3. The GRU unit's architecture.

The CNN model is initially introduced by [5] to treat the task of forms recognition and other tasks such as image classification, and character recognition. Recently, this model proved its relevance in NLP tasks and especially for tasks related to text classification. The CNN model's architecture is presented in Fig. 4. It contains mainly four types of layers. The convolution layer is a principal layer in the CNN model and it multiplies the matrix representation **M** by another called convolution matrix (or filter) to produce a feature map. The pooling layer reduces the feature map's dimensions while keeping only relevant information. The fully-connected layer refers to a neural network wherein neurons are related. The output layer predicts the adequate class.

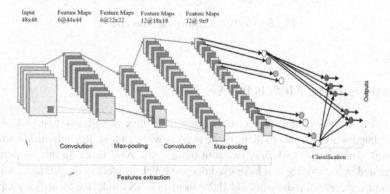

Fig. 4. The CNN model's architecture.

2.5 Bidirectional Encoder Representations from Transformers Model

The Bidirectional Encoder Representations from Transforms (BERT) model is a cutting-edge NLP model presented by Google [6]. This model is based on transformers and uses attention mechanisms to acquire contextual relationships among words within a text. A vast corpus of information is utilized to train BERT. Upon being trained on an extensive data corpus, BERT is fine-tuned to accomplish diverse NLP tasks such as sentiment analysis and named entity recognition.

One of the key advantages of BERT is that it is bidirectional, meaning that it takes into account contextual information of the target word when making predictions, as opposed to traditional models that only consider the context to the left. This allows BERT to perform better on many NLP tasks, particularly those that require an understanding of the context in which a word appears.

BERT has achieved impressive outcomes on an extensive array of NLP benchmarks and has been widely adopted in the NLP community. It has also inspired the development of several related models, including RoBERTa, which builds upon the original BERT architecture and exhibits enhanced performance on some tasks. The BERT model's architecture is presented in Fig. 5.

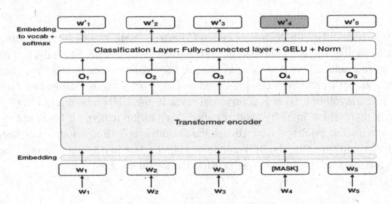

Fig. 5. The BERT model's architecture.

3 Deep Learning Models for AE Task

As we discussed above, the Aspect Extraction (AE) task is an important task in any ABSA-related work. It goals to identify the aspects within a text. Recently many studies have focused on the AE task only (without treating the ASC task). In this section, we are interested in the studies that have used deep learning models in the AE task. Among these studies, we mention the work of [7] that used an RNN model for identifying aspects existing in SemEval-2014 dataset. To achieve this work, the authors transformed, first, every word within the dataset to a word vector. These vectors were built based on Amazon Embeddings and SENNA Embeddings systems. Afterward, these vectors were used to create new context vectors that take into consideration the contextual dependencies between words. Finally, the RNN model used the constructed vectors (word embeddings vectors and context vectors) and the linguistic features' vectors to determine aspects. In the study presented in [8], an RNCRF model that combines the RNN model and the machine learning model: conditional random fields (CRF) was proposed. To implement the RNCRF model, authors constructed a dependency-tree RNN (DT-RNN) architecture. This architecture produces, for each word in the dataset, a high-level representation that takes into consideration the dependency relations between words in the dataset. These representations are included in a subsequent step in the CRF model to predict aspects.

[9] proposed a model based on LSTM for the extraction of aspects related to question answering (ASC-QA) task. In this study, authors constructed first a human-annotated benchmark dataset. After that, they proposed a Reinforced Aspect-relevant Word Selector (RAWS) model in order to select the aspect-relevant words. These selected words were incorporated in a subsequent step in the Reinforced Bidirectional Attention Network (RBAN) architecture to extract the aspect terms. This architecture treats the semantic matching problem in the QA text pair and enhances the learning algorithm. It incorporates both a bidirectional attention mechanism and a reinforcement learning (RL) component. By using a bidirectional attention mechanism, the model can identify both the aspect and its corresponding context. The RL component assists in enhancing the model's ability to comprehend the connections existing amidst the aspect and its context. This work was implemented using the LSTM model.

Other studies think that the Bi-LSTM model is more efficient than the LSTM model. For that reason, many studies used this model in their AE-related methods. Among them, we mention the work of [10] which used a multi-layers Bi-LSTM model in their work. [10] assumed that the incorporation of information about both words and clauses inside the Bi-LSTM model can enhance mainly the process of aspects detection. To realize this work, [10] segmented first the sentences into clauses. After that, they incorporated contextual vectors into the word-level aspect-specific attention layer. This layer exploits the contextual information and outputs new vectors that contain the degree of importance of each word in a given clause. These newly produced vectors were fed in the clause-level attention layer to detect which clauses are important in the dataset. Finally, the Bi-LSTM model used the vectors extracted by the word-level aspect-specific attention and clause-level attention layers to forecast the aspect terms in the dataset. This method achieved good results (68.50% F-measure). In another work, [11] enhanced the AE task using the Bi-LSTM model and Bidirectional Dependency Tree Conditional Random Field Framework (BiDTreeCRF). So, the authors constructed first a bidirectional dependency tree network (BiDTree) in order to detect the dependency relationships among the words. Afterward, they included the output of BiDTree into a Bi-LSTM model to detect the global syntactic context of each word. Finally, the outcomes produced by the preceding steps were fed into a CRF model to predict aspect terms.

[12] considered the AE as a sequence-to-sequence (seq2seq) task and proposed a Bi-GRU-based model. This model uses a seq2seq learning-based architecture that takes into consideration the meaning of sentences and labels in the extraction of aspect terms. At this level, the model takes as input word embeddings and predicts as output the label of a given word.

Other research papers preferred to combine multiple deep learning algorithms in order to perform the aspect extraction task. Among them, we mention the work of [13] that combined the Bi-GRU, CNN, and BERT models to extract the aspect terms. In this work, [13] proposed a new framework named pre-trained language embedding-based contextual summary and multi-scale transmission network (PECSMT). This framework consists of three units. The pre-training language model embedding unit generates contextualized embeddings using a BERT model. The multi-scale transmission network unit uses the multi-scale CNNs and Bi-GRU models to extract the sequential features. The contextual summary unit creates a contextualized representation of words. This model with its three units achieved good results and succeeded in the extraction of aspect terms. [14] introduced a new information-augmented neural network (IAAN) model. This model integrated informative information about the words surrounding the aspect term in order to extract the dynamic word sense. This model involves several layers. The initial layer is a contextualized embedding layer and it uses the BERT model to create contextualized word embeddings. The second layer is an encoder named MCRN and it uses the GRU model to capture the sequential data and bidirectional distant dependencies. The third layer is a decoder and it uses the GRU model to decode the encoding representations in order to predict the aspect terms.

[15] suggested a synchronous double-channel recurrent network (SDRN) model to achieve the Aspect Opinion Pair Extraction (AOPE) task. To realize this model, [15] employed first the word embeddings of BERT to learn the words' contextual semantics. Subsequently, they used this contextual information and the CRF model to detect the

Table 1. Summary of Aspect Extraction Studies.

Study	Domain	Dataset Size	Model	Available	F-measure
[7]	Restaurant	3841	RNN	yes	81.56%
	Laptop	3845		yes	77.27%
[8]	Restaurant	3841	RNN	yes	84.93%
	Laptop	3845	CRF	yes	78.42%
[9]	Products	12076	LSTM	no	64.80%
[10]	Restaurant	3841	Bi-LSTM	yes	68.50%
[11]	Restaurant	8613	Bi-LSTM	yes	84.83%
	Laptop	3845	CRF	yes	80.57%
[12]	Restaurant	2000	Bi-GRU	yes	75.14%
	Laptop	3045		yes	80.31%
[13]	Laptop	3845	BERT	yes	85.08%
	Restaurant	3841	Bi-GRU CNN	yes	88.13%
[14]	Laptop	3845	BERT	yes	87.49%
	Restaurant	3841	GRU	yes	88.40%
[15]	Laptop	3845	BERT	yes	83.67%
	Restaurant	3841	CRF	yes	89.49%

aspect and opinion terms. Table 1 contains an overview of the different AE-related studies presented.

4 Deep Learning Models for ASC Task

This part of the study gives a summary of the studies that have treated only the ASC task using deep learning models. Among them, we mention the work of [16] that proposed an RNN-based method to perform the ASC for Arabic hotels' reviews dataset. To achieve this method, [16] incorporated into the RNN model lexical, word, syntactic, morphological, and semantic features. These features enhanced significantly the effectiveness of the suggested model. [17] proposed a Target-Connection LSTM (TC-LSTM) model to detect the sentiment polarity towards the aspect terms. This model leverages the semantic relatedness between the aspect term and its context. The TC-LSTM model takes as input words embedding and aspect vectors. These aspect vectors contain information about the contextual words related to a given aspect term. This model achieved competitive results and overrode the other benchmarks, even though syntactic parsers and external sentiment lexicons were used. Similarly, to [17, 18] exploited the context of aspects and suggested an Attention-based LSTM with Aspect Embedding (ATAE- LSTM) to perform the ASC task. This model utilizes attention weights, computed using word embeddings, to capture the information associated with the aspect term. [19] proposed a hierarchical LSTM model. This model takes advantage of the interdependencies between sentences in a review to achieve the ASC task. The achieved outcomes proved the efficacy of the suggested model. Although this model didn't use hand-crafted features, it surpassed the other state-of-art models and achieved competitive results.

[20] enhanced the ASC task using linguistic regularizers and the CNN model. In this work, [20] incorporated in the CNN model two regularizers which are the Coordinating Conjunctions Regularizer (CCR) and Adversative Conjunctions Regularizer (ACR). These regularizers ameliorated the introduced model and achieved good results for the SemEval-2014 dataset. Table 2 summarizes the different studies presented for the ASC task.

Table 2. Summary of Aspect Sentiment Classification Studies.

Study	Domain	Dataset Size	Model	Available	F-measure
[16]	Hotel	340000	RNN	no	-
[17]	Twitter	-	LSTM	yes	65.9%
[18]	Restaurant	3841	LSTM	yes	-
	Laptop	3845		yes	
[19]	Multi-Domain	-	LSTM	yes	80.23%
[20]	Restaurant	3841	CNN	yes	-

5 Deep Learning Models for AE and ASC Tasks

In this section, we are interested in the studies that simultaneously treat the ABSA's two tasks: the AE and the ASC tasks. Amongst them, we mention the study of [21] which assumed that the treatment of the AE task and the ASC simultaneously is more beneficial than the treatment of each one of them separately. For that, [21] suggested a DOER (Dual crOss-sharEd RNN) framework to extract the aspects as well as the sentiment polarity. DOER consists mainly of two units: the dual RNN unit and the crOss-sharEd unit. These two units work together and enhance the AE and ASC tasks by using embeddings related to domain-specific and general-purpose.

Other studies used the LSTM model in the AE and ASC task. Amongst them, we mention the study of [22]. [22] used in their work two LSTM models. These models detect the latent relations between opinion words and aspects using a DMI (multi-hop dual memory interaction) mechanism. This mechanism is very performant and succeeded in the realization of both tasks. [23] ameliorated the work of [22] and proposed two-stacked Bi-LSTMs units. The first unit detects the unified tags (the aspect terms and their sentiment). The second unit enhances the prediction performance of the first unit. [24] proposed also a Bi-LSTM model for the identification of aspects in addition to their corresponding sentiment. This model leverages the dependency among aspects and sentiment words using a Bi-LSTM model and Biaffine score. [25] presented a Bi-LSTM-CRF model. This model incorporates first the contextualized representations of words in the Bi-LSTM in order to identify the aspects. These representations are used to detect the interactions between words. In a subsequent step, the binary classifier CRF is employed to assign a sentiment for each aspect. The model's performance was assessed on the products dataset and it demonstrated good performance in comparison to the baseline models. [26] leveraged semantic and syntactic relationships between the opinion words, and aspect terms, and suggested a model based on LSTM to tackle the AE and ASC tasks. For that, [26] modeled first the syntactic and semantic dependencies between words using Graph Neural Networks (GNN). Then, they incorporated these dependencies features and word embeddings into the LSTM model with the aim of identifying the aspects and the corresponding sentiment associated with them.

[27] adopted a CNN-based architecture to perform the ABSA task including the AE task and ASC task. In this work, [27] took advantage of the relatedness between the ABSA-related tasks and proposed an IMN network (interactive multi-task learning network). This network ensures information passing between the ABSA tasks (AE, ASC, etc.) using a common group of latent variables. This proposed method proved its efficiency in the AE and ASC tasks and achieved good results.

[28] proposed a BERT-based model to accomplish the ASC and AE tasks. This proposed method used a framework for joint learning of multiple tasks, where only one model was trained to perform these two related tasks simultaneously. The model can share information between the two tasks, allowing it to enhance the two tasks' performance. In addition, they used two independent BERT layers to extract features belonging to global and local contexts. Such features enhanced significantly the AE and ASC tasks and led to favorable outcomes for both Chinese and English reviews. In another work [29] introduced a deep contextualized relation-aware network (DCRAN) model. This BERT-based model was designed to be context-aware, taking into account

the words and phrases that appear both before and after the aspect or sentiment being identified. This model was also designed to be relation-aware, considering explicitly and implicitly the contextual information between aspects and their sentiments. This proposed method enhanced the current ABSA-related works. Also, [30] focused on the dependencies between opinion and aspect terms and suggested a BERT-based model in order to achieve the AE and ASC tasks.

[31] presented a GRU-based model and Memory Network to address the aspect extraction and aspect-based sentiment analysis tasks. Firstly, [31] pretreated the dataset. Subsequently, they augmented the GRU model with the inclusion of word vectors to extract aspect terms. Finally, they used a Memory Network to classify the sentiment of aspects. The Memory Network takes into consideration the dependence between the aspect terms.

Other studies combined several deep learning models to achieve the AE and ASC tasks. Among these studies, we mention the study of [32] which used the Bi-LSTM-CNN-based model for extracting aspects and classifying their sentiment. In this study, [32] employed a model that realizes multiple related tasks simultaneously. This model surpassed the other benchmarks related to the ABSA in English and Hindi languages. [33] introduced a method for unified aspect-based sentiment analysis. This method used a collaborative learning approach, in which the CNN model is trained to perform the AE and ASC tasks. During the training phase, this model uses the word embeddings provided by the BERT model to create useful vectors. In addition, this method takes into consideration the relationships between aspects, as sentiment towards one aspect can often be influenced by sentiment towards another. [34] presented a method based on the CNN-BERT model. This method employed an interactive architecture, in which a syntactic parser was employed to identify the syntactic dependencies between words in the text. This information was used to guide the aspect identification and classification process. The method also utilized dependency syntactic knowledge, which refers to the relationships between words in a syntactic parse tree, leading to enhanced accuracy for both identifying aspects and classifying sentiments. [35] suggested a Bi-LSTM-BERT model to identify the aspect-sentiment triplets. This model contains a neural network architecture called an Explicit Interaction Network (EIN). This architecture was designed to capture the relationships between different words in the text and then used it to identify aspects and the sentiment expressed towards them. The EIN architecture contains multiple layers that work together through explicit attention mechanisms, allowing the model's ability to concentrate on specific input parts and incorporated context from other parts of the input when making predictions.

Table 3 contains an overview of the different studies already presented for the ABSA task.

6 Discussion

This part of the study compares the deep-learning models utilized to solve ABSA-related tasks. We take into consideration the studies described in Sect. 5 (15 studies) that treat both the AE and the ASC tasks. Figure 6 presents the F-measure values achieved by the deep-learning models in the ABSAs' studies. The obtained results proved that the

BERT model achieved the highest F-measure values. This model has achieved a good performance in a wide range of ABSA-related studies for many reasons. Firstly, the BERT model underwent pre-training on a massive corpus of data, which allows it to understand the context and words' meaning in a sentence. This pre-training can be fine-tuned on specific tasks, which improves performance. Secondly, the BERT was trained in order to understand the word's context by looking at the words that come before and after it. Also, it is a fine-tunable model, which means that it can be easily fine-tuned on specific tasks, even with limited annotated data. All these advantages make the BERT model performant and suitable for the ABSA task.

Table 3. Summary of Aspect-Based Sentiment Analysis Studies.

Study	Domain	Dataset Size	Model	Available	F-measure
[21]	Twitter	3223	RNN	yes	51.37%
	Restaurant	3841		yes	72.78%
	Laptop	3845		yes	60.35%
[22]	Multi-Domain	9359	LSTM	yes	-
[23]	Laptop	2931	Bi-LSTM	yes	57.90%
	Restaurant	9738	RNN	yes	58.90%
[24]	Laptop	1487	Bi-LSTM	yes	45.05%
	Restaurant	4550			54.82%
[25]	Laptop	1435	Bi-LSTM	yes	51.33%
	Restaurant	4536	CRF	yes	61.55%
			BERT		
[26]	Laptop	1456	Bi-LSTM	yes	52.01%
	Restaurant	4533		yes	64.06%
[27]	Laptop	1903	CNN	yes	68.26%
	Restaurant	3859		yes	57.40%
[28]	Multi-Domain	41640	BERT	yes	88.14%
[29]	Restaurant	4190	BERT	yes	72.42%
[30]	Laptop	1453	BERT	yes	58.58%
	Restaurant	4536		yes	64.75%
[31]	Product	1180	GRU	no	67.00%
[32]	Product	26044	Bi-LSTM	yes	74.60%
			CNN		
[33]	Laptop	3048	CNN	yes	63.40%
	Restaurant	4359	BERT	yes	70.43%
[34]	Laptop	3048	CNN	yes	63.04%
	Restaurant	3044	BERT	yes	67.48%
[35]	Laptop	1903	Bi-LSTM	yes	61.60%
	Restaurant	5497	CNN	yes	63.77%

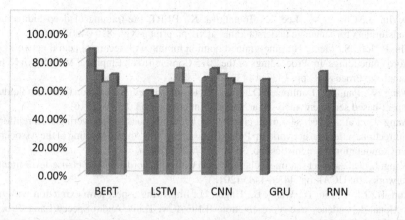

Fig. 6. The distribution of F-measure values achieved through the deep-learning models.

7   Conclusion

This survey paper gives a comprehensive review about different research that utilized deep learning models in solving the ABSA task. We first provided an overview concerning the deep learning models used in the achievement of ABSA tasks, including the RNN (Recurrent Neural Network), LSTM (Long-short term memory), etc. After that, we summarized and explained the studies that treated each one of the ABSA subtasks independently of the other subtask: the AE subtask and the ASC subtask. Also, we presented the studies that treat both tasks. Finally, we discussed the models used in the ABSA studies. The obtained results showed that the best performances have been obtained by the BERT.

Our future work intends to provide a survey with a detailed analysis of other ABSA-related studies that have used the linguistic knowledge approach, machine learning-based approach, and hybrid approach. We will also provide an overview of the dataset used in this field. In addition, we will compare different approaches by mentioning the advantages and disadvantages of each approach.

References

1. Rumelhart, D., Hinton, G.E., Williams, R.J.: Learning representations by back-propagating errors. Nature **323**, 533–536 (1986)
2. Graves, A.: Long short-term memory. In: Supervised Sequence Labeling with Recurrent Neural Networks, pp. 37–45 (2012)
3. Cho, K., Van, B., Bahdanau, D., Bengio, Y.: On the properties of neural machine translation: encoder-decoder approaches. arXiv preprint arXiv:1409.1259 (2014)
4. Hubel, D.H., Torsten, N.W.: Receptive fields, binocular interaction and functional architecture in the cat's visual cortex. J. Physiol. **160**(1), 106 (1962)
5. Fukushima, K., Sei, M.: Neocognitron: a self-organizing neural network model for a mechanism of visual pattern recognition. In: Amari, S.I., Arbib, M.A. (eds.) Competition and Cooperation in Neural Nets, pp. 267–285. Springer, Heidelberg (1982). https://doi.org/10.1007/978-3-642-46466-9_18

6. Devlin, J., Chang, M., Lee, K., Toutanova, K.: BERT: pre-training of deep bidirectional transformers for language understanding. arXiv preprint arXiv:1810.04805 (2018)

7. Liu, P., Joty, S., Meng, H.: Fine-grained opinion mining with recurrent neural networks and word embeddings. In: Proceedings of the 2015 Conference on Empirical Methods in Natural Language Processing, pp. 1433–1443 (2016)

8. Wang, W., Pan, S.J., Dahlmeier, D., Xiao, X.: Recursive neural conditional random fields for aspect-based sentiment analysis. arXiv preprint arXiv:1603.06679 (2016)

9. Wang, J., et al.: Aspect sentiment classification towards question-answering with reinforced bidirectional attention network. In: Proceedings of the 57th Annual Meeting of the Association for Computational Linguistics, pp. 3548–3557 (2019)

10. Wang, J., et al.: Aspect sentiment classification with both word-level and clause-level attention networks. In: IJCAI, pp. 4439–4445 (2018)

11. Luo, H., Li, T., Liu, B., Wang, B., Unger, H.: Improving aspect term extraction with bidirectional dependency tree representation. IEEE/ACM Trans. Audio Speech Lang. Process. **27**, 1201–1212 (2019)

12. Ma, D., Li, S., Wu, F., Xie, X., Wang, H.: Exploring sequence-to-sequence learning in aspect term extraction. In: Proceedings of the 57th Annual Meeting of the Association for Computational Linguistics, pp. 3538–3547 (2019)

13. Feng, C., Rao, Y., Nazir, A., Wu, L., He, L.: Pre-trained language embedding-based contextual summary and multi-scale transmission network for aspect extraction. Procedia Comput. Sci. **174**, 40–49 (2020)

14. Liu, N., Shen, B.: Aspect term extraction via information-augmented neural network. Complex Intell. Syst., 1–27 (2022)

15. Chen, S., Liu, J., Wang, Y., Zhang, W., Chi, Z.: Synchronous double-channel recurrent network for aspect-opinion pair extraction. In: Proceedings of the 58th Annual Meeting of the Association for Computational Linguistics, pp. 6515–6524 (2020)

16. Al-Smadi, M., Al-Ayyoub, M., Jararweh, Y., Qawasmeh, O.: Deep recurrent neural network for aspect-based sentiment analysis of Arabic hotels reviews. J. Comput. Sci., 386–393 (2018)

17. Tang, D., Qin, B., Feng, X., Liu, T.: Effective LSTMs for target-dependent sentiment classification. In: COLING, pp. 3298–3307 (2016)

18. Wang, Y., Huang, M., Zhu, X., Zhao L.: Attention-based LSTM for aspect-level sentiment classification. In: EMNLP, pp. 606–615 (2016)

19. Ruder, S., Ghaffari, P., Breslin, J.G.: A hierarchical model of reviews for aspect-based sentiment analysis. In: Conference on Empirical Methods in Natural Language Processing, ACL, pp. 999–1005 (2016)

20. Zeng, D., Dai, Y., Li, F., Wang, J., Sangaiah, A.K.: Aspect based sentiment analysis by a linguistically regularized CNN with gated mechanism. J. Intell. Fuzzy Syst. **36**(5), 3971–3980 (2019)

21. Luo, H., Li, T., Liu, B., Zhang, J.: DOER: dual cross-shared RNN for aspect term-polarity co-extraction. arXiv preprint arXiv:1906.01794 (2019)

22. Li, Z., Li, X., Wei, Y., Bing, L., Zhang, Y., Yang, Q.: Transferable end-to-end aspect-based sentiment analysis with selective adversarial learning. arXiv preprint arXiv:1910.14192 (2019)

23. Li, X., Bing, L., Li, P., Lam, W.: A unified model for opinion target extraction and target sentiment prediction. In: Proceedings of the AAAI Conference on Artificial Intelligence, pp. 6714–6721(2019)

24. Zhang, C., Li, Q., Song, D., Wang, B.: A multi-task learning framework for opinion triplet extraction. arXiv preprint arXiv:2010.01512 (2020)

25. Xu, L., Li, H., Lu, W., Bing, L.: Position-aware tagging for aspect sentiment triplet extraction. arXiv preprint arXiv:2010.02609 (2020)

26. Chen, Z., Huang, H., Liu, B., Shi, X., Jin, H.: Semantic and syntactic enhanced aspect sentiment triplet extraction. arXiv preprint arXiv:2106.03315 (2021)
27. He, R., Lee, W.S., Ng, H.T., Dahlmeier, D.: An interactive multi-task learning network for end-to-end aspect-based sentiment analysis. arXiv preprint arXiv:1906.06906 (2019)
28. Yang, H., Zeng, B., Yang, J., Song, Y., Xu, R.: A multi-task learning model for Chinese-oriented aspect polarity classification and aspect term extraction. Neurocomputing **419**, 344–356 (2021)
29. Oh, S., et al.: Deep context-and relation-aware learning for aspect-based sentiment analysis. arXiv preprint arXiv:2106.03806 (2021)
30. Huang, L., et al.: First target and opinion then polarity: enhancing target-opinion correlation for aspect sentiment triplet extraction. arXiv preprint arXiv:2102.08549 (2021)
31. Ismet, H.T., Mustaqim, T., Purwitasari, D.: Aspect based sentiment analysis of product review using memory network. Sci. J. Inf. **9**, 73–83 (2022)
32. Liu, Q., Liu, B., Zhang, Y., Kim, D.S., Gao, Z.: Improving opinion aspect extraction using semantic similarity and aspect associations. In: Thirtieth AAAI Conference on Artificial Intelligence (2016)
33. Chen, Z., Qian, T.: Relation-aware collaborative learning for unified aspect-based sentiment analysis. In: Proceedings of the 58th Annual Meeting of the Association for Computational Linguistics (2020)
34. Liang, Y., Meng, F., Zhang, J., Chen, Y., Xu, J., Zhou, J.: A dependency syntactic knowledge augmented interactive architecture for end-to-end aspect-based sentiment analysis. Neurocomputing **454**, pp 291–302 (2021)
35. Wang, P., et al.: Explicit interaction network for aspect sentiment triplet extraction. arXiv preprint arXiv:2106.11148 (2021)

A Real-Time Deep UAV Detection Framework Based on a YOLOv8 Perception Module

Wided Souid Miled[1,2]($\boxtimes$), Moulay A. Akhloufi[3], and Hana Ben Asker[1]

[1] Department of Mathematics and Computer Science, National Institute of Applied Science and Technology (INSAT), Tunis, Tunisia
{wided.miledsouid,hana.benasker}@insat.ucar.tn
[2] Electrical Engineering Department, SITI Laboratory, National School of Engineers of Tunis, University of Tunis El Manar, BP. 37 Belvédère, 1002 Tunis, Tunisia
[3] Perception, Robotics, and Intelligent Machines Research Group (PRIME), Dept of Computer Science, Université de Moncton, Moncton, NB, Canada
moulay.akhloufi@umoncton.ca

Abstract. Unmanned Aerial Vehicles (UAVs) or drones are currently gaining a lot of popularity due to the versatility of this technology and its ability to perform multiple tasks in various industries. However, arbitrary or malicious use of drones can pose a major risk for public and aviation safety. The automated detection and neutralization of malicious drones to avoid fatal incidents is therefore of primary interest for aerial security systems. Recently, deep learning based approaches for object detection have gained great attention due to their high prediction accuracy. In the proposed work, we use a deep learning object detection model based on latest versions of YOLO, i.e. v7 and v8, to detect and track one drone by another drone in a pursuit-evasion scenario. The detection accuracy achieved with the YOLOv8 model is 99% Average Precision (AP) and the inference is 107.5 FPS, proving the effectiveness of the proposed approach for real-time UAV detection. A comparative study shows that the YOLOv7 model achieves the same accuracy but with slower inference.

Keywords: Unmanned Aerial Vehicle · Drone Chasing · Object Detection · Deep Learning · YOLOv7 · YOLOv8

1 Introduction

Artificial intelligence (AI) is a research field that attracted a lot of interest in recent years, with the goal of building machines that performs tasks requiring human intelligence. The tremendous advances in AI have been made possible by various studies in this field, along with the exponential growth of both data and computational resources, allowing research communities and companies to develop and explore new ideas. Indeed, thanks to the technological evolution, the amount of data we generate daily has increased exponentially resulting in

© The Author(s), under exclusive license to Springer Nature Switzerland AG 2024
A. Bennour et al. (Eds.): ISPR 2023, CCIS 1941, pp. 184–197, 2024.
https://doi.org/10.1007/978-3-031-46338-9_14

the emergence and evolution of deep learning (DL), a subarea of AI, which revolutionized multiple fields of research, especially in the field of computer vision and image processing, thanks to convolutional neural networks (CNNs).

One of the most studied field in deep learning is autonomous drone navigation. The objective is to design and develop deep learning algorithms allowing the unmanned aerial vehicle (UAV) or drone to evolve autonomously in an unknown environment, while avoiding possible obstacles. In [1], a review on recent advanced applications of deep learning for UAVs has been performed. The authors identified and discussed three major applications: (1) feature extraction which has been widely applied thanks to the ability of deep learning to learn representations of data; (2) path planning [2] for collaborative search and rescue missions where a UAV explores and scans the environment in order to find the optimal path without any human supervision; and (3) motion control which can provide UAVs with more flexibility while navigating autonomously through unknown dynamic environments. UAV supervision, path planning, motion control and trajectory tracking systems implement complex behaviors that need to be learned and tracked. Therefore, they are commonly addressed within deep reinforcement learning (DRL) frameworks, which combine the ability of deep learning to extract information with the decision-making capabilities of reinforcement learning [3–6].

Several deep reinforcement learning developments have also been reported for tasks related to drone chasing with collaborative or competitive strategies between multiple UAVs. In the work of Çetin et al. [7], a deep reinforcement learning framework was proposed for the task of training a drone to counter another drone, while avoiding stationary and moving obstacles. In [8], Akhloufi et al. propose to combine deep reinforcement learning and deep search areas to a develop a solution to the problem of drones pursuit-evasion. In this work, DRL is first used to predict the actions to perform to keep the target in sight. Then, a deep learning object detection is used for the detection and tracking of the target drone by another UAV follower. Counter-UAV systems imply developing and testing efficient tracking and detection models, by associating the predicted objects from detection model across temporal frames. In [9], the authors propose the use of a deep object detector based on the YOLOv2 algorithm combined with a search area proposal to detect and predict the position of the target drone in a pursuit-evasion scenario. In [10], deep learning frameworks based on YOLOv2 and YOLOv3 were implemented for the real-time UAV's detection task. Both models were trained with a specific created dataset and comparative study showed that YOLOv3 outperformed YOLOv2, in terms of mean average precision and accuracy. In the work of Isaac-Medina et al. [11], a benchmark study on UAV detection and tracking is conducted with multiple datasets and across state-of-the-art object detection algorithms. This study shows that YOLOv3, overall, yields the best precision and inference time but Faster RCNN and Detection Transformer (DETR) are best suited towards detecting small UAVs. Popular deep object detection methods for the task of UAV detection are also reviewed in [12], in order to test their performance for the perception part of a UAV collaboration framework using different real-life scenes. For detection speed, YOLOv3 has the fastest speed while achieving a good average precision performance. In [13], three drone detection models are tested using a simulated dataset of drone images generated from the AirSim flight simulator. The paper demonstrates that using transfer learning techniques along with a real-time object detection algorithm, we can detect drones with high accuracy.

Deep learning techniques applied on images acquired from UAVs have also attracted a lot of attention, with different target applications. In [14], Wu et al. provide a comprehensive review of DL based UAV object detection and tracking methods by discussing the key issues and main challenges, such as image degradation, small object size, scale and real-time problems, etc. In [15], a YOLOv5 based framework with gradient boosting is developed for solving the problem of human action localization and recognition in drone videos. The proposed method achieved promising results in case of single image action detection than state-of-the-art methods. However, the speed of the drone, causing motion blurring, may affect the performance of the algorithm. In [16], authors modified YOLOv5 according to the properties of UAV aerial images to improve the detection performance of the model, while maintaining the computation speed. Experimental results on UAV aerial images show that the proposed model achieves a high detection accuracy despite the challenges of small targets, sparse distributions, dense arrangements and especially complex backgrounds. In [17], an improved YOLO algorithm called YOLODrone for detecting objects in drone images is proposed, improving detection results when compared to YOLOv3 algorithm. Recently, the authors in [18] proposed an improved object detector based on YOLOv7 [19] to detect people, ships and other objects from maritime UAV images. They proposed the use of a prediction head for detecting small objects and an attention block to find regions of interest in images, which greatly improved detection performances.

To summarize, DL and DRL are making drones more efficient and effective, both in terms of their ability to gather information and their ability to carry out complex tasks, autonomously. For UAV detection task, YOLO algorithm category has played an important role, providing a good balance between detection accuracy and computation speed.

In this paper, we propose the use of the two recent YOLO object detection models, namely YOLOv7 and YOLOv8, which outperform many previous object detectors in terms of speed and accuracy. We tested and evaluated the performance of different variants of both models on images containing one drone captured by another drone, in a pursuit-evasion scenario and within different and diversified environments. A comparative study shows that both models obtain a good performance achieving an AP50 above 99%, and that YOLOv8 has the fastest speed reaching 107.5 FPS, which proved its effectiveness for UAV real-time detection. The proposed object detector model is integrated as a perception part of the proposed deep UAV detection and tracking framework.

2   Theoretical Background

In this section, we present some basic theoretical concepts on deep learning based object detection and we give a brief overview of YOLOv7 and YOLOv8 object detection models.

2.1   Object Detection

In recent years, Object Detection has become one of the most popular and widely researched topics. It has attracted the attention of many researchers working in different areas, such as medical imaging, computer vision, robotics, and autonomous driving. Object detection focuses on localizing and recognizing objects of a particular category in images and videos, using discriminate features. The task aims to accurately obtain bounding boxes that contain and enclose the objects of interest as tightly as possible. The recent advances in the field of deep learning have dramatically improved the results of object detection because of the variety of features that powerful DL models can learn and detect [20]. State-of-the-art methods are mostly categorized into two main types: one-stage methods and two-stage methods.

Two-stage detectors have two separate stages. In the first stage, the model proposes a set of regions of interest likely to contain the desired objects, by selective search as in R-CNN [21] and Fast R-CNN [22], or by a Region Proposal Network (RPN) as in Faster R-CNN [23]. In the second stage, the model performs object classification and location refinement using bounding box regression based on characteristics retrieved from the suggested regions [24]. Recently, a novel two fully-sparse stage detector, called Sparse RCNN, have been proposed [25]. The key idea of this framework that contains a backbone network, a dynamic instance interactive head and two prediction layers, is to replace large number of regions selected from RPN with a small fixed set of learned proposal boxes. It shows better results on small objects and shorter running time compared to other detectors.

One-stage detectors contain You Only Look Once (YOLO) series and Single Shot Multibox Detector (SSD) series. In this category, object classification and bounding box regression are done directly in one step without using pregenerated region proposals as candidate object bounding boxes [26]. These detectors leverage the help of anchors and grid boxes to obtain the position of the region of detection in the image. In comparison to two-stage detectors, single-stage detectors are much faster because they predict the bounding box and the class of objects simultaneously, and are therefore better suited for real-time detection applications. On the other side, two-stage detectors are more accurate but also relatively more complex and time consuming.

2.2   YOLO Series

YOLO is a family of algorithms employed in many research studies for object detection. YOLO considers the object detection problem as a regression problem that separate the bounding boxes and their associated class probabilities. A convolutional neural network runs on the image to predict the probability of an object's bounding box and category, thus enabling end-to-end detection. Newer versions of YOLO, propose the use of deeper neural networks allowing more robust accuracy and speed.

As shown in Fig. 1, the YOLO detector is divided into three major components: the Backbone, the Neck and the Head. Images are passed through the Backbone of the network. Extracted features are combined and mixed in the Neck (FPN in this case) and then passed along to the Head that predicts the bounding boxes. The backbone is a deep neural network composed mainly of convolutional layers, aiming to extract the essential features. The neck represents additional layers inserted between the backbone and the head, mainly used to combine information from the feature maps at different levels. The prediction happens in the head, where object localization and classification tasks happen at the same time. There are three different types of loss functions that guide the detection model, namely box loss, objectness loss and classification loss.

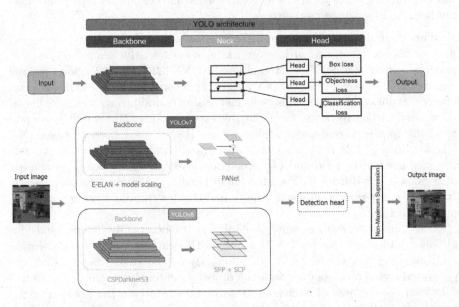

Fig. 1. The architecture of YOLO detector and its variants YOLOv7 and YOLOv8. YOLO baseline architecture is mainly composed of a backbone network, neck layer as feature pyramid network (FPN), and prediction heads. YOLOv7 Backbone integrates an E-ELAN network along with a model scaling method. YOLOv8's backbone is Darknet53. Both models include the non-maximum suppression algorithm.

2.3   YOLO Version 7

YOLOv7, proposed in [19], is considered among state-of-the-art object detection algorithms. It creates a network architecture that would predict bounding boxes more accurately than its predecessors at similar inference speeds. It shows a high accuracy with 30 FPS or higher on a V100 GPU [19]. In order to achieve these results, the authors of YOLOv7 make a number of changes to the YOLO network and new training routines. First, they noticed that the efficiency of the convolutional layers in the backbone is critical to reach an efficient inference speed. Building on previous research on this topic, they take into account the memory access cost as well as the distance that a gradient need to back-propagate to other layers. So, they integrate the Extended Efficient Layer Aggregation Network (E-ELAN), which allows the model to learn more diverse features for better learning.

Moreover, object detection models typically take into account the depth and the width of the network, as well as the resolution on which the network is trained. The authors of the YOLOv7 paper show that the model can be further optimized with a compound model scaling approach. They propose, therefore, to scale the network depth and width simultaneously while concatenating layers together, which keeps the architecture optimal while scaling for different sizes. In addition, they focused on a new re-parameterization technique that uses gradient flow propagation paths to determine how re-parameterized convolution should be combined with network modules. Finally, YOLOv7 adopts the concept of deep supervision, with a lead head that predicts the final result and an auxiliary head to assist training the middle layers.

Combined together, these improvements led to significant increases in detection accuracy and decreases in cost.

2.4   YOLO Version 8

YOLOv8, developed by Ultralytics[1], is a real-time object detection model and the latest version of the YOLO family. It introduces several improvements, including a more efficient backbone network, anchor-based object detection, and a mosaic data augmentation method making it faster, more accurate, and more efficient for object detection tasks. It is able to detect about 80 different object classes in real time, making it a valuable tool for various fields such as robotics, autonomous vehicles, and security.

The YOLOv8 architecture consists of a backbone network, a feature extractor, a detection head, anchor boxes, and improved post-processing techniques. It uses a more efficient backbone network, Darknet53, which is significantly faster and more accurate than the previous backbone used in YOLOv7. DarkNet53 is a deep network with 53 convolutional layers, that is responsible of extracting high-level features from the input image. The feature extractor takes the features generated by the backbone network and transforms them into a form that

---

[1] https://ultralytics.com/yolov8.

is suitable for object detection. Then, the detection head makes the final prediction of the object locations and class scores in the input image. The YOLOv8 architecture integrates a spatial pyramid pooling (SPP) structure and a spatial channel-wise pooling (SCP) layer, applied after the last convolutional layer, to extract multi-scale features from different regions of the feature maps and perform channel-wise feature pooling, respectively. The integration of SPP and SCP improves the representation of features and enhances the detection performance of the model. YOLOv8 includes a new anchor-free detection head that uses anchor boxes to predict the locations of objects in the input image. It also includes non-maximum suppression, an improved post-processing technique, which is used to reduce the number of false positive detections and enhance its accuracy. The architecture of YOLOv8 is designed to be fast and efficient, while still maintaining high accuracy for object detection tasks, making it a powerful tool for real-world applications.

3   UAV Detection Framework

In this section, we present the used drone detection and tracking framework, where a UAV tracker detects and follows another UAV target in a pursuit-evasion scenario.

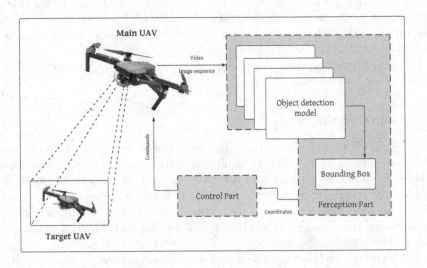

Fig. 2. The proposed UAV detection framework

As we can see in Fig. 2, this framework is divided into two parts: the perception part of the target UAV and the control navigation of the follower UAV. The detection part uses images acquired by a UAV and a deep object detector based on YOLOv7 or YOLOv8 to locate the target UAV. The position of the detected target UAV, given as a bounding box, is then sent to a high-level controller that uses a vision-based deep reinforcement learning algorithm to adjust the position of the follower UAV in order to keep the target UAV as close as possible to the center of the image in the next input frame.

4 Experiments and Results

4.1 Dataset Description

We used the dataset generated by [12]. The DJI Air 2 was used as the main UAV and the Mini 2 as the target UAV. The Air 2 has a more advanced camera and can capture panoramas. Its control unit has more functions and excellent obstacle avoidance. The Mini 2 is smaller and moves faster than the Air 2. Different complex and diversified scenes were used in order to improve the robustness of the object detection model. Seven various scenarios were acquired: indoor (factory), outdoor (factory), road, farm, swamp, lake, and forest. Four to five videos were captured for each scene, each lasting about 2 min. These videos produced 2753 images with a resolution of 1920 × 1080. We used 1663 images for training. 545 images were randomly selected for validation and 545 images for testing. The annotations are available in COCO format, including bounding box coordinates, associated file names, etc. Examples of used scenes are shown in Fig. 3.

(a)   (b)   (c)   (d)

Fig. 3. Sample images from the used dataset (a) Indoor (factory) (b) Road (c) Outdoor (factory) (d) farm.

4.2 Models

We evaluate the performance of different variants of YOLOv7 and YOLOv8, using the dataset described in Sect. 4.1. The models were trained using the NVIDIA Tesla T4 GPU. The variants of YOLOv8 include: (1) YOLO Extra Large (YOLOv8x) is the most accurate yet the slowest among them. It uses additional convolutional layers, anchor boxes, and data augmentation techniques to improve performance; (2) YOLO Large (YOLOv8l) is a lightweight version, with a smaller number of parameters and a more compact architecture compared to the standard YOLOv8. It is designed to be faster and more efficient while still maintaining reasonable accuracy; YOLO Nano (YOLOv8n) is designed to be even faster and more efficient compared to YOLOv8l. It uses a new backbone network, an improved feature extractor, and a more efficient post-processing technique to achieve faster object detection. As for YOLOv7, there are three models, namely YOLOv7-E6, YOLOv7-E6E and YOLOv7-D6. To obtain YOLOv7-E6 and YOLOv7-D6, the authors of YOLOv7 used a compound scaling method. Furthermore, the network E-ELAN is used for YOLOv7-E6, completing YOLOv7-E6E.

Table 1. Comparison of one stage and two stage detection methods.

Type	Method	Backbone	AP50	FPS	F1 score
Two-stage	Faster R-CNN [12]	R-101-FPN	0.959	18.50	–
	Libra R-CNN [12]	X-101-64x4d-FPN	0.959	15.00	-
	Cascade R-CNN [12]	R-101-FPN	0.979	17.90	–
One-stage	YOLOv7	320	0.989	36.70	–
		mstrain-416	0.989	37.70	–
		mstrain-608	0.988	33.40	–
	YOLOv3	E6E	0.992	61.80	0.985
		E6	0.987	61.75	0.978
		D6	0.990	61.77	0.984
	YOLOv8	Extra Large	0.994	26.04	**0.992**
		Large	0.994	27.01	0.991
		Nano	**0.995**	**107.53**	**0.992**

4.3   Quantitative Results

Table 1 shows the detection results of the tested YOLOv8 and YOLOv7 models compared to the results reported in [12] for YOLOv3 as a single-stage detector where three versions were used (320, MStrain-416, and MStrain-608). YOLOv3 uses DarkNet-53 as its backbone. For the neck, MStrain-416 and MStrain-608 use multi-scale training like FPN. As for two-stage methods, we reported the results of Faster RCNN, Libra RCNN, and Cascade RCNN [12]. For Faster RCNN and Cascade RCNN, the ResNet-101 was used as backbone and FPN as neck. For Libra RCNN, authors changed the backbone to ResNeXt.

All above models were trained on the dataset described in Sect. 4.1. As performance metrics, we used the Average Precision 50 (AP50) and the Frames per Second (FPS) evaluation metrics. Average Precision (AP) is defined as the area under the precision-recall curve (PR curve). 50% is the Intersection over Union

(IoU) threshold. So the AP50 is the AP value with the IoU threshold at 50%. FPS defines how fast the model processes the input and generates the desired output.

As can be seen in Table 1, YOLOv8 has more accurate predictions compared to the other methods. We particularly notice that YOLOv8n achieved the best performance both in terms of FPS and AP50, which make it an appropriate choice for UAV real-time detection. YOLOv8 is fast and more accurate than previous versions because it uses a more efficient convolutional network with FPN, which helps to better recognize patterns and objects of different sizes and improves its overall accuracy, while reducing training cost. Table 1 also show that YOLOv7 performed overall better than YOLOv3 and two-stage methods in terms of precision and detection speed. We can also notice that the detection speed of YOLOv7 is always over 61 FPS, which surpasses the detection speed of all tested YOLOv3 models in [12]. This can be explained by the introduction of the Extended Efficient Layer Aggregation (E-ELAN) in YOLOv7.

Furthermore, to evaluate the performance of YOLOv7 and YOLOv8 models to detect UAV of different sizes, we have categorized the UAVs of the dataset as small, medium or large. First, we calculated the area of each bounding box, which gives an estimation of the size of the UAV. Then, we used the k-means algorithm to group UAVs into three clusters based on size. The percentage of small size UAV in this dataset is 58.72%, medium size UAV account for 27.15% and the remaining 14.13% are large size UAV. Table 2 breaks out the results of both models by size category to demonstrate their effectiveness in detecting UAVs of different sizes. We particularly notice the strong performance of YOLOv8 for small UAV detection.

Table 2. Results of YOLOv7 and YOLOv8 methods according to UAV size category.

Size	Percentage of UAV [%]	YOLOv7 AP50	YOLOv8 AP50
Small	58.72	99.18	**99.34**
Medium	27.15	99.31	**99.51**
Large	14.13	99.52	**99.59**

4.4 Qualitative Results

Figure 4 illustrates the results given by YOLOv7-E6E. We notice that the model is very suitable for identifying medium and small size UAVs. It achieved a high accuracy for Farm, Road and Factory indoor environments. However, the confidence rate in the factory outdoor scene images is relatively low but still acceptable. This decrease can be explained by the complexity of the background in the image, which makes the drone more difficult to detect.

As shown in Fig. 5, YOLOv8 model is able to correctly detect UAVs under varying conditions, i.e., different backgrounds, different lighting environments, various scales, and several angles of view. Indeed, although the drone is relatively far in the road environment, the model manages to detect the drone with high confidence level. Moreover, in the factory outdoor scene, the complexity of the background didn't play a big role in decreasing the detection confidence level.

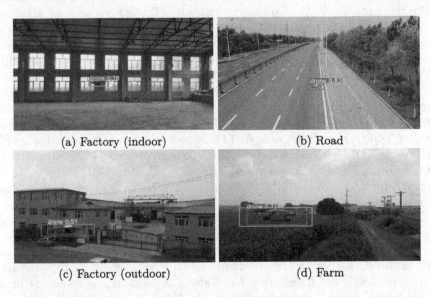

(a) Factory (indoor)   (b) Road

(c) Factory (outdoor)   (d) Farm

Fig. 4. Detection results of one stage method YOLOv7 in various scenes.

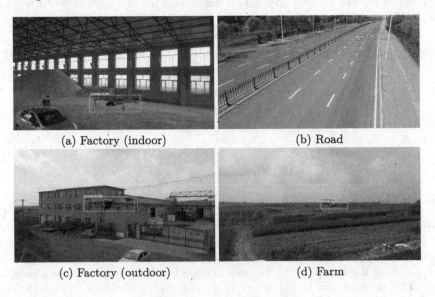

(a) Factory (indoor)   (b) Road

(c) Factory (outdoor)   (d) Farm

Fig. 5. Detection results of one stage method YOLOv8 in various scenes.

5   Conclusion

Drones are becoming increasingly autonomous and the use of artificial intelligence is playing a significant role in this direction. AI-enabled drones can make real-time decisions without human intervention. This is especially useful in urgent situations, such as disaster search and rescue missions. In this work, we used state-of-the-art deep learning object detection YOLOv8 to detect a target UAV from images captured by another UAV, using a dataset of real-life scenarios. We evaluated the performance of different variants of YOLOv8. Experimental results show that the best model achieves an AP50 of 99% and an FPS of 107.53, which proved that it is feasible and effective for UAV real-time detection. This detection module constitutes our perception part of a UAV collaboration/chasing framework. For future work, we plan to explore other object detection models such as anchor-free detectors and transformer architectures that continue to gain in popularity. Also, further work needs to focus on the control part for real-time tracking of UAVs in various challenging scenarios.

Acknowledgements. This research was enabled in part by support provided by the Natural Sciences and Engineering Research Council of Canada (NSERC), funding reference number RGPIN-2018-06233.

References

1. Sampedro, C., Rodriguez-Ramos, A., Campoy, P.: A review of deep learning methods and applications for unmanned aerial vehicles. J. Sens. **2**(1), 1–13 (2017)
2. Yan, C., Xiang, X.: A path planning algorithm for UAV based on improved Q-learning. In: 2nd International Conference on Robotics and Automation Sciences, pp. 1–5 (2018)
3. Bouhamed, O., Ghazzai, H., Besbes, H., Massoud, Y.: Autonomous UAV navigation: a DDPG-based deep reinforcement learning approach. In: 2020 IEEE International Symposium on Circuits and Systems (ISCAS). IEEE (2020)
4. Hu, Z., Gao, X., Kaifang, W., Zhai, Y., Wang, Q.: Relevant experience learning: a Deep Reinforcement Learning method for UAV Autonomous Motion Planning in complex unknown environments. Chin. J. Aeronaut. **34**(12), 187–204 (2021)
5. Kaifang, W., Bo, L., Xiaoguang, G., Zijian, H., Zhipeng, Y.: A learning-based flexible autonomous motion control method for UAV in dynamic unknown environments. J. Syst. Eng. Electron. **32**(6), 1490–1508 (2021)
6. Zaier, M., Miled, W., Akhloufi, M.-A.: Vision-based UAV tracking using deep reinforcement learning with simulated data. In: Autonomous Systems: Sensors, Processing and Security for Ground, Air, Sea and Space Vehicles and Infrastructure, vol. 12115, pp. 92–108. SPIE (2022)
7. Çetin, E., Barrado, C., Pastor, E.: Counter a drone in a complex neighborhood area by deep reinforcement learning. Sensors **20**(8), 2320 (2020)
8. Akhloufi, M.-A., Arola, S., Bonnet, A.: Drones chasing drones: reinforcement learning and deep search area proposal. Drones **3**(3), 58 (2019)

9. Arola, S., Akhloufi M.A.: UAV pursuit-evasion using deep learning and search area proposal. In: Proceedings of the IEEE International Conference on Robotics and Automation (2019)

10. Hassan, S.-A., Rahim, T., Shin, S.-Y.: Real-time UAV detection based on deep learning network. In: International Conference on Information and Communication Technology Convergence (ICTC), pp. 630–632 (2019). https://doi.org/10.1109/ICTC46691.2019.8939564

11. Isaac-Medina, B.K.S., Poyser, M., Organisciak, D., Willcocks, C.-G., Breckon, T.-P., Shum, H.-P.-H.: Unmanned aerial vehicle visual detection and tracking using deep neural networks: a performance benchmark. CoRR abs/2103.13933 (2021)

12. Qi, Z., Laplante, J.F., Akhloufi, M.-A.: Transformers and deep CNNs for unmanned aerial vehicles detection. In: Unmanned Systems Technology XXIV 12124. SPIE (2022). https://doi.org/10.1117/12.2622387

13. Çetin, E., Barrado, C., Pastor, E.: Improving real-time drone detection for counter-drone systems. Aeronaut. J. **125**(1292), 1871–1896 (2021). https://doi.org/10.1017/aer.2021.43

14. Wu, X., Li, W., Hong, D., Tao, R., Du, Q.: Deep learning for unmanned aerial vehicle based object detection and tracking: a survey. IEEE Geosci. Remote Sens. Mag. **10**(1), 91–124 (2022)

15. Ahmad, T., Cavazza, M., Matsuo, Y., Prendinger, H.: Detecting human actions in drone images using YOLOv5 and stochastic gradient boosting. Sensors **22**(18), 7020 (2022)

16. Luo, X., Wu, Y., Wang, F.: Target detection method of UAV aerial imagery based on improved YOLOv5. Remote Sensing **14**(19), 5063 (2022)

17. Sahin, O., Ozer, S.: YOLODrone: improved YOLO architecture for object detection in drone images. In: 44th International Conference on Telecommunications and Signal Processing (TSP), pp. 361–365 (2021). https://doi.org/10.1109/TSP52935.2021.9522653

18. Zhao, H., Zhang, H., Zhao, Y.: YOLOv7-sea: object detection of maritime UAV images based on improved YOLOv7. In: Proceedings of the IEEE/CVF Winter Conference on Applications of Computer Vision (WACV) Workshops, pp. 233–238 (2023)

19. Wang, C.-Y., Bochkovskiy, A., Liao, H.-Y.-M.: YOLOv7: trainable bag-of-freebies sets new state-of-the-art for real-time object detectors. arXiv (2022)

20. Zhao, Z.-Q., Zheng, P., Xu, S., Wu, X.: Object detection with deep learning: a review. arXiv (2018)

21. Girshick, R., et al.: Rich feature hierarchies for accurate object detection and semantic segmentation. In: Proceedings of the IEEE Conference on Computer Vision and Pattern Recognition (2014)

22. Girshick, R.: Fast R-CNN. In: Proceedings of the IEEE International Conference on Computer Vision (2015)

23. Ren, S., et al.: Faster R-CNN: towards real-time object detection with region proposal networks. In: Advances in Neural Information Processing Systems, vol. 28 (2015)

24. Du, L., Zhang, R., Wang, X.: Overview of two-stage object detection algorithms. J. Phys. Conf. Ser. IOP Publ. **1544**(1) (2020)

25. Sun, P., et al.: Sparse R-CNN: end-to-end object detection with learnable propos-als. In: Proceedings of the IEEE/CVF Conference on Computer Vision and Pattern Recognition, pp. 14454–14463 (2021)
26. Zhang, Y., Li, X., Wang, F., Wei, B., Li, L.: A comprehensive review of one-stage networks for object detection. In: IEEE International Conference on Signal Processing, Communications and Computing (ICSPCC), pp. 1–6 (2021). https://doi.org/10.1109/ICSPCC52875.2021.9564613
27. Long, X., et al.: PP-YOLO: an effective and efficient implementation of object detector. CoRR abs/2007.12099 (2020)

A Deep Neural Architecture Search Net-Based Wearable Object Classification System for the Visually Impaired

Aniketh Arvind[✉]

Hackley School, Tarrytown, NY 10591, USA
`aarvind@students.hackleyschool.org`

Abstract. The *World Health Organization* estimates that a staggering 2.2 billion individuals worldwide suffer from vision impairments, drastically limiting independence and quality of daily life and leading to billions of dollars in direct costs and annual productivity losses. Although the field of machine learning has made significant strides in recent years, particularly in image classification, these advances have predominantly focused on tasks that are visual in nature, which can be challenging for vision-impaired individuals. Much work has been published on obstacle avoidance and large-object detection for the visually impaired. However, little has been done to aid them in better understanding complex indoor daily-living environments. For these reasons, this study develops and presents a wearable object classification system specifically designed to assist the visually impaired in identifying small tabletop objects commonly found in their surrounding indoor environments. Through transfer learning, the system uses a pretrained neural architecture search network called NASNet-Mobile and a custom image dataset to conduct highly effective small-object classification with model accuracies of over 90.00%. The proposed transfer-learning model is subsequently deployed on a wearable wrist device for real-world applicability. This study ultimately evaluates and demonstrates the system's ability to accurately classify small tabletop objects using an eight-trial experiment that calculates the system's average precision, recall, and F1 score to be 99.30%, 97.93%, and 98.61%, respectively. Overall, this system represents a significant step forward in the development of machine learning systems that constructively assist the visually impaired while simultaneously improving their daily independence and quality of life.

Keywords: computer vision · classification · object detection · transfer learning · visually impaired · daily living

Supported by the Biorobotics Lab at the University of Wisconsin-Milwaukee.

A. Bennour et al. (Eds.): ISPR 2023, CCIS 1941, pp. 198–213, 2024.
https://doi.org/10.1007/978-3-031-46338-9_15

1 Introduction

1.1 Background

Visual impairment and blindness severely impact daily life. As of 2021, approximately 2.2 billion individuals worldwide suffer from vision impairments. Although most of these individuals are over 50, blindness and visual impairment can affect all age groups. Vision impairments also cause low productivity, decreased independence, depression, anxiety, and financial burdens. The *World Health Organization* determines that productivity losses from uncorrected myopia and presbyopia alone account for more than $244 billion and $25.4 billion in annual losses, respectively. Furthermore, eye care and education surrounding eye health are often not prioritized in developing countries leading to various preventable eye diseases, such as trachoma, cataract, age-related macular degeneration, and glaucoma. Recent studies estimate that nearly one billion people, roughly half the world's visually impaired population, are currently affected by preventable eye diseases [2].

As of 2017, in the United States alone, over 7,970,000 adults reported being visually disabled [12]. However, this figure has increased drastically over the past five years due to an increased prevalence of acute age-related eye diseases like diabetic retinopathy and primary open-angle glaucoma [8]. With the rise of vision loss and blindness in the United States, the resulting economic costs have been severe for such individuals. On average, vision loss is estimated to cost roughly $16,838 per person annually, with New York and Massachusetts reporting estimates as high as $27,314. These estimates incorporate direct costs, such as medical and nursing home care, support services (excluding hourly personal assistance), prescription costs, and indirect costs (absenteeism, reduced labor force participation, and reduced earnings, among others) [12].

1.2 Literature Review

Obstacle Avoidance for the Visually Impaired. To address the decreased independence and financial burdens of visually impaired individuals, *Gada et al. (2019)* designed a low-cost, real-time visual aid that mounts a Raspberry Pi camera module to the end of the traditional, widespread probing cane. The real-time output of the camera is fed into an object recognition system that applies a typical Haar Cascade algorithm trained on custom and open-source datasets via transfer learning. For the benefit of the visually impaired, the recognition outcome is then converted to audio feedback [5]. The developed system primarily depends on the probing cane for obstacle avoidance rather than specific object detection. Additionally, smaller objects on elevated surfaces, such as toothbrushes and cellphones, frequently encountered by visually impaired individuals daily, demand far more robust detection and recognition tools and are nearly impossible for such a system to detect.

Handheld Obstacle-Detection for the Visually Impaired. Using the PAS-CAL VOC 2007 and PASCAL VOC 2012 datasets, *Rachburee (2021)* utilized the widely-accepted YOLOv3 architecture with a pre-trained Darknet-53 approach. A smartphone camera's output serves as input for the model, primarily to allow for the detection of elevated obstacles. Although the study detected and classified various obstacles with accuracies ranging from 95% to 97%, they were exclusively large in size [10]. Small or medium-sized obstacles were presumably excluded due to the possibility of lower model accuracies and confidence. Additionally, smartphones themselves face limitations, such as battery life and costs, which hinder the efficacy of the study's proposed system.

Handheld Object-Detection for the Visually Impaired. *Denić et al. (2021)* developed a smartphone-based object detection and classification system. Using a scratch-built model consisting of various convolutional, pooling, fully connected, and softmax layers, the study conceptualizes an app that uses a smartphone camera's output as input for the model. Once an object is classified, the output is relayed to the user via audio through a Text-to-Speech (TTS) model [4]. Nevertheless, the study only presents basic classification predictions of a single object class and reveals low accuracies of around 70%.

Joshi et al. (2020) use a hand-held device to aid visually impaired users in the detection and perception of surrounding objects. The study employs a YOLOv3 framework with the Darknet-53 (53 convolutional layers) pre-trained model for transfer-learning-based training. Accuracies of 97% to 99% are achieved on medium-to-large objects, with small objects being excluded, limiting the amount of assistance provided [7]. Furthermore, handheld devices restrict vital two-hand interaction with objects, hindering ease of use.

1.3   Summary and Contribution

The limitations that have arisen thus far within this field of research can be placed into four categories: 1) predominance of obstacle avoidance over object recognition, 2) subpar accuracies, 3) lack of small-object detection, and 4) restricted user-object interaction. This research addresses all four of the above-mentioned limitations, pushing the bounds of computer vision and machine-learning applications that constructively assist the visually impaired. It does so by designing and developing a consistent, high-accuracy, NASNet-based object-classification system capable of recognizing small daily-living-related tabletop objects in indoor environments. This system manifests itself in the form of a cost-effective novel wearable arm device that enables users to interact freely with surrounding objects without inhibitions. Ultimately, this study not only offers a novel, effective solution to the decreased independence and quality of daily life within the visually impaired population, but it also helps save users valuable money, which would otherwise be spent on hourly pay for personal assistants and other similar aids.

The remaining parts of this paper are organized as follows: Sect. 2 reveals the methods used to develop the proposed system. Next, Sect. 3 highlights the

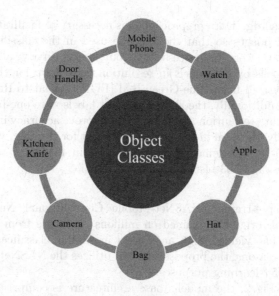

Fig. 1. The eight classes of objects commonly used in daily living, and also used for training the object-classification model.

general experimental setup for future replication. Section 4 discloses the results of the research, followed by Sect. 5, which discusses all significant findings. Lastly, Sect. 6 presents important conclusions and potential future directions for this work.

2 Methodology

2.1 Object Detection and Classification Model

This module can be divided into four sections: 1) dataset, 2) data preprocessing, 3) metrics, and 4) compilation. Throughout, this deep transfer-learning object-classification model was loaded and trained in the Google Colab environment and later saved to a Hierarchical Data Format 5 (H5) file for further use.

Dataset. The Google Open Images Dataset consists of roughly 10 million open-sourced, pre-annotated images, making it ideal for use in such applications [3]. The object detection and classification model developed in this research utilized a custom subset of data from Open Images in order to specifically target objects of daily living, which are necessary for the proposed application. Additionally, given the nature of the Open Images dataset, falsely classified images were discarded to avoid feature confusion and increase model accuracies. Moreover, the custom subset consisted of 6024 images split equally (to avoid data imbalance and model bias) into eight distinct, carefully-selected classes of objects (i.e., each class had 753 images). The selection of the object classes was based on the recommendations of a blind relative of the author and is shown in Fig. 1.

Data Preprocessing. Data preprocessing is necessary to facilitate the uniform formatting of all images so that they can be used in the classification model. In this study, data were preprocessed in four steps. Firstly, all images were resized to 300 pixels by 300 pixels in resolution, and the color channels of the images were converted from Blue-Green-Red (BGR) format to Red-Green-Blue (RGB) format. Additionally, the assigned object labels were one-hot-encoded to improve overall data resolution and potentially improve accuracy in classification predictions. Lastly, the final image dataset was split for training, validation, and testing via an 80-10-10 scheme. The split resulted in 4819 training images, 602 validation images, and 603 test images.

Metrics. The pre-trained NASNet-Mobile Convolutional Neural Network (CNN) model was originally trained on millions of images from the ImageNet dataset [1] and has yielded high accuracies in similar classification problems [6,14]. For these reasons, the proposed system utilizes the NASNet-Mobile architecture for transfer-learning purposes.

As shown in Fig. 2, the model's base architecture is comprised of two sub-states: 1) normal cell and 2) reduction cell. Both sub-states present themselves as classification cells. However, while the normal cell sub-state returns a same-dimension feature map, the reduction cell returns a feature map where the height and width are scaled down by a factor of two [14]. This dichotomy of sub-states allows the NASNet CNN to achieve state-of-the-art results.

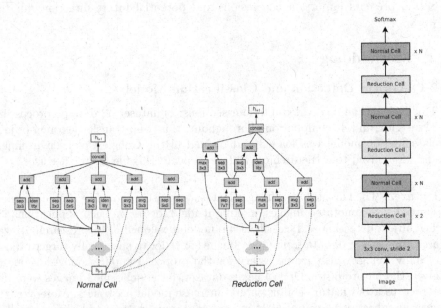

Fig. 2. NASNet image classification flow (left), normal cell and reduction cell architectures for the NASNet CNN (right) [14].

To tackle the application proposed in this research, transfer learning was utilized to add four dense layers, which classify images using convolutional layer outputs. Furthermore, two dropout layers were added to the final model to reduce and prevent the over-fitting of data during training. The architecture of the added layers with their respective parameters is illustrated in Fig. 3. Note that the first two layers in the figure are components of the base NASNet-Mobile model and are included to showcase the transition between the base model and the added layers.

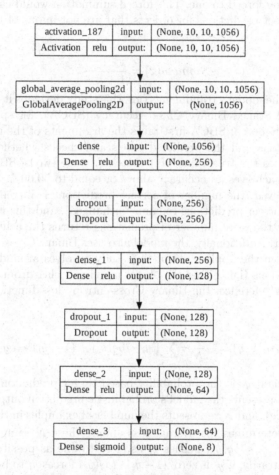

Fig. 3. The final two layers of the base NASNet-Mobile architecture with four added dense layers and two added dropout layers.

Rectified Linear Units (ReLU) were used for the first five layers, while Sigmoid was employed for the last. Furthermore, given the constraints of Eq. 1, ReLU eliminates the vanishing gradient problem and learns linear and non-linear functions within hidden layers with high efficiency.

$$R(z) = \begin{Bmatrix} z & z > 0 \\ 0 & z \leq 0 \end{Bmatrix} \tag{1}$$

Equation 2 displays the Sigmoid activation, which represents the function that calculates individual probabilities (i.e., the sum of all individual probabilities is allowed to be greater than one) for specified target classes and outputs them as a vector to determine predictions [9]. The Sigmoid activation function was chosen over the Softmax activation function to avoid the sum of all class probabilities being forced to one. This forced summation would cause the model to output incorrect predictions for objects that are not in one of the eight specified classes.

$$Sigmoid(z) = \frac{1}{1 + e^{-z}} \tag{2}$$

Concerning the model's training metrics, this study develops its own accuracy function, named Staged Binary Class Accuracy (SBCA), for specialized class predictions. Stage one of SBCA first takes the arguments of the maxima of the ground truth labels and the predicted labels and checks whether the model's prediction matches the true value. If it matches, stage two of SBCA will check if the prediction achieves an accuracy above or equal to 50.00%. If the required accuracy is achieved, the number of correct predictions is incremented (if not, the number of correct predictions remains unchanged), and this newly-obtained value divided by the total number of predictions returns the achieved accuracy as a scalar tensor. Additionally, the model also uses Binary Cross-entropy as the loss function since the application of this study revolves around various "yes-or-no" classifications (i.e., either the object matches a class from the dataset or it does not). For reference, the Binary Cross-entropy loss function is shown in Eq. 3.

$$Loss_{BCE}(\hat{y}, y) = -\frac{1}{N} * \sum_{i=1}^{N} [\, y_i * log(\hat{y}_i) + (1 - y_i) * log(\hat{y}_i)] \tag{3}$$

In this function, y_i is the probability-based target prediction at a certain i (real value), $\hat{y}_i$ represents the model's probabilistic prediction output at a certain i (predicted value), and N represents the number of examples in the input vector [9]. $\frac{1}{N} * \sum_{i=1}^{N}$ determines the average loss among these N samples. If the real value, y_i, is one, $y_i * log(\hat{y}_i)$ pushes $\hat{y}_i$ as close to one as possible, in order to reduce loss. Contrarily, if y_i is zero, $(1 - y_i) * log(\hat{y}_i)$ forces $\hat{y}_i$ to be close to zero.

Compilation. The proposed object detection and classification model was trained via transfer learning over a span of 100 epochs in batches of 32 data points. However, to avoid data over-fitting, the dropout was increased during testing, and an Early-Stopping function, monitoring validation loss with a minimum delta of 0.0001 and patience of eight epochs, was executed, thus resulting in the completion of only 50 epochs during training. Furthermore, the model was compiled with the Adam optimizer, which automatically adjusts the model's learning rate accordingly, and finds the lowest local minimum of the given loss function during training.

2.2   Wearable Object Classification and Recognition Device

The novel object classification device developed in this study revolves around Raspberry Pi hardware. The device utilizes a Raspberry Pi 4 Model B Wireless (64-bit, 4GB RAM) in conjunction with additional modules containing a light, portable battery and a mini High-Definition (HD) color camera. Overall, with respect to the above-mentioned components, the full device is incredibly cost-effective to produce. Additionally, both wired and wireless earphones can be used with this device, which is shown in Fig. 4. The application-based flowchart of the proposed system is depicted in Fig. 5.

3   Experimental Design and Setup

To establish the usefulness and accuracy of the prototype developed in this research, a trial-based experiment is designed. The various trials validate the ability of the device to classify an assortment of selected objects into one of the eight classes or to identify it as an outlier (i.e., not a member of the eight classes).

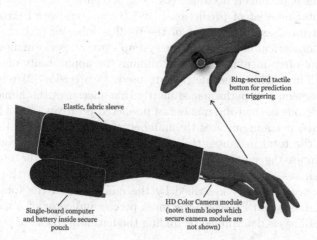

Ring-secured tactile
button for prediction
triggering

Elastic, fabric sleeve

Single-board computer
and battery inside secure
pouch

HD Color Camera module
(note: thumb loops which
secure camera module are
not shown)

Fig. 4. The proposed system (i.e., Raspberry Pi, portable battery pack, camera), which is carefully designed to provide the most optimal viewing angle for the camera module, is mounted on the user's forearm via a comfortable, elastic, and fabric sleeve. This sleeve allows visually impaired users to easily slide their dominant hand into the system without any hassle. The camera is secured to the base of the user's thenar eminence with the sleeve's thumb loops. Furthermore, a tactile button is attached to the index finger via a velcro-secured ring to trigger object predictions.

Table 1. Individual class assignments for the eight experimental trials. For each class, 10 objects that were not part of one of the object classes were tested on the device.

Trial	Object Classes
1	Mobile Phone
2	Watch
3	Apple
4	Hat
5	Bag
6	Camera
7	Kitchen Knife
8	Door Handle

The experiment consists of eight trials. As shown in Table 1, trials one to eight, respectively, are dedicated to the eight classes of objects. In all eight trials, the device is fed 90 true-positive (TP) images and 10 true-negative (TN) images in various lighting conditions. From these 100 total images, the experiment measures the number of true-positive, true-negative, false-positive (FP), and false-negative (FN) predictions made by the device. Using these resulting values, the system's average accuracy, percent error, precision, recall, and F1 score can be calculated.

Within this experiment, accuracy (i.e., the percentage of correct predictions out of the total number of predictions) and the percentage of error (i.e., the percentage of misclassifications out of the total number of predictions) act as two powerful numerical indicators of the system's overall performance. However, these alone are often insufficient in determining the applicability of a classification system. Three additional metrics are used: 1) precision, 2) recall, and 3) F1 Score. Precision is a performance metric that measures the number of true positives made out of the total number of predicted positives (i.e., TP and FP). In other words, it communicates the number of detected objects that are relevant out of the total instances the model retrieved [13]. Recall, on the other hand, determines the number of true positives made given the total number of actual positives (i.e., TP and FN). Put simply, it measures the number of relevant instances correctly identified by the model out of the total pertinent samples [13]. Lastly, an F1 score, the most precise indicator of a model's real-world applicability, is determined by finding the harmonic mean of precision and recall.

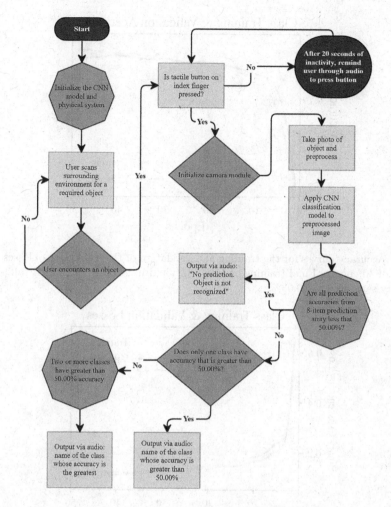

Fig. 5. Flowchart detailing the various steps needed for the proposed system

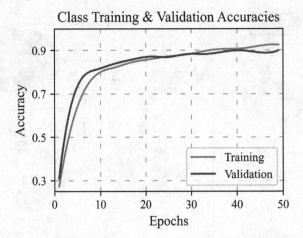

Fig. 6. Accuracy curves for the training and validation of the eight object classes over a span of 50 epochs. Final training accuracy resulted in 93.05%, while final validation accuracy saturated to 90.03%.

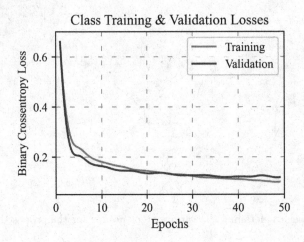

Fig. 7. Binary Crossentropy Loss curves for the training and validation of the eight object classes over a span of 50 epochs. Final training loss exacted to 0.0992, while final validation loss saturated to 0.1164.

4   Results

4.1   Performance of the Classification and Recognition Model

Through the following results, we prove that the transfer-learning-based object classification model developed in this research, which utilizes a NASNet-Mobile architecture and ImageNet weights, is ideal for the visually impaired and similar applications. Figure 6 reveals, over a period of 50 epochs, the saturation and stabilization of training and validation accuracy. With the completion of training, class training accuracy saturated to 93.05% and validation to 90.03%. Furthermore, the loss, which demonstrates the numerical difference from the desired prediction, is shown in Fig. 7. Here, losses for class training and class validation saturated to 0.0992 and 0.1164, respectively. Additionally, subsequent testing was required to discern irregularities in the model with respect to object classification and the study's specific application. This testing resulted in a final saturated accuracy and loss of 88.06% and 0.1255, respectively.

4.2 Object Classification and Recognition Performance of the Proposed Wearable Device

The prototype developed in this research uses the proposed image-classification model and Raspberry Pi hardware to allow visually impaired users to better understand their surrounding environments while interacting with vital daily objects. As established in Sect. 3, an eight-trial experiment was conducted to validate and test the accuracy of the complete system. 90 TP and 10 TN images were used in each of the eight trials to evaluate the system's performance. Table 2 presents the number of TP, TN, FP, and FN predictions the device made from the 800 total objects presented to it.

Table 2. The number of true-positive (TP), true-negative (TN), false-positive (FP), and false-negative (FN) predictions made by the system for each of the eight trials from a total sample size of 90 true images and 10 false images.

Trial	Total Samples	True Samples	False Samples	TP	TN	FP	FN
1	100	90	10	90	9	0	1
2	100	90	10	89	8	1	2
3	100	90	10	88	9	1	2
4	100	90	10	87	9	1	3
5	100	90	10	89	8	1	2
6	100	90	10	89	9	0	2
7	100	90	10	90	9	0	1
8	100	90	10	89	8	1	2

The results from Table 2 were used to calculate accuracy, percent error, precision, recall, and the F1 score of the system for each of the eight object classes. The outputs from these equations are put forward in Table 3. Here, the minimum accuracy of 96.00% is associated with the Hat object class, and the maximum accuracy of 99.00% corresponds to both the Mobile Phone and Kitchen Knife object classes. The entire system achieves an accuracy of 97.50%, thus demonstrating that objects in real-world scenarios are correctly recognized and classified into their respective classes. Furthermore, it can be seen from Table 3 that the percent error in the eight trials ranges from 1.00% (Mobile Phone and

Kitchen Knife) to 4.00% (Hat), resulting in an average of 2.50% error for the entire system. Additionally, the system attains an average precision, recall, and F1 score of 99.30%, 97.93%, and 98.61%, respectively.

Table 3. Accuracy, percent error, precision, recall, and F1 score of the proposed system within individual classes, based on the results from Table 2.

Trial	Accuracy (%)	Error (%)	Precision (%)	Recall (%)	F1 (%)
1	99.00	1.00	100.00	98.90	99.45
2	97.00	3.00	98.89	97.80	98.34
3	97.00	3.00	98.88	97.78	98.32
4	96.00	4.00	98.86	96.67	97.75
5	97.00	3.00	98.89	97.80	98.34
6	98.00	2.00	100.00	97.80	98.89
7	99.00	1.00	100.00	98.80	99.45
8	97.00	3.00	98.89	97.80	98.34

5 Discussion

The method developed and proposed in this study is designed to help visually impaired individuals better understand and interact with necessary objects encountered in their daily lives. Other studies, such as [7, 10], utilize similar methods. However, their respective systems are handheld and expensive, preventing users from tangibly interacting with their surrounding environments. This study tackles the above-mentioned limitation by engineering a cost-effective and novel wrist-mount for the classification system. This allows users to intuitively and noninvasively scan, understand, and fully interact with daily indoor objects.

Additionally, studies such as [5,7,10], and [11] focus predominantly on large outdoor objects, as opposed to smaller indoor objects, which the visually impaired population most commonly encounter. No studies in this research field stress the importance of detecting vital tabletop objects, such as mobile phones or cutlery. For these reasons, this research heavily emphasizes the detection and classification of smaller objects found on tabletops and indoor environments.

The system designed and engineered in this study was validated in two steps: 1) classification model validation and 2) prototype device testing. Regarding the validation of the transfer-learning classification model, all necessary precautions were implemented to prevent data overfitting. For example, all data points were normalized, the custom dataset was split into training, validation, and testing with random states, and multiple dropout layers were added to the model's architecture. The positive effect of these precautions and the almost-perfect fitting of the data is shown in Fig. 6 and Fig. 7 as the validation accuracy and loss closely track training accuracy and loss, respectively, throughout the latter part of the 50 epochs. It can therefore be concluded that the accuracy of 90.03% and loss of

0.1164 obtained from the model's validation indicate the system's effectiveness with respect to the classification of small tabletop objects in indoor environments. Moreover, as revealed in Sect. 4, testing of the prototype device resulted in a remarkably high average precision, recall, and F1 score of the system. Using the definitions put forth in Sect. 3, these high average values demonstrate the device's overall correctness, predictive ability, and predictive repeatability with regard to detecting specific classes of objects.

The results, shown in Sect. 4, were expected, proving the proposed system's effectiveness, usefulness, and applicability. This research provides a legitimate, high-accuracy solution for visually impaired individuals who require an immediate understanding of their daily indoor surroundings and who desire more everyday independence.

6   Conclusions and Future Work

This work aimed to design and engineer a novel, cost-effective system to accurately and efficiently detect and classify small tabletop objects and output those classification predictions through wireless audio to visually impaired users who consistently encounter such objects in their everyday lives. To realize this goal, a custom dataset containing eight common tabletop objects was curated as a subset of the Google Open Images dataset. A transfer-learning-based model was subsequently developed using a pre-trained neural architecture search net called NASNet-Mobile. The transfer-learning model achieved a high validation accuracy of over 90.00% and an accuracy of 88.06% on the test portion of the dataset. Additionally, a novel prototype device was engineered using Raspberry Pi hardware and specially designed to be mounted on the forearm or wrist for heightened intractability with encountered daily objects. To achieve a more realistic measurement of the device's performance, the averages for precision, recall, and F1 score were calculated through a validation experiment consisting of eight trials (each trial dedicated to an object class). Respectively, they amounted to 99.30%, 97.93%, and 98.61%, therefore proving the real-world applicability of the proposed system and the positive impact it could have on the 2.2 billion visually impaired individuals worldwide [2]. A future direction for this study involves incorporating additional objects in the custom dataset. Furthermore, we hope to conduct a user study where visually impaired individuals will be able to utilize the proposed system in real-world scenarios. Lastly, we plan to develop a medicine image classifier embedded within the proposed system to aid the visually impaired in discerning the various medications they may require.

Acknowledgments. The author thanks M. Rahman, M. Sunny, M. Shahria, and the Bio-Robotics Lab at the University of Wisconsin-Milwaukee for mentorship and research discussion; A. Ying for editing, research management, and discussion; and V. Dheer and S. Nadol for occasional assistance with prototype development.

References

1. Imagenet dataset (2020). https://www.image-net.org/
2. Blindness and vision impairment, October 2022. https://www.who.int/news-room/fact-sheets/detail/blindness-and-visual-impairment
3. Open images dataset v7 and extensions (2022). https://storage.googleapis.com/openimages/web/index.html
4. Denić, D., Aleksov, P., Vučković, I.: Object recognition with machine learning for people with visual impairment. In: 2021 15th International Conference on Advanced Technologies, Systems and Services in Telecommunications (TELSIKS), pp. 389–392 (2021). https://doi.org/10.1109/TELSIKS52058.2021.9606436
5. Gada, H., Gokani, V., Kashyap, A., Deshmukh, A.A.: Object recognition for the visually impaired. In: 2019 International Conference on Nascent Technologies in Engineering (ICNTE), pp. 1–5 (2019). https://doi.org/10.1109/ICNTE44896.2019.8946015
6. Howard, A.G., et al.: Mobilenets: efficient convolutional neural networks for mobile vision applications. arXiv preprint arXiv:1704.04861 (2017)
7. Joshi, R.C., Yadav, S., Dutta, M.K., Travieso-Gonzalez, C.M.: Efficient multi-object detection and smart navigation using artificial intelligence for visually impaired people. Entropy. **22**(9), 941 (2020). https://doi.org/10.3390/e22090941, http://dx.doi.org/10.3390/e22090941
8. NIDILRR (ed.): Rehabilitation Engineering Research Centers (RERC) Program: RERC on Blindness and Low Vision HHS-2021-ACL-NIDILRR-REGE-0030. National Institute on Disability, Independent Living and Rehabilitation Research (2021). https://govtribe.com/file/government-file/hhs-2021-acl-nidilrr-rege-0030-rerc-low-vision-blind-dot-pdf
9. Nwankpa, C., Ijomah, W., Gachagan, A., Marshall, S.: Activation functions: comparison of trends in practice and research for deep learning. arXiv preprint arXiv:1811.03378 (2018)
10. Rachburee, N., Punlumjeak, W.: An assistive model of obstacle detection based on deep learning: Yolov3 for visually impaired people. Int. J. Elect. Comput. Eng. (IJECE). **11**, 3434 (2021). https://doi.org/10.11591/ijece.v11i4.pp3434-3442
11. Rahman, M.A., Sadi, M.S.: IoT enabled automated object recognition for the visually impaired. Comput. Methods Prog. Biomed. Update. **1**, 100015 (2021). https://doi.org/10.1016/j.cmpbup.2021.100015, https://www.sciencedirect.com/science/article/pii/S2666990021000148
12. Rein, D.B., et al.: The economic burden of vision loss and blindness in the united states. Ophthalmology **129**(4), 369–378 (2022)
13. Shafi, A.: How to learn the definitions of precision and recall (for good), April 2022. https://towardsdatascience.com/precision-and-recall-88a3776c8007
14. Zoph, B., Vasudevan, V., Shlens, J., Le, Q.V.: Learning transferable architectures for scalable image recognition. In: 2018 IEEE/CVF Conference on Computer Vision and Pattern Recognition, pp. 8697–8710 (2018). https://doi.org/10.1109/CVPR.2018.00907

Multicarrier Waveforms Classification
with LDA and CNN for 5G

M'hamed Bilal Abidine[(⊠)]

Laboratoire d'Ingénierie des Systèmes Intelligents et Communicants (LISIC),
Faculty of Electrical Engineering, University of Science and Technology Houari Boumediene
(USTHB), B. P. 32 El Alia, 16111 Bab Ezzouar, Algiers, Algeria
abidineb@hotmail.com

Abstract. Accurate classification of multi-carrier waveforms is significant for ensuring quality signal reception, improved system throughput, and reduce the power consumption in future wireless generations as 5G. The aim of this paper is to improve the precision classification of various multicarrier waveforms. Here, we propose a novel representation of multicarrier signals in AWGN environment and use suitable networks for classification, which utilizes deep convolutional neural networks to classify OFDM-QAM, and FBMC-OQAM. LDA-based (Linear Discriminant Analysis) method is proposed in this paper to reduce the input dimensions of CNN. The results reveal that LDA-CNN is a promising candidate for wireless communication.

Keywords: Automatic modulation classification · Cognitive radio · OFDM-QAM · FBMC-OQAM · LDA · CNN

1 Introduction

5G is the next evolution of telecom networks. The new wireless applications have been facing the increasing demand of the mobile consumers, high data rates, and mobility requirements. Orthogonal frequency division multiplexing (OFDM) being the most adopted one in current wireless systems such as LTE (Long Term Evolution). Its implementation is simple to use inverse fast Fourier transform (IFFT) and FFT. Its implementation is straightforward through orthogonally both in time and frequency.

The main drawback in all multi-carrier waveforms is the High Peak-to-Average Power Ratio (PAPR). 4G mobile communications technology has two drawbacks. The first, it uses the multicarrier techniques that are not immune to fading. The second that it is not immune to inter-symbol interference [1]. Bazin [2] and Bellanger [3] have proposed various non-orthogonal multicarrier methods as in case of filter bank. The parallel filters that are applied for each sub-carrier in research project PHYDYAS.

Automatic signal recognition is an important role for Cognitive Radio systems and dynamic spectrum access, where the receiver must be able to recognize the friendly signals. AMC can be used for single carrier signal with multicarrier signal. The conventional statistical-based approaches for Automatic modulation classification (AMC) extract

© The Author(s), under exclusive license to Springer Nature Switzerland AG 2024
A. Bennour et al. (Eds.): ISPR 2023, CCIS 1941, pp. 214–222, 2024.
https://doi.org/10.1007/978-3-031-46338-9_16

hand-crafted expert features such as cyclostationarity, spectral correlation function (SCF) [6, 10, 11], and likelihood-based judgment [7–9].

Recently, many research works interest to deep learning (DL) for Automatic modulation classification (AMC) that single carrier signals can outperform expert features based techniques. The receiver needs to classify or recognize automatically the multicarrier waveforms, while increasing the data rate and decrease the power expending. This paper aims to classify the waveform between OFDMQAM, and FBMC-OQAM. We design the filters in Filterbank-based multicarrier in such a way that we reduce the interference between overlapping symbols.

Many traditional methods have been advanced for AMC using single/multicarrier signals [10]. The crafting of expert features (more commonly known as 'feature engineering') can be also avoided with this method. However, as [10, 11] mention that the classification performance of AMC using raw I/Q (multipath interference, I/Q balance) data can be affected. In [12], they proposed an approach of combining HMM classifiers with received signal cycling spectrum features to recognize OFDMQAM and OFDM-OQAM. The Principal Components Analysis (PCA) is mainly used in Automatic modulation classification (AMC) to minimize the number of dimensions, and also to eliminate the noise.

In this study, we used LDA to improve the dimensions of classification that represents the signal amplitude data, and then CNN is performed on the LDA features to classify multicarrier waveforms in dense transmission environment. We considered the AWGN channel, the offset of carriers and multi-fading channels.

This document makes the following contributions: 1) We proposed a new multicarrier waveforms classification system using LDA-CNN. LDA process is described in this work to reduce the inputs of CNN. 2) The proposed approach can classify with a better classification performance of both multicarrier waveforms which have their filter applied to each subcarrier in dense transmission environment including multi-fading channels.

2 Signals of Multicarrier Modulations

We detailed in this paragraph both modulation techniques for the specific transmit signal models. OFDM integrates easily with advanced multiple antenna systems for example massive MIMO for beamforming. To facilitate the formulation of problem, we define some parameters here: Each multicarrier signal consist of N subcarriers that they are fed by quadrature amplitude modulation (QAM) data. In this type of modulation data, the zeros are added. In the Fig. 1, we demonstrated the system block of OFDM as follow:

2.1 OFDM/QAM

Each subcarrier mapped the complex-valued symbol after the QAM modulation through IFFT, and then the results generate an OFDM/QAM symbol. Finally, the symbols are generated through P/S block in the discrete time-domain as follows:

$$s_1(k) = \sum_m \sum_{n \in \Omega} x_{m,n} P_1(k - mN) e^{j2\pi(k-mN)n/N} \tag{1}$$

where $x_{m,n}$ is the QAM data in m-th OFDM-QAM symbol.

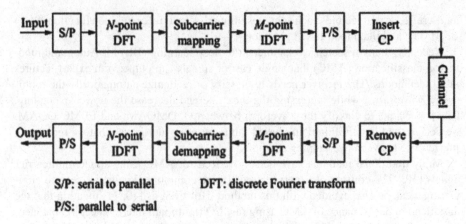

S/P: serial to parallel   DFT: discrete Fourier transform

P/S: parallel to serial

Fig. 1. System block of OFDM.

2.2 FBMC/OQAM

The FBMC-OQAM, has been greatly studied in [2, 3, 13], and it is considered as the challenger in future communication system. We can define the baseband transmit data in discrete-time domain as:

$$s_2(k) = \sum_m \sum_{n \in \Omega} a_{m,n} P_2(k - mN) \, e^{j2\pi nk/N} \, e^{j(m+n)\pi/2} \tag{2}$$

where $a_{m,n}$ is an amplitude in a complex-valued QAM symbols.

FBMC aims to improve the frequency localization using a prototype filter that covers over K multi-carrier symbols QAM. The spectral efficiency is higher in FBMC/OQAM [11]. FBMC-OQAM can very easily separate the users in frequency by using very small guard bands and no synchronization in time is required.

3 Proposed Model

This paragraph presents the LDA reduction method, the proposed system model and describes the CNN architecture. Automatic Modulation Classification (AMC) becomes important for wireless communications. It can extract the modulation features from the received IQ samples.

3.1 System Model

We presented in this paragraph, the structure of our proposed model as shown in Fig. 2. The system inputs are the received baseband data. Then the amplitude data are reduced by LDA, we extract the output sequence of LDA. We should normalize the power to proceed on sequence of amplitude. In the last step, and we obtained the classification results using a low dimensional LDA based on data enters to CNN. These features can then be used as inputs to neural networks to achieve high classification accuracy. CNN can directly classify modulation from the LDA features.

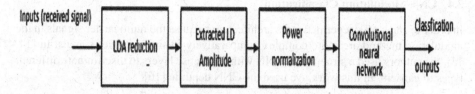

Fig. 2. System block of our proposed multicarrier classification system.

In the proposed system, the CNN does not need the step of feature extraction because it describes automatically their dimensions within each FFT period of two type's OFDM/QAM, and FBMC/OQAM waveforms.

3.2 Linear Discriminant Analysis (LDA)

The LDA method try to find directions that are efficient for discrimination between classes. The aim of LDA is to maximize the between-class Scatter matrix S_B while minimize the within-class Scatter matrix S_W. Let C_i be the class containing the state binary vectors $\vec{x}$ corresponding to the ith activity class. LDA consists to resolve the eigenvalue problem:

$$L = Eig\left(S_W^{-1} S_B\right) \tag{3}$$

$$S_B = \sum_{i=1}^{N} N_i(\overline{m}_i - \overline{m})(\overline{m}_i - \overline{m})^{T} \tag{4}$$

$$S_W = \sum_{i=1}^{N} \sum_{j=1}^{N_i} (x_j - \overline{m}_i)(x_j - \overline{m}_i)^{T} \tag{5}$$

where, N corresponds to classes, $\overline{m}_i$ is the average value of the samples in class C_i.

3.3 Power Normalization

In this section, we discuss the power normalization of fading channel model. Physically, radio channels correspond to passive circuits and follow the energy conservation law. We should carry out power normalization because the effects of channel can mystify the CNN algorithm and have an effect on the classification accuracy. Channel effects can present residuary features. To deal this, we normalized the power by modulus to conduct the equation as:

$$\tilde{r}_{m,k}^{normalized} = \left|\tilde{r}_{m,k}\right| / \sqrt{\frac{1}{k} \sum_{l=1}^{K} \left|\tilde{r}_{m,k}\right|^2} \tag{6}$$

3.4   CNN Modulation Classification

In order to carry off deeper networks architecture because the multicarrier signals in the modulation method are more complex comparatively to single-carrier signals. In [14, 15], the authors used a structure of CNN with few dense layers to discriminate different types of carriers. In this work, we used the CNN depth in [16].

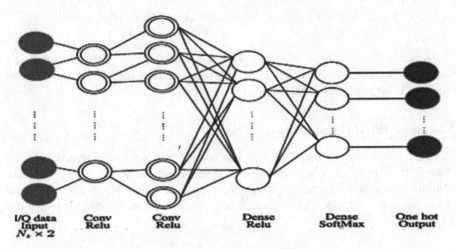

Fig. 3. Deep neural networks architecture.

Furthermore, we used in this paper the input data as an amplitude sequence [15] instead of using two-dimensional raw I/Q sequence. To obtain optimal classification accuracy, we have adapted parameters in each layer. The input data for CNN corresponds to receive data of one-channel amplitude sequence. The proposed five-layer CNN is visualized in Fig. 3 [17]. In addition, in avoid of over fitting in CNN, the parameters are all set to 0.5 between the two adjacent layers.

4   Experimental Results

In our study, we detailed the simulations performed in MALTAB of the proposed approach using two type of modulation form multi-magnitude sequence. Furthermore, we compared in this section and discussed diverse architectures of networks.

4.1   Parameter Settings

In Fig. 4, we demonstrate that the choose signal magnitudes of one spread signal whether for OFDM/QAM or for FBMC/OQAM that used as CNN input data. The FFT point of both waveforms is 512, and the number of sub-channel frequencies is 20. Each sub-bands contains Nc = 16 sub-carries with the subcarrier spacing $\Delta f = 15$ kHz. The CP length = Nc/4. We used the best parameters that maximize the accuracy. For the FBMC/OQAM [18] with K = 2 corresponding to the factor of overlap.

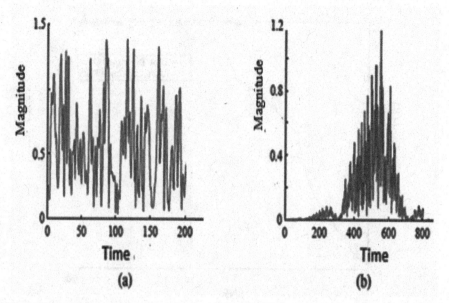

Fig. 4. Time domain representation of magnitude symbol of both multicarrier signals. (a) OFDM-QAM, and (b) FBMC [19].

4.2 Performance of LDA

We present here the performance in terms of accuracy of LDA. We used the same parameters settings of multi-carrier signal as in the previous section by considering the AWGN model. We discussed in this section the improvement of LDA to reduce the data inputs using the raw amplitude data. The effect of LDA on the CNN classifier is also analyzed. The normalization step is required to normalize the power by considering the amplitude gains. After reducing neural network model will be applied to classification task.

In the Fig. 5, we noticed that LDA gives a better separation between classes and has the strong noise removing ability when SNR ranges in −5.5 dB to 13 dB. Furthermore, when SNR surpasses 7 dB, the accuracy of CNN classifier performing LDA reduction data accomplishes close to 100%. Accuracy of CNN classifier that used raw amplitude data accomplishes close to 100% when SNR exceeds 12 dB.

4.3 Classification Performance

Our suggested method can obtain better performances in considering the used channel in Fig. 6. In our experimentations, we normalized by a uniform distribution the carrier frequency offset between the transmitter and receiver carrier by 0.5. Then, we plot the classification performance of CNN by using the raw I/Q data as inputs of CNN. Before to use the CNN classifier.

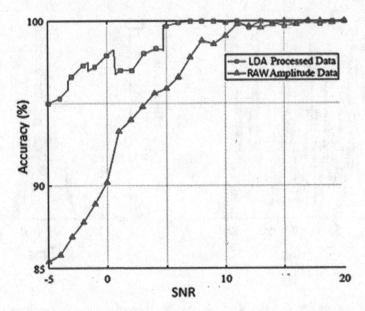

Fig. 5. Classification accuracy vs. SNR (dB) of LDA processed data and raw data.

The idea of this work is to add the LDA reduction of raw amplitude features to provide more informative features for CNN. However, when the signal power is more

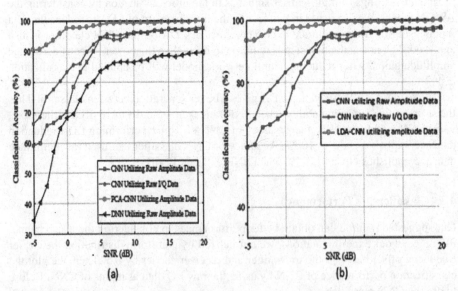

Fig. 6. Classification accuracy vs. SNR (dB) of neural networks. (a) PCA-CNN [19] (b) Proposed classification model.

than the power of noise, of neural networks obtains better performances which are close to CNN. The suggested system in Fig. 6 (b) that uses LDA reduction in the range of SNR in -5.2 dB to 20 dB accomplishes about average 97.7% accuracy. In Fig. 6 (a), the classification accuracy of PCA-CNN [19] is less well compared to LDA-CNN in (b).

In this work, the neural networks training complexity in terms of amount of data is reduced with two processes. The first, when we use amplitude as inputs instead of the raw I/Q, and the second when we use LDA reduction in the proposed system for the neural networks. We conclude that the proposed system economize the computing resources of CNN. In future work, it would be interesting to perform the PCA-LDA space [20] combined with CNN to classify the multicarrier waveforms.

5 Conclusion

Automatic modulation classification is a crucial factor in identifying the spectrum utility by classifying the modulation type of the signal. In this study, we performed the automatic classification of two multi-carrier waveforms (OFDM/QAM, and FBMC/OQAM) for future generations 5G. We conducted a new architecture system based on LDA and deep CNN classifier. LDA aims to reduce the dimension of time-domain signal amplitude series from the raw I/Q input data. CNN implicitly learned and classified correctly the features used by the specific model in the AWGN channel. The comparison between our proposed system using LDA features with CNN, and PCA-CNN using raw amplitude data gives the advantage to the proposed in terms of better performance.

References

1. Al-Jawhar, Y.A., Ramli, K.N., Taher, M.A., Shah, N.S.M., Mostafa, S.A., Khalaf, B.A.: Improving PAPR performance of filtered OFDM for 5G communications using PTS. ETRI J. **43**(2), 209–220 (2021)
2. Bazin, A.: Massive MIMO for 5G scenarios with OFDM and FBMC/OQAM waveforms. Doctoral dissertation, INSA de Rennes (2018)
3. Bellanger, M.G.: Specification and design of a prototype filter for filter bank based multicarrier transmission. In: Proceedings IEEE International Conference on Acoustics, Speech, and Signal Processing. Proceedings (Cat. No.01CH37221), SaltLake City, UT, May 2001, vol. 4, pp. 2417–2420 (2001)
4. Chang, D.C., Shih, P.K.: Cumulants-based modulation classification technique in multipath fading channels. IET Commun. **9**(6), 828–835 (2015)
5. Su, W.: Feature space analysis of modulation classification using very high-order statistics. IEEE Commun. Lett. **17**(9), 1688–1691 (2013)
6. Bouchou, M., Wang, H., Lakhdari, M.E.H.: Automatic digital modulation recognition based on stacked sparse autoencoder. In: Proceedings IEEE International Conference on Communication Technology (ICCT), Chengdu, China, October 2017, pp. 28–32 (2017)
7. Dulek, B., Ozdemir, O., Varshney, P.K., Su, W.: Distributed maximum likelihood classification of linear modulations over nonidentical flat blockfading Gaussian channels. IEEE Trans. Wireless Commun. **14**(2), 724–737 (2015)
8. Yuan, Y., Zhao, P., Wang, B., Wu, B.: Hybrid maximum likelihood modulation classification for continuous phase modulations. IEEE Commun. Lett. **20**(3), 450–453 (2016)

9. Ozdemir, O., Li, R., Varshney, P.K.: Hybrid maximum likelihood modulation classification using multiple radios. IEEE Commun. Lett. **17**(10), 1889–1892 (2013)

10. An, S., Jang, M., Yoon, D.: Classification of single-and multi-carrier signals using CNN based deep learning. In: 2021 7th IEEE International Conference on Network Intelligence and Digital Content (IC-NIDC), pp. 196–199. IEEE, November 2021

11. Hauser, S.C., Headley, W.C., Michaels, A.J.: Signal detection effects on deep neural networks utilizing raw IQ for modulation classification. In: Proceedings IEEE Military Communications Conference (MILCOM), Baltimore, MD, October 2017, pp. 121–127 (2017)

12. Moles-Cases, V., Zaidi, A.A., Chen, X., Oechtering, T.J., Baldemair, R.: A comparison of OFDM, QAM-FBMC, and OQAM-FBMC waveforms subject to phase noise. In: 2017 IEEE International Conference on Communications (ICC), pp. 1–6. IEEE, May 2017

13. Ramadhan, A.J.: Implementation of 5G FBMC PHYDYAS prototype filter. Int. J. Appl. Eng. Res. **12**(23), 13476–13481 (2017)

14. Jackisch, F.: Spectral efficiency gains through the use of FBMC/OQAM for DOCSIS systems. In: 2018 IEEE International Conference on Consumer Electronics (ICCE), pp. 1–4. IEEE, January 2018

15. O'Shea, T.J., Corgan, J., Clancy, T.C.: Convolutional radio modulation recognition networks. In: Jayne, C., Iliadis, L. (eds.) Engineering Applications of Neural Networks, pp. 213–226. Springer, Cham (2016). https://doi.org/10.1007/978-3-319-44188-7_16

16. Zhang, R., Yin, Z., Wu, Z., Zhou, S.: A novel automatic modulation classification method using attention mechanism and hybrid parallel neural network. Appl. Sci. **11**(3), 1327 (2021)

17. Scherer, D., Behnke, S.: Evaluation of pooling operations in convolutional architectures for object recognition. In: Proceedings IEEE International Conference on Artificial Neural Networks, Thessaloniki, Greece, pp. 92–101 (2010)

18. Viholainen, A., Bellanger, M., Huchard, M.: Prototype filter and structure optimization. Document D **5** (2009). www.ict-phydyas.org

19. Duan, S., Chen, K., Yu, X., Qian, M.: Automatic multicarrier waveform classification via PÇA and convolutional neural networks. IEEE Access **6**, 51365–51373 (2018)

20. Abidine, B.M.H., Fergani, L., Fergani, B., Oussalah, M.: The joint use of sequence features combination and modified weighted SVM for improving daily activity recognition. Pattern Anal. Appl. **21**(1), 119–138 (2018)

Weeds Detection Using Mask R-CNN and Yolov5

Merzoug Soltane[1]([⊠]) [iD] and Mohamed Ridda Laouar[2] [iD]

[1] LAVIA Laboratory Larbi, Tebessi University, Tebessa, Algeria
soltane.merzoug@univ-tebessa.dz
[2] Larbi Tebessi University, Tebessa, Algeria

Abstract. Agriculture have seen a decrease in yields crops due to weeds, so they were encouraged to use pesticides and chemical treatments to eliminate them and obtain the results despite their environmental damage, another hand the new IA technology can be give best solution help farmers to detect and elimination weeds. The detection of objects is crucial in many computer vision applications, and our focus is on detecting weeds, which pose a significant challenge in agriculture due to their negative impact on harvest yield and quality. In this study, we evaluate two state-of-the-art convolutional neural network-based object detectors for weed detection under real-world conditions without staging. Our evaluation considers both speed and accuracy metrics using a dataset of images captured using a smartphone camera, the performance of weed detection is evaluated according to a predefined geographic location in Tebessa, an Algerian eastern region. Additionally, we compare the performance of these models with and without additional training using examples from different databases.

Keywords: object detection · weeds detection · R-CNN · YOLO

1 Introduction

According to Food and Agriculture Organization, the global food demand will continue to grow, leading the food production to jump up to 70% over the few coming decades, in order to fulfil the nurture of the predicted population in 2050 (1). Studies count about 2,50,000 plant species, including up to 250 species associated with weeds. Weeds could bring significant damage to the agriculture production.

Data processing (Machine learning or deep learning) is booming, affecting many fields and disciplines in the research or professional world. The agricultural sector is no exception to this reality, especially with intelligent precision agriculture. The availability and ease of obtaining more and more data to process, in particular images made easy thanks to the technology of new means of capture such as cameras, drones, and especially smartphones, has greatly contributed to the development of means that allow it to be exploited. Various methods and techniques have been explored in order to process the captured images in order to be able to extract usable data to improve quantitative and qualitative throughput.

Nevertheless, picking single one object detection method is related obviously to the task that it should handle. There is a large spectrum of detectors that can treat

A. Bennour et al. (Eds.): ISPR 2023, CCIS 1941, pp. 223–233, 2024.
https://doi.org/10.1007/978-3-031-46338-9_17

speed, loss, precision, and others can provide details of detection results with some surrounding boxes. Furthermore, some cutting-edge technologies have offered the ability of the detector to provide Hash masks at the pixel level.

In the present study, our focus is directed toward detection of weeds in agricultural field scenes. Previous experiments have shown that while modern detectors perform relatively well in identifying various objects in generic image datasets, they may not perform as well in detecting weeds. In the next section, we will provide an overview of related works that are relevant to our project. We will then describe the dataset we used to train and test our detectors, as well as our contributions in Sect. 3. In Sect. 4, we present the results and discuss our findings, followed by a conclusion and future perspectives.

2 Problem Statement

Weeds are a new research problem in the agricultural field that wants to get help from computing to detect unwanted weed growth with other crops/plants. Usually in agriculture when farmers grow something due to land ownership and availability of small seeds there is additional growth of weeds which reduces the actual crop yield for cultivation like they affect the growth of cultivated plants. Therefore, weed detection is an accuracy problem Define the weed area so that specific areas can be targeted for spraying Minimal spraying on other plants of interest. In recent years, with the dwindling global population growth and current and natural resources, the precision of agriculture has declined. It attracts more and more researchers' attention. Image processing approaches can be applied to solve this problem.

Data processing (machine learning or deep learning) thrives, affecting many areas and disciplines of research or the professional world. The agricultural sector does not escape this reality because it is more and more important to human life.

3 Related Work

Numerous studies were led to enhance and implement superior methods to identify and manage weeds. The primary and prevalent technique employed in research is utilizing artificial intelligence for detection, particularly with image processing and deep learning. This method has displayed promising outcomes in agriculture. Latha et al. [2] put forward a K-means clustering approach to recognize weeds in a corn field using inter-row and inter-plant techniques, resulting in varying outcomes with each application.

In their study, Michel et al. [3] combined image processing techniques and aspect ratio component labeling methods to compare image height and width for effective weed detection and control. Similarly, Radhika and Roopa [4] introduced a machine vision system for identifying weeds in vegetable crops, which used image filtering to extract color and area features while considering the illuminance of the detection area to ensure efficient analysis. Sandeep et al. [5] used a combination of segmentation methods, convolutional neural networks (CNN), and image processing techniques to separate weeds and crops in a carrot orchard. They collected RGB weed data, stored it in a database, and implemented their method using the Python programming language.

Radhika and Roopa [4] developed a machine vision system for weed detection in vegetable crops that uses image filtering to extract color and area features while considering the illuminance of the detection area to ensure efficient analysis. Sandeep et al. [5] combined segmentation methods with convolutional neural networks (CNN) and image processing techniques to differentiate weeds and crops in carrot plantations. They stored a set of RGB weed data in a database and implemented their method using the Python programming language.

Peyman et al. [6] proposed a scattering transform as an energy contrast-based method for weed detection in high-density crops. They evaluated the use of artificial intelligence techniques for weed classification. Adel and Abdou-labbas [7] combined multiple shape features with machine learning techniques, including support vector machines and artificial neural networks, for accurate weed detection in sugar beet fields. The results showed that the artificial neural network correctly classified 92.5 percent of the weeds.

Grinblat et al. [8] used a Deep Convolutional Neural Network to classify leaf veins based on morphological patterns to identify species of legumes. Femi Temitope Johnson and Joel Akerele [9] provided an overview of the application of Artificial Intelligence (AI) for weed detection, and their results showed a detection accuracy of 90% using MATLAB. Similarly, the Supporting Vector Machine (SVM) classification method also achieved a 90% accuracy rate for weed detection.

Femi Temitope Johnson and Joel Akerele. [9] Provides an overview of the application of artificial intelligence (AI) as the best techno-efficient approach for weed detection Execution results made with MATLAB, It gave a detection accuracy of (90%). Similarly, Supporting Vector Machine classification (SVM) Approved for weed detection also gave (90%).

Weeds have had a significant impact on agriculture globally, resulting in increasing crop losses. While technology has advanced in ML and DL, AI has also played a crucial role in detecting and eliminating weeds through traditional or machine methods, as well as educating farmers to reduce the use of harmful chemicals and pesticides. Among the various techniques used for weed detection, DL offers promising data-driven solutions. Although k-means clustering methods using inter-row and inter-plant techniques have been used, they struggle to cluster data with varying sizes and densities, and do not provide clear results. Therefore, DL has emerged as the most appropriate technique to meet the needs of farmers, as it offers better recognition accuracy and learning function from images sent from sensors. Other technologies such as k-means and SVM with manually extracted features have been studied, but they did not meet the needs of the farmers. The vision is to continue developing the application to provide real-time analysis of fields observed from different perspectives.

4 Dataset

Our dataset contains 202 images over 858 × 644 saved in JPG format, and we divided them into 80% which are 183 images used for training, and 20% which are 19 images for validation (Fig. 1).

Fig. 1. Our Data set tebessa aera

5 Contribution

In the very beginning of this experiment, the performance of weed detection is evaluated according to a predefined geographic location in Tebessa, an Algerian eastern region.

The detections performed are based on a series of images taken by the Region-Based Convolutional Neural Networks detector in standard configuration and YOLO v5 in small configuration. Both detectors capture the same predefined weights for MS-COCO. Both models are then additionally trained on our training set and performance is re-evaluated. The detectors were trained and tested using Google Colab with generated Python code in its environment. The Google Colab environment is considered a suitable and powerful tool, especially for machine learning, data analysis, and educational analysis.

The recognition speed results of the above experiments are in favor of Yolov5, which is nearly 20 times faster than Mask R-CNN. in our paper, we use the dataset contains 202 images of 858 × 644 format saved in JPG format, of which 183 images were used for training and 19 images for validation. Weed detection tools consist of weed images from public sources.

The specified part of the dataset is bounded in polygon format and annotated by VGG Annotator, where it is downloaded as JSON file for R-CNN mask, and for yolov5 it is extracted by roboflow and downloaded by file in yolov5 Pytorch format.

To see a well-defined result, we make two detection models and we compare between their results and find the best one from their prediction and their results. In the next section we detail architecture. We start with the Mask RCNN 2017 model.

Mask R-CNN adds a mask prediction branch to the last stage of Faster RCNN. First, two networks, backbone (Res Net 101) and RPN are designed. These networks operate once per frame to yield a set of region maps. Why are the proposals for the zone within the framework of the selective zones that still hold object.

In the second step, the network predicts the box and the class of objects for each of the proposed regions obtained in step 1. Each proposed region can be of different size while the fully connected layers of step 1 are fixed for make predictors. The Size of these regions proposes an overhaul of the use of pools in the ROI (Figs. 2 and 3).

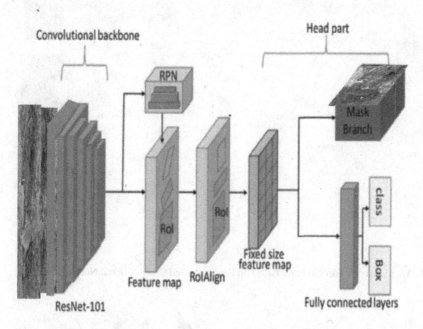

Fig. 2. Weed detection architecture based on Region-Based Convolutional Neural Networks

Dependency, the FPN uses multi-layer convolution architecture features and one can choose one or a combination of them. This FPN hierarchy gives a better prediction. The yolov5 detector is based on CNN are mainly applicable to recommender systems. YOLO (You Only Look Once) models are used for high performance object detection.

YOLO divides the image into a grid system, and each grid detects objects by itself. They can be used to discover objects in real time based on data streams. It requires very few computational resources (Fig. 4).

Yolov5 utilizes a network architecture consisting of three components: (1) Backbone: CSP Darknet, (2) Neck: PANet, and (3) Head: Yolo Layer. The input data is initially processed by CSP Darknet for feature extraction, and then these features are passed to PANet for feature fusion. The primary role of the backbone model is to extract essential features from the input image. In YOLO v5, CSP Darknet (Cross-Phase Partial Networks) serves as the backbone to extract valuable information from the input image.

The neck module plays a crucial role in creating distinct pyramid structures. These distinct pyramids aid in successful generalization when it comes to object measurement. They enable the model to recognize objects of varying sizes and scales effectively.

Distinct pyramids are particularly beneficial in facilitating efficient model performance on unseen data. Other models, such as FPN, Bi FPN, and PANet, employ diverse

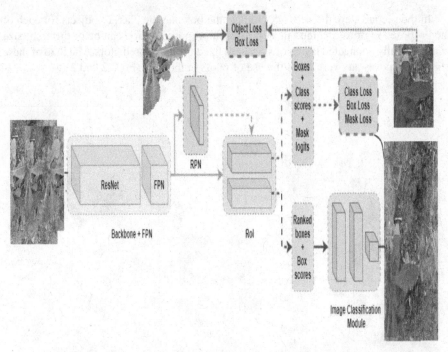

Fig. 3. Weed detection architecture based on Region-Based Convolutional Neural Networks

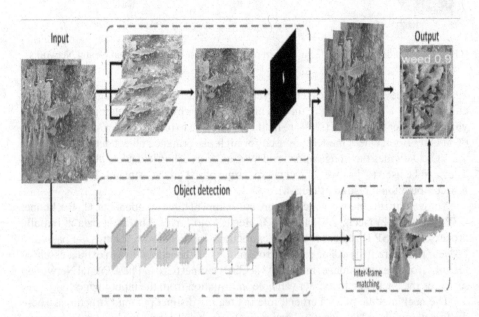

Fig. 4. YOLOV5 architecture

approaches to feature hierarchy. The model head is primarily responsible for the final detection stage. It utilizes bounding boxes to generate output vectors containing class probabilities, object scores, and bounding box coordinates.

6 Result and Discussion

For our model, we started with the MASK RCNN model with our data set and the number of epochs 40. After analyzing the results obtained, we make the following observations: According to the numbers, the error rate of learning and validation decreases with the number of periods until stability, so it is normal that the rate of learning and validation increases with the number of epochs and this reflects that each epoch learns more information. During the training, several metrics of RPN [rpn_class, rpn_bbox] and ROI area (rpn_rois) obtained by RPN, [output_rois] obtained by rpn_rois and the label within the suite were obtained target and hides the detection results of two maskrcnn subclasses [loss, mrcnn_class, mrcnn_bbox, mrcnn_mask] In addition to loss metrics [rpn_class_loss, rpn_bbox_loss, class_loss bbox_loss, mask_loss] (Fig. 5).

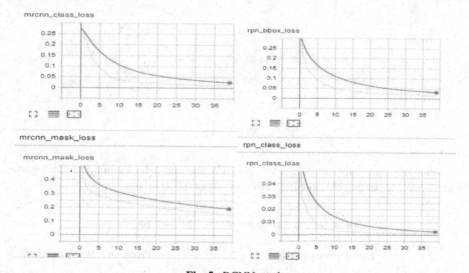

Fig. 5. RCNN result

We used a dataset consisting of 201 images in which 183 images were used for training and 19 images for validation. It took about 80 min for 40 epochs, each period containing 100 (batch) training pulses.

A loss rate of around 0.20 so our model still has decent accuracy despite our small training data set. Tensorboard is used for data visualization (Fig. 6).

On the other hand, the YOLO loss function consists of three parts:

- Box_loss: bounding box regression loss (rms).
- Obj_loss: Confidence in the existence of an object is a loss of objectivity (Binary Cross Entropy).

- Cls_loss: loss of classification (via entropy).

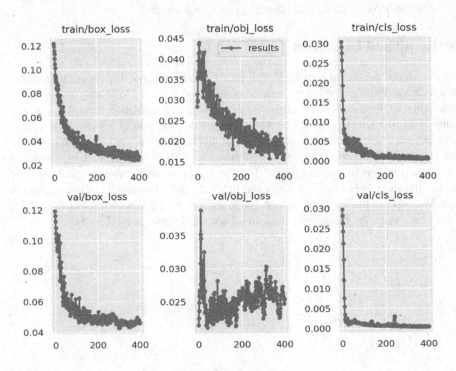

Fig. 6. YOLO v5 result

Since our data only contains a category (Weed) and a background, the category identification error and classification error are about 0.000 (Table 1).

Table 1. Comparison between R-CNN and Yolov5.

	Taux d'erreur	Taux d'erreur de classe	Taux d'erreur de box	Précision	Précision-recall
MASK-RCNN	0.19	0.02	0.03	0.777	0.700
YOLO V5	0.03	0.000	0.04	0.89	0.77

The performance comparison of Mask R-CNN and YOLOv5 aims to produce the best weed detection and recognition models. Based on experiments, YOLOv5 outperformed Mask R-CNN with an accuracy score of 0.89 and faster training times. However, R-CNN Mask function adds a mask to the final detection, which shows better pattern segmentation with a mask as well as the recognition rate is better than yolov5. We tested our model in several samples among these (Figs. 7, 8 and 9).

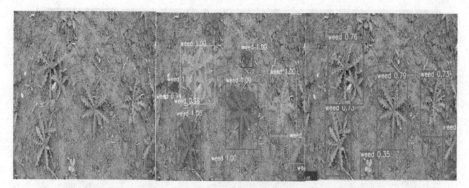

Fig. 7. Experimental result R-CNN

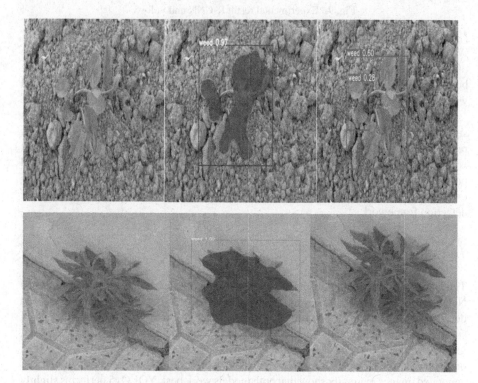

Fig. 8. Experimental result YOLO

To show the accuracy of our model, we experimented on an area where there is a plant next to the weeds, and we found that our model is very encouraging and its application will be very successful in the agricultural sector.

The initial performance of YOLO and Mask R-CNN with pre-trained weights in the weed detection task was unsatisfactory when using our custom sample dataset. However, once we trained the models on our dataset, their performance improved significantly.

Fig. 9. Experimental result R-CNN and yolov5

We observed a notable increase in true positive detections, but this came at the expense of additional false positives. This improvement can be attributed to the use of bounding boxes to represent objects, which, in some cases, did not tightly enclose the detected weeds.

While YOLO exhibited faster detection speed at higher resolutions, it still had a significant advantage over Mask R-CNN. Although Mask R-CNN did not improve its recall as much as YOLO, it had fewer false positives, resulting in higher precision. It is important to note that Mask R-CNN excelled in accurately identifying objects and provided supplementary segmentation information. By incorporating additional training examples of weeds, both methods achieved results that can be integrated into a weed detection framework.

7   Conclusion

This paper proposed a newly designed deep learning model, founded on Convolutional Neural Network (CNN), that was deployed for weed detection. Our proposed weed detection system is an R-CNN mask and model based on yolov5 implemented using some Google Colab libraries such as Keras, NumPy, etc. To train on the herbal database, which includes200 images and test on new images from the new dataset, we are testing on several samples. Our model turns out to be a success, because experiments on 200 annotated images manually show that both models work best. YOLOv5 performs slightly better than Mask R-CNN with 2% and 19% error rate Mask R-CNN, but Mask R-CNN's predictions are more accurate than Yolo's because Mask R-CNN defines the goal more clearly and precisely.

References

1. 2050: 2,3 milliards de bouches de plus à nourrir. ONU (2009)
2. Latha, A., Poojith, B.V., Kumar, G.V.: Image processing in agriculture. Int. J. Innov. Res. Electr. Electron. Instrum. Control Eng. **2**(6), 1562–1565 (2014)
3. Viegas, M.A.E., Kurian, A., Rebello, V.J., Gaunker, N.M.: Weed detection using image processing. Int. J. Sci. Res. Dev. **4**(11), 660–662 (2017)
4. Radhika Shetty, D.S., Roopa, G.K.: Weed detection using image filtering in vegetables crops. IOSR J. Comput. Eng. **21**(1), 61–64 (2019)
5. Sandeep Kumar, K., Rajeswari, Usha, B.N.: Convolution neural network based weed detection in horticulture plantation. Int. J. Sci. Res. Rev. **7**(6), 41–47 (2018)
6. Rasti, P., Ahmad, A., Samiei, S., Belin, E., Rousseau, D.: Supervised image classification by scattering transform with application to weed detection in culture crops of high density. Remote Sens. J. **11**(249), 1–16 (2019)
7. Bakhshipoura, A., Jafarib, A.: Evaluation of support vector machine and artificial neural networks in weed detection using shape features. Comput. Electron. Agric. **145**, 153–160 (2018)
8. Grinblat, G.L., Uzal, L.C., Larese, M.G., Granitto, P.M.: Deep learning for plant identification using vein morphological patterns. Comput. Electron. Agric. **127**, 418–424 (2016)
9. Johnson, F.T., Akerele, J.: Artificial Intelligence for Weed Detection-A Techno-Efficient Approach (2020)

Boruta-AttLSTM: A Novel Deep Learning Architecture for Soil Moisture Prediction

Bamory Ahmed Toru Koné[1]([⊠]) [iD], Bassem Bouaziz[2] [iD], Rima Grati[3] [iD],
and Khouloud Boukadi[1] [iD]

[1] Faculty of Economics and Management of Sfax, University of Sfax, Sfax, Tunisia
konebamory18@gmail.com
[2] Higher Institute of Computer Science and Multimedia, University of Sfax, Sfax, Tunisia
[3] College of Technological Innovation, Zayed University, Dubai, United Arab Emirates

Abstract. Water scarcity is worsening due to poor water management in irrigated areas, which directly impacts global food safety. Furthermore, effective irrigation scheduling necessitates predicting future soil moisture content, representing soil water availability. For this purpose, the current study proposes a novel data-driven architecture based on deep learning algorithms to predict soil volumetric water content. The proposed architecture combines the time-processing ability of Long Short-Term Memory with the attention mechanism's ability to process long sequences. The suggested architecture's resulting model is compared to a 2-layer LSTM in terms of MSE, MAE, RMSE, and R^2 score. This study also examines the relationships between various climate and soil parameters and targets soil moisture. The relevance of input features is considered by the feature selection strategy using their computed shapley values. The findings of this study suggest that attention mechanisms can increase the performance and generalizability of regular LSTMs.

Keywords: Soil Moisture Prediction · Irrigation Scheduling · Attention Mechanism · Boruta · Shap

1 Introduction

Global water scarcity is a growing problem caused by a variety of factors, including global warming and wasteful water use. As global temperatures rise, the earth's climate is changing, leading to droughts, reduced rainfall, and increased evaporation rates. These changes can make it harder for crops to grow and for people to access clean water, leading to food insecurity and water scarcity. In addition to global warming, Many regions of the world use water inefficiently, leading to the depletion of water sources and increased competition for limited water resources. Wasteful water use can include practices like over-irrigation, inefficient water distribution systems, and water-intensive agricultural practices. To address these challenges, it's important to take action to improve water use efficiency for irrigation technologies and practices.

Given that irrigated agriculture produces more than 40% of all food produced globally despite accounting for only 17% of all cultivated land [1], efficient water use in such

A. Bennour et al. (Eds.): ISPR 2023, CCIS 1941, pp. 234–246, 2024.
https://doi.org/10.1007/978-3-031-46338-9_18

environment could significantly reduce water scarcity while increasing food production. In irrigated agriculture, precise crop water requirements based on irrigation scheduling (IS) are required for efficient water use [2]. Unfortunately, most irrigation scheduling studies ignore incoming soil moisture, which is a critical component of effective irrigation scheduling. Furthermore, understanding incoming soil moisture is essential for effective irrigation scheduling because it reveals the state of water in the soil, providing an overview of water stress/availability. Forecasting such a critical parameter is therefore desirable for proactive, long-term decisions in an efficient IS [3] and water management [4]. This is because future soil moisture knowledge could capture otherwise difficult-to-record unexpected land-water fluctuations, avoiding inefficient irrigation decisions. For this purpose, researchers have extensively used machine learning (ML) to predict such a critical parameter. ML algorithms have already proven their ability to solve problems in a wide range of fields, such as risk assessment, manufacturing, psychiatry, machine fault diagnosis, agriculture, and so on.

Moreover, soil moisture prediction is a time series prediction scenario because data from sensors is recorded at time intervals, which is a critical parameter in this scenario. ML algorithms such as Long Short-Term Memory (LSTM) [5, 6] and neural networks have been extensively studied to address time series forecasting in general, and have achieved significant results due to their ability to learn from data sequences (i.e., time series data). However, standard LSTM suffers from vanishing/exploding gradients for long data sequences. In other words, they have difficulty remembering information from earlier time steps. As a result, it is most likely that valuable patterns will be lost while processing long data sequences with standard LSTM. As a result, when processing long input sequences, these types of networks may be limited. Although many studies have been using LSTM to forecast soil moisture based on daily observations, it's sometimes preferable to have longer prediction timesteps, such as weekly or even monthly input data, for better irrigation scheduling. This is because relevant information can be learned from earlier days, preventing the model from being completely dependent on the previous day's data.

As a result, in this paper, we propose a novel architecture to deal with longer observations for soil moisture prediction based on the attention mechanism and LSTM models. The objective is to handle longer observation sequences without losing valuable information from previous observations. In contrast to most studies in the literature, the current study considers input data from the previous seven days in order to eventually capture valuable patterns. In this study, we additionally address some of the shortcomings of traditional LSTM with regard to lengthier input sequences. Hence, our proposed architecture takes advantage of LSTM's native temporal processing ability while also addressing its lack of sufficient memory by adding attention layers to it. Furthermore, the current study aims to evaluate the most impacting features of soil moisture prediction to avoid using potentially harmful parameters and thus reduce model complexity. A new feature selection technique based on Shapley value computation, BorutaShap [7], is being investigated. The remainder of this paper is structured as follows. Section 2 provides an overview of the current state of soil moisture prediction research. Sections 3 and 4 explain the methodology used in this study and evaluate the performance of the deep learning models, respectively. Conclusions and recommendations for further work are outlined in Sect. 5.

2 Related Works

Several data-driven algorithms for predicting future soil moisture values have been evaluated by researchers. Feed-forward neural networks (FNN), Random Forest (RF), and Support Vector Machine (SVM) have been compared in [8]. Furthermore, [9] investigated the applicability of deep regression neural networks. Based on thermal properties, fuzzy logic (FL) has also been evaluated to forecast soil moisture content. However, due to the time-series structure of soil-moisture data, traditional ML algorithms, and neural networks frequently encounter challenges such as substantial data pre-processing and/or insufficient accuracy.

To address these issues, some studies used a hybridization technique, in which a standalone model is combined with an optimization algorithm. For example, in [3], the authors combined standalone ensemble learning machines (ELM) and RF models with the EEMD (Ensemble Empirical Mode Decomposition) optimization algorithm to forecast upper- and lower-layer soil moisture. Similarly, in [10], the grey wolf optimization was combined with adaptive neuro-fuzzy inference systems (ANFIS). The resulting hybrid model was compared to standalone artificial neural network (ANN), support vector regression (SVR), and ANFIS. Although hybridized models can produce reliable and accurate predictions [10], the models used in these studies still require significant data pre-processing because they are not well adapted to time series prediction.

Machine learning models based on the Recurrent Neural Networks (RNNs) architecture require substantially less pre-processing because they are organically suited for sequential data. For this purpose, modern works primarily focus on these types of models. For example, [11] demonstrated the effectiveness of an RNN-based model, namely Long Short-Term Memory (LSTM), in predicting soil moisture. In their work, the authors built an LSTM that performed similarly to a feed-forward ANN while requiring fewer data pre-processing. Furthermore, [12] used LSTM to forecast the soil water index, outperforming MLP and ANFIS models. Moreover, the RNN-based gated recurrent unit (GRU) is combined with convolutional networks (CNN) to predict soil water content in the maize root zone [13].

However, the vast majority of these studies assume that the previous day's observations are sufficient to predict the soil moisture value for the following day. As a result, their predictive models are directly trained on a small timestep of input data. The goal of this research is to account for additional patterns obtained from older timesteps when predicting soil moisture, as well as to investigate the relationship between target soil moisture and older input data observations. Furthermore, because standard RNN models struggle with long input sequences, we propose a novel attention LSTM architecture in this paper.

3 Material and Methods

3.1 Overall Architecture

The overall approach used in this study is depicted in Fig. 1 and described further below. We fed our models with Cosmic-ray soil moisture monitoring (COSMOS) data [14]. COSMOS data is gathered from 51 sites across the United Kingdom that record various hydro-meteorological and soil characteristics. The dataset includes sub-hourly hydrometeorological and soil observations from October 2013 to December 2019. Meteorological data include, for example, radiation (shortwave, longwave, and net), precipitation, atmospheric pressure, air temperature, wind speed and direction, and humidity. Soil observations consist of measurements of soil heat flux, soil temperature, and volumetric water content (VWC) at various depths.

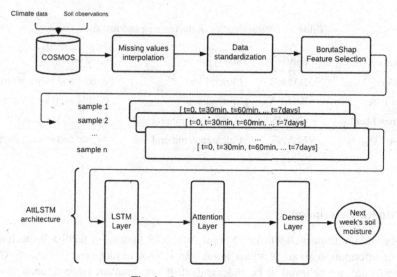

Fig. 1. Overall Architecture

Furthermore, missing values were interpolated to prevent the models from performing significantly worse. Interpolation is the process of calculating missing values for an observation based on its preceding values. The sequential nature of this interpolation technique fits the temporal nature of time-series data. The dataset was then divided into training, validation, and test data. Because we're working with time-series data, the split was performed sequentially to preserve the data's temporal dynamics. As a result, we used Balruderry data from 2014 to 2017 as the training set and data from 2018 and 2019 as the validation and test sets. In addition, data collected at locations different than the training site were considered additional test sets. The soil type and land cover of each location are shown in Table 1. The dataset was then transformed into samples/sequences of weekly observations to predict soil moisture. Each sample contains sub-hourly climate and soil observations corresponding to weekly data. As a result, each sequence has a length of 336 (2 * 24 * 7) timesteps.

Moreover, we standardized data as an additional preprocessing step to deal with the wide range of input parameters. This ensures that the mean value of each input parameter is 0 and the standard deviation is 1, allowing neural network-based models to learn more effectively from data. Following that, we examine the relevant input features to the target soil moisture, eventually discarding unnecessary features from the prediction. Feature selection is desirable to avoid poor model performance while reducing model complexity. The novel BorutaShap method [7], which is based on Shapley value computation, was used in this study to select features. For each prediction, each feature is assigned an importance value, known as a Shapley value. The Boruta algorithm then selects the relevant features for the overall prediction based on these values.

Finally, we built and trained two deep learning models to make predictions on these sequences, described in the following subsection: standard LSTM and attention-augmented LSTM.

Table 1. Characteristics of the training and test sites

Site name	Label	Soil type	Land cover
Balruderry	BALRD	Mineral soil	Arable and Horticulture
Bunny Park	BUNNY	Mineral soil	Arable and Horticulture
Chinmey Meadows	CHIMN	Calcareous mineral soil	Improved grassland
Elmsett	ELMST	Calcareous mineral soil	Arable and Horticulture

3.2 Models Description

LSTMs are a variant of Recurrent Neural Networks that use a feedback mechanism to store information about previous inputs. In RNNs, unlike other forms of ANNs, the inputs are not believed to be independent of one another. Instead, each input is handled depending on the feedback supplied by the processing of preceding inputs. This ability is critical when dealing with sequential problems involving data dependencies. However, LSTM is limited by the size of information that they can remember as they gradually "forget" information contained in earlier periods. To address this shortcoming, the attention mechanism was introduced [15–17]. The attention mechanism was first introduced in natural language processing and has since been adapted to all sequence-processing tasks such as time series prediction. Its unique feature is the ability to process long sequential data more efficiently by learning to focus on relevant parts of the input sequence.

In this study, rather than building a model entirely of LSTM layers, we add an attention layer on top of the initial LSTM layer, as shown in Fig. 1. Using the LSTM layer, the resulting model first learns the temporal dynamics of the input data. The produced time-aware input is then fed into the attention layer, which learns the relevant sequences of this input data without losing the temporal order of the sequences. As a result, rather than processing all sequences of input data equally, our architecture prioritizes relevant ones for each prediction. Figure 2 depicts a simulation of an input passing through the attention layer, with the most relevant sequences colored blue, less relevant sequences colored green, and irrelevant sequences colored red.

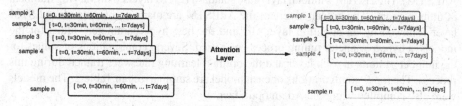

Fig. 2. Simulated attention layer. (Color figure online)

4 Results

The regression metrics used to evaluate the performance of the models are described below, along with the respective formula used to compute them:

- Mean Squared Error (MSE) represents the mean of the square of the individual prediction errors.

$$MSE = \frac{\sum_1^n (y_{pred} - y_{obs})^2}{n}$$

- Mean Absolute Error (MAE) represents the mean of the absolute values of the individual prediction errors on overall instances.

$$MAE = \frac{\sum_1^n |y_{pred} - y_{obs}|}{n}$$

- Root Mean Squared Error (RMSE) represents the square root of the mean of the square of the individual prediction errors.

$$RMSE = \sqrt{\frac{\sum_1^n (y_{pred} - y_{obs})^2}{n}}$$

- Coefficient of determination (R^2) is a goodness-of-fit measure for models based on the proportion of explained variance.

$$R^2 = 1 - \frac{\sum (y_{pred} - y_{obs})^2}{\sum (y_{pred} - \underline{y_{obs}})}$$

where y_{pred} is the predicted value, y_{obs} is the observed value, n is the number of instances, and the prediction error represents the difference between the predicted and the observed values.

4.1 Finding the Right Hyperparameters

During this study, two types of models were developed and compared: a 2-layer LSTM (referred to as 2LSTM throughout this paper) and an attention LSTM (referred to as AttLSTM). The 2LSTM's hidden layers are standard LSTM layers with varying numbers of units/neurons. The hidden layers of the AttLSTM are made up of a standard LSTM layer followed by an attention layer. To find the best hyperparameter values for the models, a hyperparameter tuning process was used. Several combinations of values for the number of units in each layer and the optimal learning rate were trained during this process. The top-3 configurations of each model are summarized in Table 2. The models were also compiled using the Adam optimizer.

Table 2. Models' configurations during Hyperparameter tuning

Model Name	Network Structure	Learning Rate
2LSTM	64 - 64	0.0003
	64 - 64	0.0001
	256 - 64	0.0001
AttLSTM	32 - 32	0.003
	64 - 128	0.001
	32 - 16	0.0003

Compared to the AttLSTM, which has 32 neurons in its first and second hidden layers, the best variant of the 2LSTM has 64 neurons in both. Table 3 compares the highest performing models of the hyperparameter tuning process in terms of the performance metrics listed above. Based on these results, we can see that AttLSTM outperforms 2LSTM in terms of all evaluation metrics in the training site. Figure 3 depicts the evolution of predicted soil moisture (using both models) versus actual soil moisture. We can observe from this figure, that AttLSTM (lower plot) better follows the evolution of targets soil moisture than 2LSTM (upper plot).

Table 3. Performance of tuned models on test data from the training site

Model	MSE (%)	MAE (%)	RMSE (%)	R^2
2LSTM	0.3555	0.3897	0.5963	0.9493
AttLSTM	0.1704	0.2543	0.4128	0.9756

4.2 Multi-site Evaluation

The evolution of soil moisture values can vary significantly between sites with different soil properties since the soil characteristics determine the water availability. For this

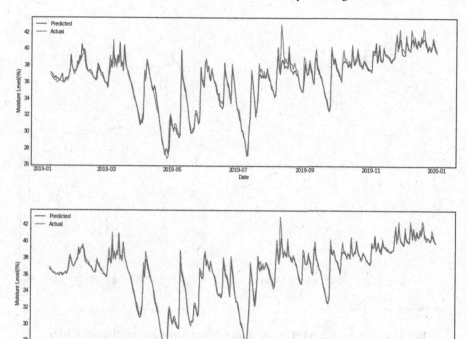

Fig. 3. Evolution of predicted and actual soil moisture in training site

purpose, it is desirable to know the variations in model performance with variations in soil characteristics. As a result, we also assess the models' generalization ability across three sites with characteristics that differ from the training site, namely Bunny Park, Chinmey Meadows, and Elmsett (see Table 1). Figure 4 and Fig. 5 illustrate the evolution of soil moisture (actual vs predicted) on each test site for the 2LSTM and AttLSTM models, respectively. We can see that the 2LSTM model barely follows the evolution of target soil moisture, especially at the Bunny and Elmsett sites, even though the Bunny site has the same characteristics as the Balrd training data.

In contrast, despite having different soil characteristics, AttLSTM produced significant results across all test sites. We can see that the distance between predicted and target soil moisture for AttLSTM is smaller than for 2LSTM. Even though relatively lesser accuracy is noticeable at Bunny site, AttLSTM still ouperforms 2LSTM. Overall, this demonstrates the ability of the attention model to generalize across sites. This ability to generalize can be traced back to the attention mechanism's ability to identify relevant sequences for each prediction scenario. Furthermore, Table 4 summarizes the statistical evaluation of the models in terms of MSE, MAE, RMSE, and R^2.

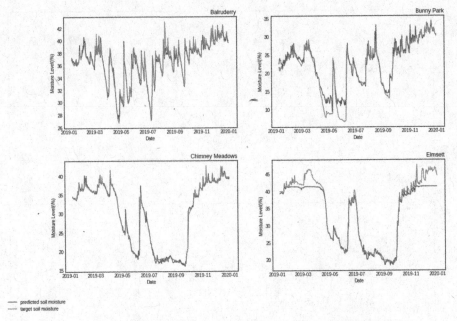

Fig. 4. Evolution of 2LSTM predicted vs actual soil moisture on evaluation sites

Table 4. Statistical evaluation of models on evaluation sites

Test site	Model	MSE (%)	MAE (%)	RMSE (%)	R^2
BUNNY	2LSTM	2.9828	1.0864	1.7271	0.9023
	AttLSTM	1.4190	0.7162	1.1912	0.9589
CHIMN	2LSTM	0.7114	0.4299	0.8434	0.9708
	AttLSTM	0.1442	0.2535	0.3798	0.9940
ELMST	2LSTM	3.8219	1.3060	1.9550	0.9125
	AttLSTM	2.4356	0.9421	1.5606	0.9446

4.3 Feature Selection

Finally, the BorutaShap model feature selection technique resulted in high model performance with fewer features. For instance, the mean squared error (MSE) for the Bunny Park, Chimney Meadows, and Elmsett sites overall decreased from 1.4190, 0.1442, and 2.4356 to 0.3386, 0.1630, and 1.5805, respectively. Furthermore, the R^2 score for the same test sites overall increased from 0.9589, 0.9940, and 0.9446 to 0.9819, 0.9934, and 0.9646, respectively. Because the R^2 score measures how well the model fits, it implies that the features chosen were indeed important for the model. Table 5 provides additional information on other performance metrics for each test site while focusing solely on the positive/negative effects of BorutaShap feature selection. Negative changes

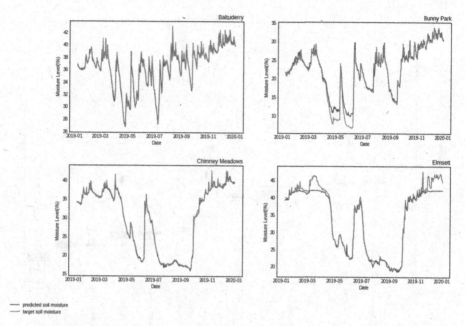

Fig. 5. Evolution of AttLSTM predicted vs actual soil moisture on evaluation sites

in MSE, MAE, and RMSE values are better because they represent a reduction in error rate; positive changes in R^2 are better.

Table 5. Effects of feature selection on evaluation sites

Test site	MSE (%)	MAE (%)	RMSE (%)	R^2
BUNNY	−1.0804	−0.3223	−0.6093	+0.0230
CHIMN	+0.0188	−0.0163	+0.0240	−0.0006
ELMST	−0.8551	−0.2654	−0.3034	+0.0200

The selected features are soil temperature, solar radiation, relative humidity, precipitation, air pressure, and past soil moisture (Fig. 6). This is especially useful when working with deep learning models, which frequently necessitate extensive training. Hence, Shapley values proven to be an outstanding predictor of feature contribution to prediction. In this study, we demonstrated the usefulness of BorutaShap for identifying pertinent soil moisture prediction features.

Furthermore, significant performance improvements are noticeable at the Bunny and Elmset sites, where the models previously did not perform effectively for a short length of time. This demonstrates that BorutaShap can identify relevant features for soil moisture prediction. Figure displays the reevaluation of AttLSTM model at the Bunny location after feature selection (Fig. 7).

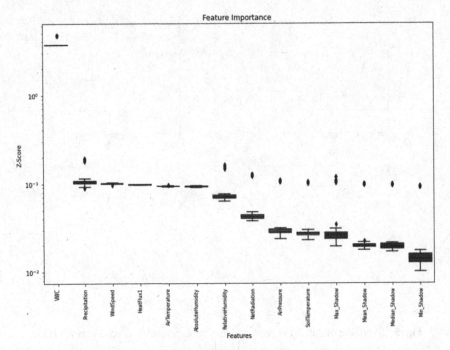

Fig. 6. BorutaShap box plot

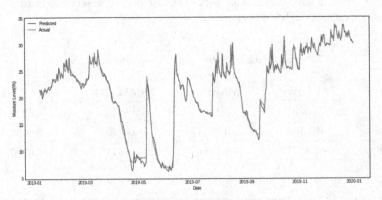

Fig. 7. Evaluation of AttLSTM at Bunny site after feature selection

5 Conclusion

Future soil moisture information is essential for more proactive irrigation scheduling. Furthermore, a specific soil moisture value can be influenced by older climates and soil observations. Additionally, the value of soil moisture on a given day may be determined not only by the previous day's gathered value, but also by a couple of days prior to that. For this purpose, our study took a different approach consisting of extending the history size of the input data while simultaneously building a model capable of learning key timesteps to the prediction. As a result, we proposed in this study a novel architecture that combines the time-processing ability of Long Short-Term Memory with the ability of the attention mechanism to focus on relevant prediction sequences. The results demonstrated that LSTM could predict agricultural time series even when dealing with long input sequences. Furthermore, the results demonstrated that by coupling these models with an attention mechanism, the overall model can consider intermediary information processed between sequences. Besides that, the study demonstrated that the novel architecture could maintain high prediction performances even under adverse conditions. Hence, LSTM combined with an attention mechanism can natively reduce overfitting. Moreover, the study investigated the effect of various climate and soil parameters on target soil moisture. As a result, the relationship between input and output features was investigated. This resulted in the determination and selection of relevant features for the prediction based on the computation of Shapley values. Furthermore, feature selection led in a significant improvement in model performance. Therefore, the architecture proposed in this study has demonstrated the ability to identify relevant predictors as well as relevant timesteps for soil moisture prediction.

In addition, because deep learning models are commonly perceived as black boxes, the current work could be enhanced by explaining the model's predictions. Hence, additional information regarding the direct effects of the selected features might be provided. Collecting more input data may further enhance the performance of the model built for this study.

References

1. Fereres, E., García-Vila, M.: Irrigation management for efficient crop production. In: Christou, P., Savin, R., Costa-Pierce, B.A., Misztal, I., Whitelaw, C.B.A. (eds.) Encyclopedia of Sustainability Science and Technology, pp. 1–17. Springer, New York, NY (2018). https://doi.org/10.1007/978-1-4614-5797-8_162
2. Jones, H.G.: Irrigation scheduling: advantages and pitfalls of plant-based methods. J. Exp. Bot. **55**(407), 2427–2436 (2004). https://doi.org/10.1093/jxb/erh213
3. Prasad, R., Deo, R.C., Li, Y., Maraseni, T.: Soil moisture forecasting by a hybrid machine learning technique: ELM integrated with ensemble empirical mode decomposition. Geoderma **330**, 136–161 (2018). https://doi.org/10.1016/j.geoderma.2018.05.035
4. Sanuade, O.A., Adetokunbo, P., Oladunjoye, M.A., Olaojo, A.A.: Predicting moisture content of soil from thermal properties using artificial neural network. Arab. J. Geosci. **11**(18), 566 (2018). https://doi.org/10.1007/s12517-018-3917-4
5. Hochreiter, S., Schmidhuber, J.: Long short-term memory. Neural Comput. **9**(8), 1735–1780 (1997). https://doi.org/10.1162/neco.1997.9.8.1735
6. Sutskever, I., Vinyals, O., Le, Q.V.: Sequence to sequence learning with neural networks. In: Proceedings of the 27th International Conference on Neural Information Processing Systems - Volume 2, Montreal, Canada, pp. 3104–3112 (2014)
7. Kursa, M.B., Rudnicki, W.R.: Feature selection with the boruta package. J. Stat. Softw. **36**(11), 1–13 (2010). https://doi.org/10.18637/jss.v036.i11
8. Dubois, A., Teytaud, F., Verel, S.: Short term soil moisture forecasts for potato crop farming: a machine learning approach. Comput. Electron. Agric. **180**, 105902 (2021). https://doi.org/10.1016/j.compag.2020.105902

9. Cai, Y., Zheng, W., Zhang, X., Zhangzhong, L., Xue, X.: Research on soil moisture prediction model based on deep learning. PLOS ONE **14**(4), e0214508 (2019). https://doi.org/10.1371/journal.pone.0214508

10. Maroufpoor, S., Maroufpoor, E., Bozorg-Haddad, O., Shiri, J., Mundher Yaseen, Z.: Soil moisture simulation using hybrid artificial intelligent model: hybridization of adaptive neuro fuzzy inference system with grey wolf optimizer algorithm. J. Hydrol. **575**, 544–556 (2019). https://doi.org/10.1016/j.jhydrol.2019.05.045

11. Adeyemi, O., Grove, I., Peets, S., Domun, Y., Norton, T.: Dynamic neural network modelling of soil moisture content for predictive irrigation scheduling. Sensors **18**(10) (2018). Art. no 10. https://doi.org/10.3390/s18103408

12. Marini, A., Termite, L.F., Garinei, A., Marconi, M., Biondi, L.: Neural network models for soil moisture forecasting from remote sensed measurements. ACTA IMEKO **9**(2) (2020). Art. no 2. https://doi.org/10.21014/acta_imeko.v9i2.797

13. Yu, J., Zhang, X., Xu, L., Dong, J., Zhangzhong, L.: A hybrid CNN-GRU model for predicting soil moisture in maize root zone. Agric. Water Manag. **245**, 106649 (2021). https://doi.org/10.1016/j.agwat.2020.106649

14. Stanley, S., et al.: Daily and sub-daily hydrometeorological and soil data (2013–2019) [COSMOS-UK]. NERC Environ. Inf. Data Centre (2021). https://doi.org/10.5285/b5c190e4-e35d-40ea-8fbe-598da03a1185

15. Bahdanau, D., Cho, K., Bengio, Y.: Neural machine translation by jointly learning to align and translate. CoRR, September 2014, Consulté le: 11 février 2023. https://www.semanticscholar.org/paper/Neural-Machine-Translation-by-Jointly-Learning-to-Bahdanau-Cho/fa72afa9b2cbc8f0d7b05d52548906610ffbb9c5

16. Luong, T., Pham, H., Manning, C.D.: Effective approaches to attention-based neural machine translation. In: Proceedings of the 2015 Conference on Empirical Methods in Natural Language Processing, Lisbon, Portugal, September 2015, pp. 1412–1421 (2015). https://doi.org/10.18653/v1/D15-1166

17. Vaswani, A., et al.: Attention is all you need. In: Advances in Neural Information Processing Systems, vol. 30 (2017). Consulté le: 11 février 2023. [En ligne]. https://papers.nips.cc/paper/2017/hash/3f5ee243547dee91fbd053c1c4a845aa-Abstract.html

Graph Autoencoder with Community Neighborhood Network

Ahmet Tüzen[1,2(✉)] and Yusuf Yaslan[2]

[1] Aselsan Inc., Ankara, Turkey
atuzen@aselsan.com.tr
[2] Istanbul Technical University, Istanbul, Turkey
{tuzen,yyaslan}@itu.edu.tr

Abstract. Neighborhood information can be extracted from graph data structure. The neighborhood is valuable because similar objects tend to be connected. Graph neural networks (GNN) represent the neighborhood in layers depending on their proximity. Graph autoencoders (GAE) learn the lower dimensional representation of graph and reconstruct it afterward. The performance of the GAE might be enhanced with the behavior of GNNs. However, utilizing the neighborhood information is challenging. Far neighbors are capable of building redundantly complex networks due to their decreasing similarity. Yet, less neighborhood models are closer to GAE. Restricting the neighborhood within the same community enriches the GNN. In this work, we propose a new unsupervised model that combines GNN and GAE to improve the representation learning of graphs. We examine the outcomes of the model under different neighborhood configurations and hyperparameters. We also prove that the model is applicable to varying sizes and types of graphs within different categories on both synthetic and published datasets. The outcome of the community neighborhood network is resistant to overfitting with fewer learnable parameters.

Keywords: graph autoencoder · graph neural network · neighborhood network · graph representation learning

1 Introduction

The graph data structure is one of the widespread methods to represent data, especially for some network types like social networks, citation networks, transaction networks, ...etc. In fact, many real-world scenarios can be represented with graphs because many objects have a strong relationship with other objects that are similar to each other. Such relationships can offer additional information that certain objects may not have, and that information can be exploited to enhance the performance of existing machine learning applications.

Yet, real-world scenarios are harder to work with because the number of objects is enormous (i.e., high-dimensional) and often have limited relationship (i.e., sparse) with each other. These difficulties require careful analysis of the content, and one solution may fail on another graph.

© The Author(s), under exclusive license to Springer Nature Switzerland AG 2024
A. Bennour et al. (Eds.): ISPR 2023, CCIS 1941, pp. 247–261, 2024.
https://doi.org/10.1007/978-3-031-46338-9_19

Graph representation learning is one method for handling large-scale graphs by mapping high-dimensional data to low-dimensional representations. It is important to use effective representation methods to handle large graphs in applications. Graph embedding is widely adopted approach for graph representation learning and achieved decent success in graph applications such as link prediction, node classification and anomaly detection. Adopting deep learning models (e.g., deep neural networks) for network embedding captures solid information. Graph embedding and network embedding terms will be used interchangeably for the rest of the paper.

Autoencoders for graph or graph autoencoders (GAE), are feed forward neural networks that reconstruct graph information (e.g., link structure) by first encoding the input to a lower dimension (latent space) and then rebuilds the input by decoding the lower dimension output in an unsupervised manner. However, GAE may generate poor outcomes without considering the graph structure through neighborhood [21]. To overcome this problem, GAE can be enlarged with graph neural networks (GNNs) which can extract valuable neighborhood information [10]. The neighborhood can mimic the layers of the neural network [13] where the neighbors can contribute depending on their distance. In this work, our aim is to enhance the performance of GAE by utilizing neighborhood information. We will examine this problem from different aspects, such as the structure of the graph, network types, and hyperparameters. Our contributions are as follows:

- Our proposed approach combines two frameworks, GAE and GNN, and we examine the performance of the model under various conditions. Such conditions are, different networks based on neighborhood information (e.g., number of hop neighbors/layers, usage of neighborhood information), different types and sizes of graphs, different behaviors of the network (e.g., weight sharing between nodes), and different hyperparameters. The outcomes of this work may provide a basis for future studies while designing and determining optimal parameters.
- A new neighbor selection technique, community hop, is proposed. To the best of our knowledge, this work is the first to use such a method. Also, community hop is compared to other neighbor selection techniques and its strengths are analyzed.
- The outcome of our work is a robust model that is resistant to overfit. To show its superiority, we used both reconstruction loss and link prediction as performance metrics. Furthermore, we tested our model with publicly available datasets and achieved significant success.

2   Related Work

Graph/network representation learning, or network embedding has been a widely studied research area. Machine learning tasks on large-scale graphs can be challenging, and methods such as those proposed in [26] have been successful in generating high-dimensional graph representations. Other studies are focused on different aspects like structure [22,24], neighborhood [11,20], local information [12] and many more [3,7]. Reconstructing graph structure with embedded data [9] also unlocks other topics. And autoencoders or GAE commonly utilized for this task [4,14,29].

Even though traditional methods make novel contributions [5], deep models are also adopted to strengthen the applications [1]. Neural networks including GNN [10,25] are a decent approach on graph embedding [6,30]. The properties of neural networks successfully applied to graph features [16,17].

Similar to our neighbor selection methods, randomly dropout models are also more resistant to overfitting [18,23].

3 Problem Statement and Solution Approach

A static plain graph is defined by $G = \{V, E\}$, where V is the node/vertex set, and E is the edge/link set. The graph G is structured by $V = \{v_1, v_2, \cdots, v_n\}$, where n denotes the number of nodes, and $E = \{e_1, e_2, \cdots, e_m\} \subseteq V \times V$, represents the edges between nodes, where m denotes the total number of edges. Each edge $e_k = (v_i, v_j)$ is defined with a pair of ordered nodes, where v_i is the starting node and v_j is the destination node. In undirected graphs, the order of pairs is not important, i.e., $\forall e_k \in E, e_k = (v_i, v_j) = (v_j, v_i)$. Each edge in the edge set has a weight $w_{i,j} > 0$. For unweighted graphs, $\forall w_{i,j} = 1$. The adjacency matrix A of a graph G is a $n \times n$ matrix that can restore the graph structure. If there is a link between nodes v_i and v_j, then $A_{i,j} = w_{i,j}$, otherwise, $A_{i,j} = 0$. Thus, the link structure of a node can be rebuilt by the corresponding row of the adjacency matrix.

The nodes of a graph represent data objects. The objects can vary depending on the problem. Edges are relationships between objects and are specific to domain as well. In reality, many graphs have thousands or even millions of nodes, which are sparse by nature. Therefore, it is challenging to work with these sparse and large matrixes/vectors. One solution is to learn an embedding function such that $f : v_i \rightarrow h_i \in R^m$, where v_i is i^{th} row of the adjacency matrix A, and m is the dimension of the embedding space such that $m \ll n$. After the embedding, the A greatly reduced to size $n \times m$, which is smaller. To acquire input matrix A again, another function $g : h_i \rightarrow v'_i \in R^n$ can be learned. Both operations can be completed with GAE by encoder and decoder parts. The aim of GAE is reconstruct graph while minimizing difference between input vector and reconstructed vector.

Our goal is to reconstruct the graph by combining node structure and neighborhood information. For this purpose, two deep models are defined as shown in Fig. 1. The core network is GAE, which contains two encoder and two decoder layers. The layered architecture of the core network remains the same throughout all the experiments. We use the hop neighborhood as layers in GNN, and the k-hop is defined as the k-step distanced nodes from the starting node. In other words, k^{th} layer of the GNN is k-hop neighbors of the starting node. During forward propagation each row vector at the kth layer is multiplied with learnable

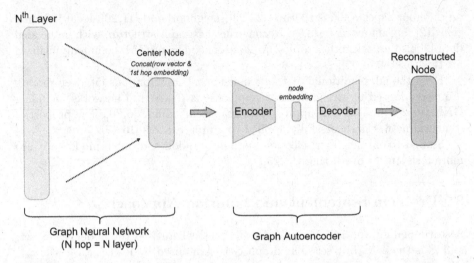

Fig. 1. The proposed model consists of two deep learning models, GNN and GAE. GNN is used to combine both the neighborhood and node information, while GAE is responsible for embedding this information into a latent space before reconstructing the input node.

weights of appropriate dimension, and the weighted sum is passed through a non-linear activation function. The result vectors of the kth layer have a dimension of size $1 \times h$ for any of the layers in GNN, where, h is the embedding dimension. The result vectors are then concatenated with the row vectors of the next layer. This dimension reduction is applied because the impact of high-hop neighbors wanted to be lower than the closer neighbors. The last layer of the GNN passed as the input of the GAE.

However, working with the neighborhood information in the model is not an easy task [2]. This is because, assuming the graph has uniformly distributed neighbors. The number of nodes increases exponentially on each layer. This exponential increase in layers can significantly slow down the model training process. While using all neighbors in the network may be a decent approach for small graphs, it is not a realistic solution for larger graphs.

The exponential increase neighbor problem can be handled with neighbor choosing, e.g., instead of using all neighbors some of the neighbors can be picked. In this paper, different neighbor selection methods are examined for each network, and their performance is analyzed based on the number of picked neighbors. The paper also proposes a new neighbor selection method called community hop. Instead of considering all possible neighbor set, community hop considers neighbors that can reach the starting node without using the link between the neighbor and the starting node in a given number of steps. The name community hop is given because that selected neighborhood can form a community. And this approach is inspired by the idea that communities contain similar objects, and these communities may provide better information extraction. A toy example of

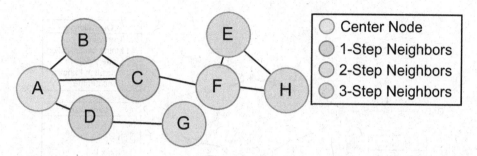

Fig. 2. The community hop with a restriction of two steps is explained in this toy example. The first-hop neighbors of node A are B, C, and D. B is community hop, because in two steps, it can reach node A without using the link between them (through C), and similarly, C is community hop as well. However, D cannot be considered a community hop because there is no path from it to get back to the A in two steps without using the link between A and D. The link structure between A, B and C forms a triangle and those three appear like a community. It can be seen that F, E and H are also community hops of each other.

community hop is given in Fig. 2 for the sake of clarity. The number of steps is restricted because in large graph, the starting node can eventually be reached. Community hop is suggested for graphs with nodes that have a decent number of neighbors and varying classes.

The forward propagation of the model begins with calculating weighted sums, starting from the outmost layer. If the weights are shared (as an input for the model), the same weights are used in the particular layer, which significantly reduces the number of learnable weights. Then, non-linear activation functions are applied to all the weighted sums. Next, the output of any previous layer is concatenated with the next layer in the GNN. The GAE is designed with two encoder and two decoder layers. The first encoder layer maps the input to double the embedding size $(2h)$. Obviously, the embedding dimension should be chosen to be less than half of the input size to achieve a meaningful embedding operation. Reconstruction loss is computed via loss function between the input graph and the output of the GAE. Then, the model parameters are updated with backpropagation. The pseudo-code of this process is given in Algorithm 1 and the forward propagation for GNN, and GAE is shown in Fig. 3 and Fig. 4 respectively for the 2-hop 2 neighbors model.

However, a graph with an extreme number of nodes may require more preprocessing (i.e., pre-embed, labeling the edges) because of computation time concerns. To achieve this, the adjacency matrix is transformed to another matrix by sampling fixed number of neighbors. If the number of neighbors for any node is lower than the given fixed amount, the rest are filled with zero to indicate that there are no neighbors. And if it is the opposite, neighbors are randomly selected equal to the fixed amount. Then unique consecutive IDs are assigned to each node (integers greater than zero). Since the integer number in the transformed matrix is arbitrary, reconstruct the nodes directly from this graph is

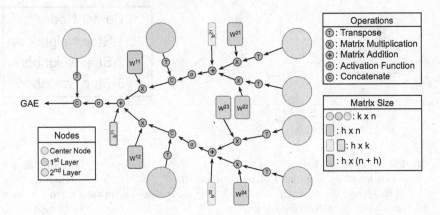

Fig. 3. Forward propagation of the GNN in the 2-hop 2 neighbors model begins with the second layer (blue nodes). In this illustration, h is the embedding dimension, n is the length of the row vector, and k is the batch size. If weights are shared, all W^{NX} are same for every N. In this example, if weights are not shared, there are $(2h \times (3n + h))$ learnable weights. On the other hand, $(h \times (2n + h))$ learnable weights for shared case. (Color figure online)

Algorithm 1. Reconstruction Loss Experiment for 2-Hop Model

Input: Graph: G,
　　　　　Activation function: σ,
　　　　　Bias usage: $bool_{bias}$, Weight sharing: $bool_{weight}$,
　　　　　Embedding dimension: h,
　　　　　Batch number: k, Epoch: e
Output: G', Reconstruction Loss

1 **Initialize:** Weight matrixes (Xavier initialization, size is depending on $bool_{weight}$), biases (zero, if $bool_{bias}$ is true; else, no bias will be used)
2 $G \leftarrow Preprocess(shuffle(G))$
3 **for** $e \leftarrow 1, 2, \ldots$ **do**
4 　**for** $k \leftarrow 1, 2, \ldots$ **do**
5 　　Sample neighbors in the batch: (nb_{L2}, nb_{L1})
6 　　$[sum_{L2}] = \sigma(MatMul(Weight_{L2}, nb_{L2}.T) + bias)$
7 　　$R_{L1} = Concatenate(nb_{L1}.T, [sum_{L2}])$
8 　　$[sum_{L1}] = \sigma(MatMul(Weight_{L1}, R_{L1}.T) + bias)$
9 　　$R_i = Concatenate(G[k].T, [sum_{L1}])$
10 　　$R_{encL2} = \sigma(MatMul(Weight_{encL2}, R_i))$
11 　　$R_{encL1} = \sigma(MatMul(Weight_{encL1}, R_{encL2}))$
12 　　$R_{decL1} = \sigma(MatMul(Weight_{decL1}, R_{encL1}))$
13 　　$G[k]' = MatMul(Weight_{decL2}, R_{decL1})$
14 　Compute reconstruction loss, backpropogate and update parameters

15 $//Matmul$: matrix multiplication, T: Transpose, Ln: Layer n, enc: encoder,
16 $//dec$: decoder

---

extremely challenging. The impact of high ID numbers overscale low ones and reconstruction loss as an evaluation metric becomes ineffective. As a solution, the node IDs are converted to binary format. Therefore, large graphs can be pre-embedded without losing the node structure and results in faster computations.

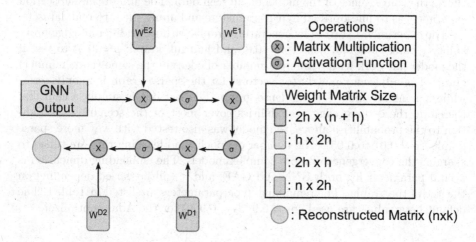

Fig. 4. Forward propagation of the GAE. The GNN output is $(n + h) \times h$ in size. h is the embedding dimension and n is the length of the row vector. The output of the GAE is a reconstructed matrix, which is used to compute the reconstruction loss and update the parameters.

In this work, we experimented with 10 different network structures, which are named as follows: NoHop, 1HopFull, 1HopFixed, 1ComHopFull, 1ComHop-Fixed, 2HopFull, 2HopFixed, 2ComHopFull, 2ComHopFixed, 3HopFixed. The first number indicates the number of layers and neighborhoods, and the second part of the name implies the behavior of the model in terms of neighborhood information.

4   Experimental Evaluation

In this section, we introduce the model, experimental setups, and evaluation metrics, as well as list graph properties and hyperparameters. Then the results are presented with comparisons, and behavior of the models are analyzed. The model is not only experimented with synthetic graphs but also experimented through real-world datasets. The model is implemented using the Python programming language and the PyTorch [19] platform; with PyTorch Geometric [8] used as a package for the baseline models. The experiments and evaluation of the model are performed in two subsections: synthetic and published dataset experiments.

4.1   Synthetic Graph Experiments

Synthetic graphs of different sizes are generated with varying numbers of edges in between graphs. As an evaluation metric for synthetic graphs, reconstruction loss is being used to analyze how combining autoencoder and graph neural network affects the convergence of the model on all seen data. The aim of this subsection experiment is to find optimal hyperparameters and apply them to real datasets.

The experiments started by generating graphs with three different edge probability (sparse: $p = 0.3$, dense: $p = 0.8$, and half full graph: $p = 0.5$) to test if the model can work with a different number of edges in the same (node number) graph. The adjacency matrix (or vectors) for the sparse graph is mainly zeros, while it is mainly ones for the dense graph. The last one is simply a coin flip; therefore, these three edge probabilities cover most of the scenarios. In addition to the probabilities above, the model was also tested with way more sparse graphs ($p = 0.01$ to 0.05). The number of epochs and learning rate are tuned to examine the convergence of more complex models. The embedding dimension is a vital parameter for both GNN and GAE, and it will be varied depending on the size of the graphs. The remaining hyperparameters are listed in Table 1. The default beta values are used ($\beta_1 = 0.9, \beta_2 = 0.999$) for the Adam optimizer.

Table 1. Experimental Setup

Hyperparameter	Experiments
Activation Function	Rectified Linear Unit (ReLU),
	Leaky Rectified Linear Unit (Leaky ReLU)
Optimizer	Adam, Stochastic gradient descent (SGD)
Neighborhood information	Full, fixed number (different numbers)
Weights	Shared, Not Shared

The experiments begin with preprocessing after generating synthetic graphs. Preprocessing involves obtaining the adjacency matrix (graphs may represented with other structures due to storage concerns), slicing the adjacency matrix into row vectors, and acquiring necessary neighborhood information. Inputs are then taken for the model, such as the length of the inner dimension vector (the output of encoder), the usage of bias, the value of the negative slope for leaky ReLU activation, the usage of neighbors (hop, community hop or none), the number of neighbors (an integer; depending on how sparse the graph is), the number of layers for the neural network, and the weight type (shared or not shared in the same layer). The weights are initialized with Xavier initialization, and bias vectors are initialized with zeros. Additionally, the activation function remains the same throughout the model e.g., there is no case where ReLU and Leaky ReLU are combined for the model.

The outcome of the synthetic graph experiments is presented in Fig. 5. The observations are listed below with the corresponding case given in parentheses:

- The convergences of the models are as follows from lower to higher: fixed size of neighbor networks (community hop converges better than hop picking (Fig. 5-C)), full neighbor networks, and no neighbor networks, respectively (Fig. 5-A). For the GNN part of the networks, 2-layer models are preferred over 3-layer models due to the computation time, even though 3-layer models may perform better in some cases. Consequently, more complex networks tend to converge better, which is particularly noticeable in lower dimensional embeddings. Additionally, community hop has shown better outcome hop-neighbor.
- More complex networks converge slower and better. (Fig. 5-A)
- Smaller reconstruction losses are achieved with a fixed size network. Interestingly, better results are obtained with a small number of neighbors (2–5) for each node compared to choosing a high number of neighbors. The neighbor picking operation similar to dropout. As a result, small number neighbor network creates more robust model. (Fig. 5-B)
- The hardest graph to reconstruct is the probability of edge 0.5, since any edge is simply a coin flip, and this pattern is difficult to learn. For the sparse and dense graphs, the models produced similar reconstruction losses. (Fig. 5-D)
- Leaky ReLU activation function converged better (with varying negative slope value). (Fig. 5-E)
- Even though using different weights in GNN achieves better convergence compared to weight sharing cases, the difference is considerably small. For the rest of the work, the tradeoff is taken to have a lesser number of trainable weights. (Fig. 5-F)
- Adam optimizer generated better reconstruction losses. However, this result should not be generalized since with better hyperparameter tuning, any optimizer can outperform another. Therefore, the choice of optimizer should depend on the field and be carefully tuned by domain experts.

The observations above hold for all sizes of generated graphs.

4.2   Published Dataset Experiments

After completing the experiments on synthetic graphs, the proposed model is tested on real-world datasets. In addition to reconstruction loss, link prediction is used as a performance measure. The F1 score is calculated as an evaluation metric in this section. There are a total of four published datasets used in this section, belonging to three different network categories, which are listed in Table 2.

To accomplish the link prediction task, some adjustments are made on the model. First, the graph is randomly split into two parts, the train and the test. In every epoch, the train and test data are shuffled. Then the model is set to training mode and trained in that particular epoch, and the reconstruction loss for train is calculated. Next, the model is set to test mode, and all the learning for the GNN and the GAE is stopped temporarily. The reconstruction loss for test and accuracy is calculated, and the loop continues with the next epoch.

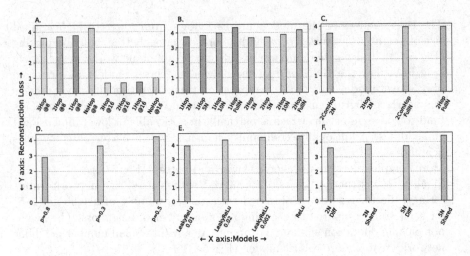

Fig. 5. The results of the experiments on synthetic graphs are summarized.

Each element of the output of GAE is converted to links (binary) depending on the determined threshold value (0.5 in experiments). The hyperparameters for the published dataset are fixed to compare the performance of models. The activation function is Leaky ReLU with negative slope of value 0.01 with Adam optimizer. The learning rate is set to a relatively small value (0.0001).

The Citeseer dataset is a citation network where each node represents a document while edges are citation between papers. The edges are directed by nature (from the citing paper to the cited paper). The papers in the dataset are classified into six classes based on the research field. Nodes in the same classes tend to have similar citation patterns, with most of the links are structured within the community and only a few popular nodes cited by the majority. Additionally, links between communities are expected to be low or non-existent (i.e., some papers are not cited yet). As a directed graph, the dataset has row vectors full of zeros, and some papers may be cited by the majority, which was not the case in previous experiments. In addition to these challenges, predicting links between nodes across different classes makes the link prediction task for this dataset even more challenging.

Table 2. Published datasets

Dataset	Number of nodes	Number of edges
Citeseer	3312	4732
Amazon	1418	3695
Disney	124	335
Mooc	5034	10022

The Amazon Fail network is a co-purchase network, and the Disney dataset is extracted from the Amazon network. The edges represent the co-purchase relationships between products, and these networks are also directed. In addition, in purchase networks, the presence of randomness is high, e.g., anomalies. Additionally, these networks have weak edges because the products are distinct with their attributes, which makes link prediction challenging in a way that is not introduced by other networks.

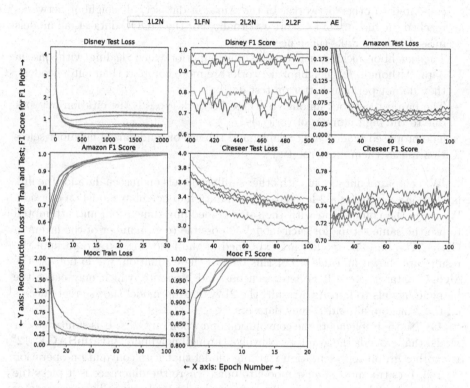

Fig. 6. Published dataset experiments results. *AE*: Autoencoder, 1*L2N*: 1 Layer 2 Neighbors, 1*LFN*: 1 Layer Full Neighbors, 2*L2N*: 2 Layer 2 Neighbors, 2*LFN*: 2 Layer Full Neighbors.

The MOOC or MOOC Student Dropout dataset is a collection of students who enrolled the course but dropped out before completion. The edges represent the actions and behavior of students. Unlike the previous networks, MOOC is a bipartite graph. Additionally, the dataset is imbalanced and noisy, making the link prediction task even more challenging.

The outcome of the experiments using published datasets under different neighbor networks is shown in Fig. 6.

- Complex models converge slower. The model with only core network (AE) converged quickly compared to other models.
- One of the key findings of our paper is that the neighbor picking models are resistant to overfitting, which can be easily observed in Disney experiments. Community-hop networks are more resistance to overfit compared to full neighbor and hop neighbor networks. Also, it achieved the best test losses even with an unrealistic number of epochs.
- The model achieved the highest F1 score in the Amazon and MOOC datasets compared to other networks. In the Amazon dataset, all neighbor networks reached the maximum F1 score eventually. For the MOOC dataset, all models missed only 11 links (false negatives).
- The neighbor picking networks exhibit superior when dealing with unseen data. Although, full neighbor networks are more complex than autoencoders, they do not perform better on test data.
- The hardest network among the three network types is the citation network due to the high number of varieties.
- The proposed models are scalable, and the use of randomized neighbor selection makes them more robust.

We compared our model with other similar models on unseen data for the link prediction task, using a 2-layered 2-community hop neighbor model on our side. We make the comparison with the same embedding dimension, and attempted to use the same setups for each model as possible (e.g., number of the layers in GNN). The baseline models are explained in the following paragraph, and the results are shown in Table 3. For this part of the evaluation, we do not use the MOOC dataset, since it is heterogeneous and bipartite, which may cause the baseline models to fail. As a result, the 2L2ComHop model shows great success on the AmazonFail and Disney datasets.

GCN [15] implements the convolution operation on GNN. It is similar to our model that embeds the graph by convolve through neighbors. **GraphSAGE** [12] is another graph representation learning model that uses convolution operation as well. It is the most similar model to ours, where the difference is it picks the neighbors based on random walks in a predetermined steps. Then aggregates the local neighborhood to embed the graph. Another GNN model, **GAT** [28] is a combination of attention mechanism [27] and GNN. The **VGAE** and **GAE** [14] is implementation of variational autoencoder and autoencoder on graphs respectively. Unlike previous models, these two models reconstructs the graph. For comparison, we implemented two types of VGAE and GAE; with addition of GCN or not.

Table 3. The F1 score comparison of link prediction on unseen data of three real-world datasets.

Model	Citeseer	Disney	AmazonFail
GCN	0.723	0.733	0.689
GraphSAGE	0.729	**0.813**	0.709
GAT	0.709	0.788	0.696
VGAE	**0.764**	0.698	0.710
VGAE-GCN	**0.769**	0.779	**0.760**
GAE	0.752	0.746	0.687
GAE-GCN	0.747	0.782	0.704
2L2ComHop	0.763	**0.991**	1

5 Conclusion

In this work, we proposed combining two deep models, GAE and GNN, for reconstructing the graph structure in an unsupervised manner. The GNN extracts neighborhood information from the graph and enhances the performance of GAE. We tested the models on synthetic data and real-world networks. The proposed framework succeeds through in different sizes and types of graphs. Using neighborhood information randomly instead of all of it achieved better results for both seen and unseen data, with fewer learnable parameters. As a result of this work, such neighbor networks are resistant to overfitting and applicable to many networks.

References

1. Al-Rfou, R., Perozzi, B., Zelle, D.: DDGK: learning graph representations for deep divergence graph kernels. In: The World Wide Web Conference, pp. 37–48 (2019)
2. Alon, U., Yahav, E.: On the bottleneck of graph neural networks and its practical implications. In: International Conference on Learning Representations (2021)
3. Bandyopadhyay, S., Lokesh, N., Murty, M.N.: Outlier aware network embedding for attributed networks. In: Proceedings of the AAAI Conference on Artificial Intelligence, vol. 33, pp. 12–19 (2019)
4. Bandyopadhyay, S., Vivek, S.V., Murty, M.: Outlier resistant unsupervised deep architectures for attributed network embedding. In: Proceedings of the 13th International Conference on Web Search and Data Mining, pp. 25–33 (2020)
5. Cao, S., Lu, W., Xu, Q.: GraRep: learning graph representations with global structural information. In: Proceedings of the 24th ACM International on Conference on Information and Knowledge Management, pp. 891–900 (2015)
6. Cao, S., Lu, W., Xu, Q.: Deep neural networks for learning graph representations. In: Proceedings of the AAAI Conference on Artificial Intelligence, vol. 30 (2016)

7. Chen, H., Perozzi, B., Hu, Y., Skiena, S.: HARP: hierarchical representation learning for networks. In: Proceedings of the AAAI Conference on Artificial Intelligence, vol. 32 (2018)
8. Fey, M., Lenssen, J.E.: Fast graph representation learning with PyTorch Geometric. In: ICLR Workshop on Representation Learning on Graphs and Manifolds (2019)
9. Gao, H., Huang, H.: Deep attributed network embedding. In: Twenty-Seventh International Joint Conference on Artificial Intelligence (IJCAI)) (2018)
10. Gori, M., Monfardini, G., Scarselli, F.: A new model for learning in graph domains. In: Proceedings. 2005 IEEE International Joint Conference on Neural Networks, vol. 2, pp. 729–734 (2005)
11. Grover, A., Leskovec, J.: node2vec: scalable feature learning for networks. In: Proceedings of the 22nd ACM SIGKDD International Conference on Knowledge Discovery and Data Mining, pp. 855–864 (2016)
12. Hamilton, W., Ying, Z., Leskovec, J.: Inductive representation learning on large graphs. In: Advances in Neural Information Processing Systems, vol. 30 (2017)
13. Hamilton, W., Ying, R., Leskovec, J.: Representation learning on graphs: methods and applications. IEEE Data Eng. Bull. **40**(3), 52–74 (2017)
14. Kipf, T.N., Welling, M.: Variational graph auto-encoders. In: NIPS Workshop on Bayesian Deep Learning (2016)
15. Kipf, T.N., Welling, M.: Semi-supervised classification with graph convolutional networks. In: International Conference on Learning Representations (2018)
16. Lyu, T., Zhang, Y., Zhang, Y.: Enhancing the network embedding quality with structural similarity. In: Proceedings of the 2017 ACM on Conference on Information and Knowledge Management, pp. 147–156 (2017)
17. Niepert, M., Ahmed, M., Kutzkov, K.: Learning convolutional neural networks for graphs. In: International Conference on Machine Learning, pp. 2014–2023. PMLR (2016)
18. Papp, P.A., Martinkus, K., Faber, L., Wattenhofer, R.: DropGNN: random dropouts increase the expressiveness of graph neural networks. Adv. Neural. Inf. Process. Syst. **34**, 21997–22009 (2021)
19. Paszke, A., et al.: PyTorch: an imperative style, high-performance deep learning library. In: Advances in Neural Information Processing Systems, vol. 32 (2019)
20. Perozzi, B., Al-Rfou, R., Skiena, S.: DeepWalk: online learning of social representations. In: Proceedings of the 20th ACM SIGKDD International Conference on Knowledge Discovery and Data Mining, pp. 701–710 (2014)
21. Ren, Y., Liu, B., Huang, C., Dai, P., Bo, L., Zhang, J.: Heterogeneous deep graph infomax. ArXiv abs/1911.08538 (2019)
22. Ribeiro, L.F., Saverese, P.H., Figueiredo, D.R.: struc2vec: learning node representations from structural identity. In: Proceedings of the 23rd ACM SIGKDD International Conference on Knowledge Discovery and Data Mining, pp. 385–394 (2017)
23. Rong, Y., Huang, W., Xu, T., Huang, J.: DropEdge: towards deep graph convolutional networks on node classification. In: International Conference on Learning Representations (2020)
24. Rossi, R.A., Ahmed, N.K., Koh, E., Kim, S., Rao, A., Abbasi-Yadkori, Y.: A structural graph representation learning framework. In: Proceedings of the 13th International Conference on Web Search and Data Mining, pp. 483–491 (2020)
25. Scarselli, F., Gori, M., Tsoi, A.C., Hagenbuchner, M., Monfardini, G.: The graph neural network model. IEEE Trans. Neural Netw. **20**(1), 61–80 (2008)

26. Tang, J., Qu, M., Wang, M., Zhang, M., Yan, J., Mei, Q.: LINE: large-scale information network embedding. In: Proceedings of the 24th International Conference on World Wide Web, pp. 1067–1077 (2015)
27. Vaswani, A., et al.: Attention is all you need. In: Advances in Neural Information Processing Systems, vol. 30 (2017)
28. Veličković, P., Cucurull, G., Casanova, A., Romero, A., Liò, P., Bengio, Y.: Graph attention networks. In: International Conference on Learning Representations (2019)
29. Wang, D., Cui, P., Zhu, W.: Structural deep network embedding. In: Proceedings of the 22nd ACM SIGKDD International Conference on Knowledge Discovery and Data Mining, pp. 1225–1234 (2016)
30. Xu, K., Hu, W., Leskovec, J., Jegelka, S.: How powerful are graph neural networks? In: International Conference on Learning Representations (2019)

Question-Aware Deep Learning Model for Arabic Machine Reading Comprehension

Marwa Al-Harbi[1], Rasha Obeidat[1]($\boxtimes$) (iD), Mahmoud Al-Ayyoub[2], and Luay Alawneh[3]

[1] Department of Computer Science, Jordan University of Science, Irbid, Jordan
maalharbi18@cit.just.edu.jo, rmobeidat@just.edu.jo
[2] Department of Information Technology, Ajman University, Ajman, UAE
maalshbool@just.edu.jo
[3] Department of Software Engineering, Jordan University of Science, Irbid, Jordan
lmalawneh@just.edu.jo

Abstract. Machine reading comprehension is one of the most long-standing challenges in artificial intelligence. It is a subtask question-answering system that aims to find a text span in a reading context that answers a question. It is the core of several applications, such as customer service systems and chatbots. However, However, few studies have targeted machine reading comprehension of Arabic text. Besides, The existing models generate a context representation independently from the question. As an analogy to some people's comprehension styles who believe that by reading the question, they gain a better understanding of the reading passage, we present a question-aware neural machine reading comprehension model for Arabic. Our model extracts a representation of the context by incorporating the question using several bidirectional attention units to achieve various levels of question-centered context understanding. Our experimental evaluation shows that our model outperforms strong machine reading comprehension baselines, including DrQA and QaNET models applied to Arabic, by a significant margin with an F1 score of 59.6%.

Keywords: Arabic Machine Reading Comprehension · Deep learning · Question Answering · Bi-directional Attention · Arabic MRC datasets · Question-Aware MRC

1 Introduction

Machine Reading Comprehension (MRC) is a Natural Language Processing (NLP) task that aims to understand and answer questions based on a given text. It is considered important as it imitates human reading and comprehension capabilities [18]. It is an essential testbed for evaluating machine natural language understanding. Besides, it is the core building block of several applications, including smart assistants [22] and chatbots [28], as it enables them to respond better to user questions and return more precise answers.

A. Bennour et al. (Eds.): ISPR 2023, CCIS 1941, pp. 262–276, 2024.
https://doi.org/10.1007/978-3-031-46338-9_20

Based on the answer type, the reading comprehension tasks can be divided into four categories: cloze test, multiple choice, free answering and span extraction MRC [18]. The cloze test, called gap-filling, is the MRC setup where the answer is a single word or entity missing from the sentence (the question). Multiple-choice MRC involves questions that come with a passage and a list of candidate answers, and the task is choosing the choice that correctly answers the question. In free answering MRC, the machine needs to reason over multiple passages to summarize the answer, which can be part of the context or be predicted by the machine (i.e., not part of the context). Extractive (or Span-based) MRC, the focus of this study, answers a question based on a given context. The task can be formulated as a supervised task with dataset examples in the form of (context, question, answer)-triples; the objective is to learn a model that receives the context and the question as input and produces a continuous span from the context as the answer.

The field of MRC has progressed rapidly since the release of the Stanford Question Answering Dataset (SQuAD) [26], Where strong Deep Learning models have been successively proposed. Seo et al. [27] proposed a deep architecture with a bi-directional attention flow network to learn the interactions between the question and the context. Chen et al. [9] trained a multi-layer recurrent neural network model to detect answers. R-NET [14] is an end-to-end gated attention-based Recurrent Neural Network (RNN) to obtain a query-aware context representation. The network also uses self-matching attention to encode the important information from the whole passage. Yu et al. [32] proposed an MRC architecture composed of convolutions and self-attention units to model local and global interactions. On the other hand, the lack of Arabic annotated datasets matching the SQuAD size and quality slowed the development of Arabic-specific MRC models or even applying the models proposed for English to Arabic.

In human reading comprehension, some people prefer to read the question before reading the passage to get a question-focused understanding of the reading passage. O'Reilly et al. show that reading purpose affects how students process reading texts [23]. Motivated by this idea, we propose a **Question-Aw**are extractive **MRC** (QAw-MRC) for the Arabic Language that uses several bi-directional attention blocks to incorporate the question in creating question-aware context embeddings on various levels of understanding. The generated context embedding helps in extracting a more precise answer span. We train and evaluate the model on Arabic Question Answering Dataset, a large Arabic reading comprehension manually created through language experts and available on GitHub. We compared and experimentally evaluated our approach to two strong baseline methods with a similar architectural flavor to ours, namely DrQA [9] and QANet [32]. Our results show the superiority of our model compared to the baselines, with an F1 score of 59.6%. We chose not to compare our model with a BERT-based one to cancel the influence of transfer learning achieved by using pre-trained BERT models for a more fair comparison.

2   Related Work

There has been extensive research on English MRC with impressive systems, and some of them even surpassed human-level performance. Both large-scale datasets and neural models drive this success. Arabic MRC could not match the pace due to the language challenge and the lack of available large, high-quality, general-purposed MRC datasets. This section overviews the main datasets and approaches previously evaluated for Arabic MRC.

Arabic MRC Datasets. The availability of massive Benchmark datasets is a major driver in the English MRC progress [27]. On the other hand, Existing datasets for Arabic suffer from several drawbacks. The manually-built Arabic MRC datasets are small to train neural models such as [20]. An example is ARCD [21], a native Arabic RC dataset created similarly to SQuAD v1.1. It contains 1,395 examples. In contrast, the available dataset large datasets such as AAQAD [7], Arabic-SQuAD [21], Arabic-SQuAD2.0 [1], and Arabic-WikiReading are noisy as they are fully or partially translated or collected by distant supervision [3]. AAQAD consists of 17K+ questions mapped and partially machine-translated from SQuAD. Arabic-SQuAD consists of 48k+ questions obtained from SQuAD and fully translated by machine. Human-translated datasets have also been released, such as MLQA [17] and *XQuAD* [6]. For example *XQuAD* is a cross-lingual dataset translated into ten languages by professional translators. It has relatively good quality since it is translated from SQuAD by professional translators; however, since questions are sourced from a text originally written in English, they do not reflect the Arabic culture. Some other datasets are domain-specific, such as Hadith [2] and QRCD [20], or only cover limited types of questions (e.g., KaifLematha [3]). Besides the general-purpose Arabic MRC dataset, there are a few domain-specific datasets proposed to serve specific domains, including the Arabic Poem Comprehensive Dataset (APCD) [31] and Qur'anic Reading Comprehension Dataset (QRCD) [20].

More recently, a notable multilingual dataset called TyDiQA Gold-P [11] has been made available. It consists of a large number of question-answer-passage pairs written in several languages, including Arabic. It consists of 14k+ examples, representing a valuable source for elevating the research on Arabic MR. However, different question tools are not evenly covered in TyDiQA. Besides, It also has HTML artifacts in the passages, answers training instances, and text in other languages [5]. In this work, an MRC dataset for Arabic is available on Github, encompassing 16k+ examples sourced by language experts from Wikipedia articles. Table 1 summarizes the existing Arabic RC datasets.

Arabic MRC Methods. A limited number of studies on Arabic Reading Comprehension (RC) have been proposed, and most of them utilize the transfer learning obtained by using Language Models (LMs) such as BERT [13], RoBERTa [19] and ELECTRA [12], mainly to deal with the scarcity of large-scale MRC datasets as shown in see Table 2. Fine-tuning language models is adopted as a practical solution for learning from small datasets.

Various language models have been trained using large amounts of training data and contributed heavily to the improvement of many Arabic Natu-

Table 1. A Summary of the available Arabic MRC datasets.

Dataset	Source	Questions Source	Size
Arabic-SQuAD [21]	SQuAD	Machine-translated	48k+
Arabic-SQuADv2.0 [1, 24]	SQuADv2.0	Machine-translated	100k
ARCD [21]	Arabic Wikipedia	Crowd-workers	1.3K+
AAQAD [7]	Arabic Wikipedia and SQuAD	Automated	17k+
Hadith [2]	Sahih al-Bukhari	Field experts	3k+
TyDiQA Gold-P [11]	Arabic Wikipedia	Crowd-workers	14.8k+
QRCD [20]	Tanzil project	Crowd-workers	1k+
XQuAD [6]	SQuAD	Professional translators	1.1k
Arabic WikiReading [3]	Arabic Wikipedia	Distant supervision	98k+
KaifLematha [3]	Arabic Wikipedia	crowd-workers	6.7K+
ArQuAD	Arabic Wikipedia	Language expert	16k+

ral language understanding tasks, including MRC. AraBERT [4] is a BERT-based model pretrained on 200m+ Arabic MSA sentences gathered from various resources, including large Arabic corpora and news websites, in addition to the Arabic Wikipedia. mBERT[1] is a multilingual BERT-based model pretrained on more than one hundred languages in Wikipedia, including Arabic. AraBERT achieved an F1 score of 62.7% on the ARCD dataset compared to 61.3% by the mBERT [4]. Table 2 summarizes some of the MRC models evaluated for Arabic along with the type of the adopted approach, the utilized dataset and the best-achieved results. AraELECTRA [5] is an ELECTRA-based model [5] pretrained on the same dataset as AraBERTv0.2. It achieved F1 scores of 86.68% and 71.22% on TyDiQA Gold-P and ARCD, respectively. Pandya et al. [24] fined-tuned xlm-roberta-base-arabic is XML-RoBERTa model pre-trained using Arabic unsupervised data, using a combination of XQuAD and MLQA testing set. Their model achieved an F1 score of 58.07%. T5 [25] is a Text-to-Text Transfer Transformer built on converting all text-based language problems into a text-to-text format. mT5 [30] is a multilingual variant of T5, pretrained on a huge dataset covering 101 languages. A version of mT5 called mT5-XXL achieved an F1 score of 80.3% on XQuAD, surpassing strong pre-trained language models such as mBERT, which achieved an F1 score of 61.5%. mT5-XXL also attained an F1 score of 71.7% on the MLQA dataset.

Neural models that are composed of different types of neural networks, such as CNN, RNN, and LSTM, along with different attention mechanisms, have rarely been trained and evaluated using the Arabic MRC dataset. DrQA [16] and BiDAF [27] are Among the few neural architectures that have been trained and evaluated for the Arabic MRC. Mozannar et al. [21] train the QANet model on the translated ARCD and SQuAD datasets achieving F1 scores of 38.6% and 44.4%, respectively. Atef et al. [7] trained the BiDAF model using the AAQAD dataset and got an F1 score of 32%.

---

[1] https://huggingface.co/bert-base-multilingual-cased.

It is not deniable that the pre-trained language model significantly surpassed the conventional neural models. However, the conventional models facilitate modeling different human-reading comprehension styles, such as reading the question before reading the context and vice versa. To reduce the gap, we propose a neural model that uses several bi-directional attention blocks to incorporate the question in creating question-aware context embeddings on various levels of understanding. Then we prove the superiority of our model compared to DrQA and QANet experimentally.

Table 2. A comparison between the previously proposed Arabic MRC systems overview. All the reported results use F1 scores.

Model	Method	Dataset	Result
QANet [21]	neural RC model	ARCD	38.6%
QANet [21]	neural RC model	Arabic-SQuAD	44.4%
BiDAF [7]	neural RC model	AAQAD	32.0%
AraBERT [4]	Pretrained LM	ARCD	63.7%
mBERT [4]	Pretrained LM	ARCD	61.3%
mBERT [30]	Pretrained LM	XQuAD	61.5%
AraELECTRA [5]	Pretrained LM	ARCD	71.2%
Ar-XLM-RoBERta [24]	Pretrained LM	XQuAD, MLQA	58.1%
AraELECTRA [5]	Pretrained LM	TyDiQA Gold-P	86.6%
mT5-XXL [29]	Pretrained LM	XQuAD	80.3%
mT5-XXL [29]	Pretrained LM	MLQA	71.7%

3   Question-Aware Arabic MRC

This section describes our **Question-Aw**are Arabic **MRC** (QAw-MRC) neural model that receives the question and the context as inputs and then predicts the start and the end offsets of the text span within the context that answers the question. Similar to BiDAF [27], our model is composed of multiple hierarchical stages for modeling the representations of the context passage at different levels of granularity, including word level, contextual level and model level representations. However, QAw-MRC differs from BiDAF that it also creates a multi-level representation of the question and passes it to the output layer along with context representation. In addition, we combine question-aware and question-independent context representations to obtain a multi-perspective summary of the context.

The architecture of QAw-MRC is depicted in Fig. 1. Given a context c consisting of n tokens $\{c_1, c_2, ..., c_n\}$, and a question q composed of m tokens $\{q_1, q_2, ..., q_m\}$, the goal of the model is to predict a substring of the context

$(a_{start},\ a_{end})$ that best answers the question. Our model is composed of the following components:

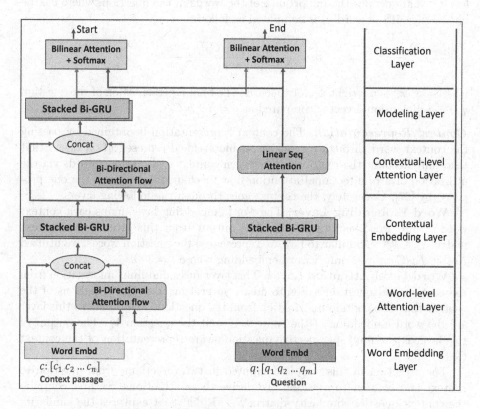

Fig. 1. The Architecture of QAw-MRC model.

Question Representation. Using our model, question q goes through three consecutive layers to obtain a final question representation passed to the output layer, namely, the word embedding Layer, contextual embedding Layer and linear sequence attention layer.

Word Embedding Layer. The word embedding layer maps each question word q_j into word vector $v_q^j \in \mathbb{R}^l$. The output is the question embedding matrix $V_q \in \mathbb{R}^{m \times l}$. To obtain the embedding for each word, we used 300-dimensional pretrained Arabic *FastText* word embedding trained a large corpus of Arabic, including Arabic Wikipedia [8].

Contextual Embedding Layer. Since RNN is a natural solution to model variable length sequences (e.g., questions), we utilize a stack of three bidirectional GRUs (BiGRUs) [10] to model the temporal interactions between words in the given question. We concatenate the outputs of the six GRU to obtain $C_q \in \mathbb{R}^{d \times m}$. Note that d-dimensional vectors in C_q are obtained by concatenating the outputs of the three forward and three backward GRUs.

Linear Sequence Attention. To obtain a final question encoding, we employ a linear sequence attention unit [9] to compute the importance vector $b \in \mathbb{R}^m$ that encodes the importance of the words in the question, where b_j, the importance of the word q_j is computed as follows:

$$b_j = \frac{\exp(w^T.C_{q_j})}{\sum_{k=1,...l} \exp(w^T.C_{q_k})}$$

where $w \in \mathbb{R}^d$ is a weight vector to learn. The final representation of the question $\hat{q}$ is a d-dimensional vector computed as $\hat{q} = \sum_j b_j C_{q_j}$.

Context Representation. The context representation is obtained by passing the context word through a sequence of hierarchical representation layers that takes into account the relationship between context and question words via the utilization of a bi-directional attention flow mechanism similar to the one proposed by [27]. Concretely, the context goes through the following layers.

Word Embedding Layer. The word embedding layer maps each context word c_i into word vector $v_c^i \in \mathbb{R}^l$. The output from this layer is the context matrix $V_c \in \mathbb{R}^{n \times l}$. Similar to how we represented the question words, we utilized Arabic *FastText* pretrained word embedding where $l = 300$.

Word Level Attention Layer. This layer fuses and links information from the context and question words to create several matrix representations of the context that also contain information from the question. The inputs to this layer are the word embedding of the context V_c and the question V_q. The output of this layer is the multi-perspective question-aware representation of the context $G \in$.

The attention in this step is computed in two directions, from question to context and context to question. To derive these attentions, we first compute the context-question similarity matrix $S \in \mathbb{R}^{n \times m}$ that captures the similarity between each context word c_i and each question word q_j. Specifically, S_{ij} is computed as the dot products between nonlinear mappings of context and question word embeddings as follows (α is a single dense layer with the Leaky ReLU nonlinearity function):

$$S_{ij} = \alpha(V_{c_i}).\alpha(V_{q_j})$$

The word level similarity matrix S gives a relatively high attention score to two semantically similar words, such as "planet" and "earth." The similarity matrix S serves as an input to compute two context representation matrices, U and H.

- **Context-to-Question Word-Level Attention.** This attention is identical to the question-aligned embedding originally proposed by Chen et al. [9] computed from S and used as a part of the context representation. As illustrated in Fig. 2, row-wise softmax is applied S so that the attention scores for each word in the context sum up to 1. The result is the matrix $\hat{S} \in \mathbb{R}^{n \times m}$. Every row in $\hat{S}$ signifies the relative importance of question words to the context word c_i. Thus, the question word q_j that is more related to the context word

c_i gets a higher score, and that happens if the similarity between the word embeddings of c_i and q_j is high (e.g., words with similar meanings).

We use $\hat{S}$ to compute $U \in \mathbb{R}^{n \times l}$, a context-attended summary of the query vectors that will act as a question-based representation of the context. In other words, each word in the context is represented by summing up all the question vectors after being weighted by the similarities between these vectors and the context word vector. Formally, $U_i \in \mathbb{R}^l$, the question-based representation of the context word c_i is calculated as:

$$U_i = \sum_j \hat{S}_{ij} V_{q_j}$$

- **Question-to-Context Word-Level Attention.** To compute question-to-context attention, as depicted in Fig. 3, we first take the maximum score across rows of the similarity matrix S, then apply softmax on the resultant max column vector to get the attention distribution vector called $z \in \mathbb{R}^n$. Then, we use z to compute the weighted sum of the context matrix V_c. The intuition behind this step is determining which context word is most similar to either of the question words, hence crucial for answering the question. Concretely, the attended context vector $h_i \in \mathbb{R}^l$ is computed as:

$$h = \sum_{i=1,\ldots,n} z_i V_{c_i}$$

h^T is tiled m times across the row, forming $H \in \mathbb{R}^{n \times l}$. H encapsulates the information about the most important context words with respect to the question.

Finally, the context word embedding, context-to-question attention, and question-to-context attention are combined to yield G, where each row vector can be considered as the question-aware context word embedding of each context word. We define $G \in \mathbb{R}^{n \times 4l}$ by:

$$G = [V_c; U; V_c \circ U; V_c \circ H]$$

where $\circ$ is element-wise multiplication, $[;]$ is Matrix concatenation across rows. V_c is the question-independent representation of the context. $V_c \circ U$ is the context representation that summarizes the most relevant words from the question attended by the context words. $V_c \circ H$ is the context representation that summarizes the most relevant information from the context with respect to the most related words in the question.

Contextual Embedding Layer. This layer comprises three stacked bidirectional GRUs, where the input is $G \in \mathbb{R}^{n \times 4l}$ from the previous step. The output of this layer captures the interaction among the context words conditioned on the question. We concatenate the outputs of the six GRU to obtain $C_c \in \mathbb{R}^{d \times n}$. Note that d-dimensional vectors in C_c are obtained by concatenating the outputs of the three forward and three backward GRUs.

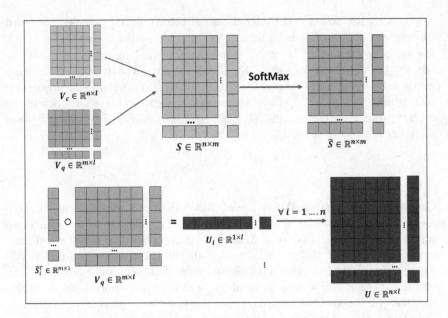

Fig. 2. Context-to-question word-level attention.

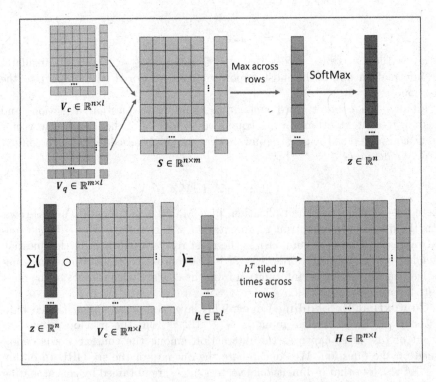

Fig. 3. Question-to-context word-level attention.

Contextual Level Attention Layer. After applying multiple layers of GRU to extract different levels of understanding of each context word, we conduct another attention from question to context and from context to question. The input of this layer is the contextual context representation C_c and the contextual question representation C_q. The computation in this layer is the same as the Word-Level Attention Layer. Assume that the $\hat{U} \in \mathbb{R}^{n \times d}$ and $\hat{H} \in \mathbb{R}^{n \times d}$ are the *contextual-level* context representations obtained from applying *contextual-level* context-to-question and question-to-context attention units, similar to what we computed at the word level, where d is the dimensionality of the vector that obtained by concatenating the output of the six GRUs from the previous layer. $\hat{G} \in \mathbb{R}^{n \times 4d}$, the output of this layer is computed as:

$$\hat{G} = [C_c; \hat{U}; C_c \circ \hat{U}; C_c \circ \hat{H}]$$

Modeling Layer. This layer contains three stacked bidirectional GRUs. The input is the matrix $\hat{G}$ from the previous layer. This output matrix $\hat{C} \in \mathbb{R}^{n \times d}$ can be considered a more comprehensive and contextually aware representation of the context. Note that d-dimensional vectors in $\hat{C}$ are obtained by concatenating the outputs of the six GRUs. This new representation is then used as an input to the output layer.

Output Layer. Given the question and the reading context, the MRC model is required to find a span of tokens that answer the question. We adopt the output layer from Chen et al. [9]. We take $\hat{q}$ and $\hat{C} = [\hat{c}_1, \hat{c}_2, ...\hat{c}_n]$, the final encoding of the question q and the reading context c, as the input to two independent classifiers that we train to predict the span *start* and *end* offsets. Specifically, we compute the similarity between context word c_i and the question q using two bilinear functions as follows:

$$P_{start}(i) = exp(\hat{c}_i W_s \hat{q})$$
$$P_{end}(i) = exp(\hat{c}_i W_e \hat{q})$$

The prediction is performed by choosing the best span from r to token k such that $r \leq k \leq r + 15$ and $P_{start}(r) \times P_{end}(k)$ is maximized. Our final training objective is to minimize the cross-entropy loss function.

4 Experimental Setup

4.1 Dataset

To train and evaluate our proposed model, we used a large dataset for the Arabic language called Arabic Question Answering Dataset, available at[2]. The dataset contains 16,020 questions, 12,816 examples for training, 1,596 for development and 1,608 for testing. Each example comprises a passage extracted from an

---

[2] https://github.com/RashaMObeidat/ArQuAD.

Arabic Wikipedia article and associated questions posed by language experts. The answer to each question is a text segment from the corresponding reading passage. Table 3 illustrates the distribution of different Arabic question tools used in the dataset. It shows that the dataset covers various Arabic question tools. The data set average question and answer length are 8.1 and 3.4, respectively.

Table 3. Question tools distribution in ArQuAD sorted by frequency.

Ques.Tool (Ar)	Ques.Tool (Ar)	Frq%	Example
ما, ماذا	What	31.8	ماذا فعل القائد صلاح الدين بالغزاة ؟
من	Who	13.7	من قاد الحملة الفرنسية؟
متى	When	12.7	متى بدأت حرب الاستنزاف؟
كم	How much/many	11.1	كم عدد سور القرآن الكريم ؟
أين	Where	7.6	أين ولد كرستيانو رونالدو؟
في اي	In which/what	6.3	في أي سنة أنشئت جامعة الدول العربية؟
أخرى	Others	5.9	على من بعث سيدنا محمد صلى الله عليه وسلم ؟
لماذا	Why	4.6	لماذا سميت البتراء بالمدينة الوردية ؟
كيف	How	2.6	كيف أثرت رسالة الاسلام على..؟
إلى أي	To which/what	1.5	الى أي حزب ينتمي حسني مبارك؟
أي	Which	1.2	أي الدول العربية اكثر سكانا؟
هل، أ	Yes/No	1.0	هل ولد ابو بكر الصديق في مكة؟

4.2 Baselines

Our model was trained and compared against two baselines using the same Arabic MRC dataset. **Document Reader (DrQA):** This model proposed by [9] is an end-to-end open-domain question answering originally proposed and applied to SQuaD. By this model, The question and the passage tokens are passed through an embedding layer, followed by aligned question embedding. The passage and question are then passed through stacked Bi-LSTM. The question is also passed through a linear attention layer to calculate the importance of each question token. Finally, both question and passage representations go through two bilinear attention layers to predict the start and end position of the answer. We used the DrQA model implementation by[3]. **Combining Local Convolution With Global Self-Attention (QANet):** This model proposed by [32] is an end-to-end RC model that uses convolutions and self-attentions

---

[3] https://github.com/kushalj001/pytorch-question-answering.

instead of recurrent as a base for the encoders. The model consists of the following layers: input embedding layer, embedding encoder layer (i.e., convolution-layer, self-attention-layer, feed-forward-layer), context-query attention layer (i.e., the similarity between each pair of context and query tokens), model encoder layer, and output layer. We used the QANet model implementation by[4].

4.3   Evaluation Metrics

Following the majority of extractive machine reading comprehension tasks, we used the Exact Match (EM) and the F1 score (macro-averaged) evaluation metrics [33]. While EM measures the percentage of predictions that exactly matches the target answer, the macro-averaged F1 score measures the average overlap between the predictions and the target answer. The predictions and target answers are handled as a bag of words to calculate the F1 scores.

4.4   Hyper-parameter Tuning

We fine-tuned the model on the ArQuAD development set. We use a 3-layer bidirectional GRU with $h = 128$ hidden units for context and question encoding. We apply the Arabic spaCy tokenizer[5] for tokenization. We use *Adamax* for optimization [15] with a constant learning rate $= 0,002$. Dropout with $p = 0.3$ is applied to word embeddings and all the hidden units of GRU. Lastly, all the training examples were divided into mini-batches of 32 examples. Table 4 presents the Hyper-parameter values used for the baseline models. We trained all models for 4 epochs as we observed performance degradation starting from epoch 5.

Table 4. Details of baselines hyper-parameter tuning.

Parameter	QANet ‡	DrQA ‡
Optimizer	Adam	Adamax
Learning rate	1e-3	2e-3
Hidden size	96	128
Batch	32	32
Dropout	0.1	0.3
Word Embd	FastText	FastText
Epochs	4	4

---

[4] https://github.com/andy840314/QANet-pytorch-.
[5] https://github.com/explosion/spaCy.

5   Results and Discussion

In this section, we present the experimental outcomes of the proposed model's training and assessment in contrast to two baseline models, DrQA and QaNET. The testing and development EM and F1 scores of all the models under consideration are reported in Table 5. The results show that our model significantly and consistently outperformed both baselines in terms of EM and F1 scores. This demonstrates the advantages of combining question-aware and question-independent representations for context and also highlights the importance of directly utilizing the question representation as a direct input to the model.

Table 5. Testing and development EM and F1 score of DrQA, QANet and QAw-MRC.

Method	Development set		Testing set	
	EM	F1	EM	F1
DrQA	29.3%	48.0%	31.6%	52.0%
QANet	18.4%	33.7%	19.1%	35.1%
QAw-MRC (ours)	**33.7%**	**55.4%**	**36.4 %**	**59.6 %**

We further analyzed the performance of the three models by hand-analyzing 100 randomly selected samples from the development set. We investigate the performance based on the answer types, including common nouns, adjectives, verb phrases, date, and other numeric and named entities (person, location, and organization). Our analysis reveals that all models performed well on the date and other numeric classes because these questions have a few plausible candidates, and the answer usually is a single token. On the other hand, all models are less capable of extracting the named entity answers of the types of person, location, and organization. Although these answers mostly consist of a single token, they have many more plausible candidates, making them more challenging to identify. The three models encountered the most difficulty in obtaining answers of the resting types, including common nouns, adjective phrases, and verb phrases.

6   Conclusion

This paper proposes a question-aware neural machine reading comprehension model for The Arabic Language. Our model employs several bidirectional attention units to incorporate the question to encode the context and to achieve various levels of question-centered context understanding. Besides, Our model extracts a representation of the question and utilizes it as a direct input to the output layer. We trained our model and evaluated it using an Arabic MRC dataset against two well-known baselines, DrQA and QaNET. Experimental results show that our model outperforms the baselines by a significant margin with an F1 score of 59.6%. Future work involves adding components to handle the unique characteristics of the Arabic language, such as diacritic restoration, which enables a better understanding of the meaning of words. Besides, we plan to encode the types of named entities in questions and answers to achieve an understanding of questions and answers, thus improving the model performance.

References

1. AraElectra for question answering on Arabic-SQuADv2. https://huggingface.co/ZeyadAhmed/AraElectra-Arabic-SQuADv2-QA. Accessed 13 Apr 2010
2. Abdi, A., Hasan, S., Arshi, M., Shamsuddin, S.M., Idris, N.: A question answering system in hadith using linguistic knowledge. Comput. Speech Lang. **60**, 101023 (2020)
3. Albilali, E., Al-Twairesh, N., Hosny, M.: Constructing Arabic reading comprehension datasets: Arabic WikiReading and KaifLematha. Lang. Resour. Eval., 1–36 (2022)
4. Antoun, W., Baly, F., Hajj, H.: AraBERT: transformer-based model for Arabic language understanding. In: Proceedings of the 4th Workshop on Open-Source Arabic Corpora and Processing Tools, with a Shared Task on Offensive Language Detection, pp. 9–15 (2020)
5. Antoun, W., Baly, F., Hajj, H.: AraELECTRA: pre-training text discriminators for Arabic language understanding. arXiv preprint arXiv:2012.15516 (2020)
6. Artetxe, M., Ruder, S., Yogatama, D.: On the cross-lingual transferability of monolingual representations. In: Proceedings of the 58th Annual Meeting of the Association for Computational Linguistics, pp. 4623–4637 (2020)
7. Atef, A., Mattar, B., Sherif, S., Elrefai, E., Torki, M.: AQAD: 17,000+ Arabic questions for machine comprehension of text. In: 2020 IEEE/ACS 17th International Conference on Computer Systems and Applications (AICCSA), pp. 1–6. IEEE (2020)
8. Bojanowski, P., Grave, E., Joulin, A., Mikolov, T.: Enriching word vectors with subword information. Trans. Assoc. Comput. Linguist. **5**, 135–146 (2017)
9. Chen, D., Fisch, A., Weston, J., Bordes, A.: Reading Wikipedia to answer open-domain questions. In: Association for Computational Linguistics (ACL) (2017)
10. Cho, K., et al.: Learning phrase representations using RNN encoder-decoder for statistical machine translation. arXiv preprint arXiv:1406.1078 (2014)
11. Clark, J.H., et al.: TyDi QA: a benchmark for information-seeking question answering in typologically diverse languages. Trans. Assoc. Comput. Linguist. **8**, 454–470 (2020)
12. Clark, K., Luong, M.T., Le, Q.V., Manning, C.D.: ELECTRA: pre-training text encoders as discriminators rather than generators. arXiv preprint arXiv:2003.10555 (2020)
13. Devlin, J., Chang, M.W., Lee, K., Toutanova, K.: BERT: pre-training of deep bidirectional transformers for language understanding. arXiv preprint arXiv:1810.04805 (2018)
14. Group, N.L.C.: R-Net: machine reading comprehension with self-matching networks, May 2017. https://www.microsoft.com/en-us/research/publication/mcr/
15. Kingma, D.P., Ba, J.: Adam: a method for stochastic optimization. arXiv preprint arXiv:1412.6980 (2014)

16. Lee, K., Salant, S., Kwiatkowski, T., Parikh, A., Das, D., Berant, J.: Learning recurrent span representations for extractive question answering. arXiv preprint arXiv:1611.01436 (2016)

17. Lewis, P., Oguz, B., Rinott, R., Riedel, S., Schwenk, H.: MLQA: evaluating cross-lingual extractive question answering. In: Proceedings of the 58th Annual Meeting of the Association for Computational Linguistics, pp. 7315–7330 (2020)

18. Liu, S., Zhang, X., Zhang, S., Wang, H., Zhang, W.: Neural machine reading comprehension: methods and trends. Appl. Sci. **9**(18), 3698 (2019)

19. Liu, Y., et al.: RoBERTa: a robustly optimized BERT pretraining approach. arXiv preprint arXiv:1907.11692 (2019)

20. Malhas, R., Elsayed, T.: Arabic machine reading comprehension on the holy Qur'an using CL-AraBERT. Inf. Process. Manag. **59**(6), 103068 (2022)

21. Mozannar, H., Hajal, K.E., Maamary, E., Hajj, H.: Neural Arabic question answering. arXiv preprint arXiv:1906.05394 (2019)

22. Nguyen, T., et al.: MS MARCO: a human generated machine reading comprehension dataset. In: CoCo@ NIPS (2016)

23. O'Reilly, T., Feng, D.G., Sabatini, D.J., Wang, D.Z., Gorin, D.J.: How do people read the passages during a reading comprehension test? The effect of reading purpose on text processing behavior. Educ. Assess. **23**(4), 277–295 (2018)

24. Pandya, H.A., Ardeshna, B., Bhatt, D., Brijesh, S.: Cascading adaptors to leverage English data to improve performance of question answering for low-resource languages. arXiv preprint arXiv:2112.09866 (2021)

25. Raffel, C., et al.: Exploring the limits of transfer learning with a unified text-to-text transformer. J. Mach. Learn. Res. **21**(1), 5485–5551 (2020)

26. Rajpurkar, P., Zhang, J., Lopyrev, K., Liang, P.: SQuAD: 100,000+ questions for machine comprehension of text. arXiv preprint arXiv:1606.05250 (2016)

27. Seo, M., Kembhavi, A., Farhadi, A., Hajishirzi, H.: Bidirectional attention flow for machine comprehension. arXiv preprint arXiv:1611.01603 (2016)

28. Shawar, A., Atwell, E.: An Arabic chatbot giving answers from the Qur'an. In: Proceedings of TALN04: XI Conference sur le Traitement Automatique des Langues Naturelles, vol. 2, pp. 197–202. ATALA (2004)

29. Xue, L., et al.: mT5: a massively multilingual pre-trained text-to-text transformer. arXiv preprint arXiv:2010.11934 (2020)

30. Xue, L., et al.: mT5: a massively multilingual pre-trained text-to-text transformer. In: Proceedings of the 2021 Conference of the North American Chapter of the Association for Computational Linguistics: Human Language Technologies, pp. 483–498 (2021)

31. Yousef, W.A., et al.: Poem comprehensive dataset (PCD) (2018). https://hci-lab.github.io/ArabicPoetry-1-Private/PCD

32. Yu, A.W., et al.: QANet: combining local convolution with global self-attention for reading comprehension. arXiv preprint arXiv:1804.09541 (2018)

33. Zeng, C., Li, S., Li, Q., Hu, J., Hu, J.: A survey on machine reading comprehension-tasks, evaluation metrics and benchmark datasets. Appl. Sci. **10**(21), 7640 (2020)

A Hybrid Deep Learning Scheme
for Intrusion Detection in the Internet
of Things

Asadullah Momand[1], Sana Ullah Jan[2]([⊠]), and Naeem Ramzan[1]

[1] School of computing Engineering and Physical Sciences, University of the West of
Scotland, Paisley PA1 2BE, UK
[2] School of Computing Edinburgh Napier University, Edinburgh EH10 5DT, UK
`S.jan@napier.ac.uk`

Abstract. The Internet of Things (IoT) is the connection of smart
devices and objects to the internet, allowing them to share and analyze
data, communicate with each other, and be controlled remotely. Several IoT devices are designed to collect, process, and store confidential
data in order to perform their intended function. This information can
be sensitive such as location, health, military, financial information, and
biometric data. The efficient implementation of IoT networks has become
increasingly reliant on security. In IoT networks, several researchers used
intrusion detection systems (IDS) for the identification of cyberattacks
where machine learning (ML) and deep learning (DL) are significant
components. The existing IDS still needs improvements for the detection of multiclass detection to identify each category of attack separately.
To improve the detection performance of IDS, this study proposes a
hybrid scheme of convolutional neural networks (CNN) and gated recurrent units (GRU). The proposed hybrid scheme integrates two CNN layers and three GRU layers. The proposed scheme was assessed using the
IoTID20 dataset.

Keywords: Convolutional Neural Networks · Gated Recurrent Units ·
Internet of Things · Intrusion Detection

1 Introduction

The Internet of Things (IoT) is a network of physical objects, devices, and sensors that are connected to each other and to the internet, allowing them to
share data and communicate with each other. These objects are capable of collecting, storing, and transmitting information, creating a vast amount of data
that can be analyzed and utilized in a wide range of applications. The IoT is
an expanding area that has immense potential to revolutionize our lives and
the way we work [1,2]. As more devices become connected to the internet, they
are able to communicate and share data in real-time, creating new opportunities for automation, optimization, and innovation. The increasing use of IoT

© The Author(s), under exclusive license to Springer Nature Switzerland AG 2024
A. Bennour et al. (Eds.): ISPR 2023, CCIS 1941, pp. 277–287, 2024.
https://doi.org/10.1007/978-3-031-46338-9_21

technologies across a wide range of industries is driving growth and creating new possibilities for businesses and individuals to benefit from this transformational technology. Figure 1 demonstrates that the projected number of devices within the IoT is 75.44 billion in 2025. IoT allows the monitoring and control of machines remotely, resulting in increased efficiency, productivity, and safety [3]. IoT has many potential applications across a wide range of industries including healthcare, manufacturing, agriculture, energy management, smart cities, and militaries [4,5].

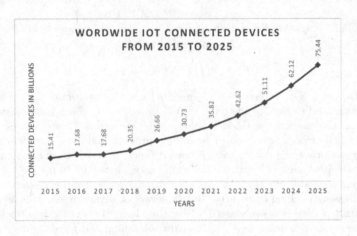

Fig. 1. The expected increase in IoT till 2025 [6]

Many IoT devices are designed to collect, process, and store confidential data in order to perform their intended function [7]. This information can be sensitive such as location, health, military, financial information, and biometric data. The efficient implementation of IoT networks has become increasingly reliant on security [8–10]. In IoT networks, several researchers used IDS for the identification of cyberattacks where ML and DL are significant components [11–13]. DL is a subfield of ML that involves the use of artificial neural networks with many layers to model and solve complex problems. Moreover, DL techniques enable systematic training of nonlinear models on large datasets [14]. This is why DL is a powerful tool for detecting network intrusions because of its ability to learn from large amounts of data and to generalize new types of attacks [15,16]. The existing IDS still needs improvements for the detection of multiclass detection to identify each category of attack separately. Moreover, the existing methods are facing a dearth of efficient and effective intrusion detection mechanisms for IoT networks, particularly in regard to their scalability and capability to adapt to dynamic network conditions. To improve the detection performance of IDS, this study proposes a hybrid scheme of CNN and GRU. CNN is used to highlight the salient features in the input and GRU is used for the detection of intrusions using these salient features [17]. The proposed scheme was assessed using the

IoTID20 dataset. Furthermore, the results were compared with advanced DL and ML models to prove the effectiveness of the proposed architecture. Outlined below are the key contributions of this study:

- In this study, a hybrid architecture of CNN-GRU is designed to improve the performance of existing IDS.
- Evaluated the performance of the proposed approach on both binary and multiclass attacks identification of the IoTID20 dataset.
- A comparison was made between the performance of both the novel approach and the ML and DL models to prove their effectiveness.

The structure of the remaining paper is outlined as follows: Sect. 2 provides a summary of the related literature, while Sect. 3 presents the proposed model and the steps involved in its implementation. In Sect. 4, the outcomes are discussed. Finally, Sect. 5 offers a summary of the paper.

2   Related Works

Several ML/DL-based frameworks for the detection of network intrusions have been proposed and examined. The most recent and useful articles are mentioned below.

Dina et al. [18] proposed a new loss function called focal loss to overcome the imbalance problems of datasets for intrusion detection in IoT. They implement the proposed function with two DL models, including a feed-forward neural network (FFNN) and a CNN. The authors used three datasets for the evaluation of the proposed system. Elaziz et al. [19] combined swarm intelligence models with deep neural networks (DNN) and designed a new model for intrusion detection in IoT cloud systems. Capuchin search was used to select the most salient features. They evaluated the presented model with NSL-KDD, BoT-IoT, KDD99, and CIC2017 datasets. Ullah et al. [20] presented a deep convolutional neural network (DCNN) to identify cyberattacks in IoT. Authors considered binary, multiclass, and subcategories of multiclass attacks in the networks. They used the IoTID20 dataset to evaluate the proposed DCNN model.

Awajan [21] proposed a four-layer fully connected DL-based model for finding malicious attacks in IoT. They explored a dataset from real-time IoT network flows and evaluated the proposed model with this dataset. Ullah et al. [22] designed a hybrid DL scheme to find cyber attacks in the Internet of Vehicles (IoV). The hybrid scheme consists of long short-term memory (LSTM) and GRU for the detection of intrusions in both inter and intra-vehicular networks. They utilized two public datasets (combined DDoS and car-hacking) for the evaluation of the proposed model. Khan et al. [23] presented a deep neural network (DNN) for the identification of intrusions in MQTT-enabled IoT networks. Experiments were carried out to explore three distinct scenarios including uniflow, biflow, and packet flow. The MQTT-IoT-IDS2020 dataset was used for the evaluation of the model. A detailed analysis of the related literature is provided in Table 1, where the accuracy, precision, recall, and Matthew's Correlation Coefficient scores are abbreviated as ACC, PR, RE, and MCC respectively.

3 Proposed Methodology

This section will provide a detailed description of the dataset, preprocessing steps, and proposed CNN-GRU architecture.

3.1 Dataset

The proposed scheme was assessed using the IoTID20 dataset. This dataset was generated from real-time IoT network cyberattacks. The different attacks were performed on that network and collected data accordingly. This dataset contains 585,342 instances of attacked data and 40,073 normal data.

Table 1. A comparative analysis of the literature on intrusion detection in IoT.

Author	Year	Technique	Dataset	Evaluation Metrics	Results in %age
Dina et al.	2023	FFNN, CNN	Bot-IoT, WUSTL-IIoT-2021, WUSTL-EHMS-2020	ACC, PR, F1-score, MCC	24, 39, 39, 60
Elaziz et al.	2023	CNN-CapSA	NSL-KDD, BoT-IoT, KDD99, and CIC2017	ACC, PR, RE, F1-score	99.917, 91.609, 92.044, 90.093
Ullah et al.	2022	DCNN	IoTID20	ACC, PR, RE, F1-score	98.12, 97.13, 97.83, 97.46
Awajan	2023	DNN	Network Generated	ACC, PR, RE, F1-score	93.74, 93.71, 93.82, 93.47
Ullah et al.	2022	LSTM-GRU	Combined DDoS, Car-Hacking	ACC, PR, RE, F1-score	99.51, 99.51, 99.60, 99.52
Khan et al.	2022	DNN	MQTT-IoT-IDS2020	ACC, PR, RE, F1-score	97.084, 94.761, 86.435, 90.605
Proposed study	2023	CNN-GRU	IoTID20	ACC, PR, RE, F1-score	98.41, 97.53, 97.99, 97.75

3.2 Preprocessing Techniques

Preprocessing is a necessary step in ML/DL approaches. It transforms data into a suitable form for ML/DL algorithms to process. Some values are missing in the utilized dataset. We excluded all incomplete data from the dataset to clean it up. The label encoder method has been employed to convert categorical attributes to numerical. Because the categories in the utilized dataset have an inherent order, and the label encoder assigns a unique integer value to each category in the data based on the order of appearance. On the other hand, one-hot encoding creates a binary vector for each category, which requires more memory and processing power. The dataset used has 83 features out of which some features are insignificant that may reduce the performance of the model. To select the most salient attributes, we used an extra tree classifier method to rank the features according to their impact on the output class. Attributes with a value greater than 0.001 were selected for the experiment. In this experiment, 62 features were chosen, and the rest of the features were discarded.

The values of the utilized dataset are in distinct ranges, for instance, some attribute values are between 0–100 and some are between 0–1000 which affects the training of the model. To address this issue, We utilized the approach of min-max normalization to standardize the data within the range of 0 to 1. Divided the normalized data 80% and 20% into the train and test sets, respectively, to evaluate the proposed scheme.

3.3   CNN-GRU Architecture

The proposed model consists of two DL algorithms including CNN and GRU, shown in Fig. 2. The architecture of the proposed scheme includes two pairs of convolutional layers and corresponding max-pooling layers, followed by three GRU layers. The CNN layers can extract spatial features from the input data, while the GRU layers can capture temporal dependencies, making the model more effective in capturing complex patterns in the data. A group of filters is associated with the convolutional layer (also known as kernels) applied to the input data. Each filter performs a mathematical operation on the input, extracting features. The pooling layer reduces the size of the feature maps produced by the convolutional layer [24]. This helps to simplify the data and make it more manageable for subsequent layers. Moreover, the reduction of features can ultimately decrease the model's response time.

GRU is a type of Recurrent Neural Network (RNN) used to retain or forget information from previous time steps [25]. This is achieved through the use of two gating mechanisms: the update gate and the reset gate. The update gate is responsible for deciding which prior hidden state is relevant and should be carried forward, whereas the reset gate governs the extent to which the previous hidden state is irrelevant and needs to be disregarded [26]. The GRU can selectively remember or forget information by using these gates, allowing it to model long-term dependencies in sequential data more effectively than traditional RNNs [27].

The proposed hybrid architecture contains two pairs of convolutional and max-pooling layers in series. The convolutional layers in this model employ 62 and 31 filters respectively, with a kernel size of 3. The max-pooling layers use a pooling size of 4 and 2, corresponding to the two convolutional layers. The GRU layer receives the output of the max-pooling layer as its input. Three layers of GRU are used to predict cyber attacks and maintain long-term records. The GRU layers in this model have 62 and 31 features in each time step for the first and second layers, respectively. The final GRU layer generates output for the given task.

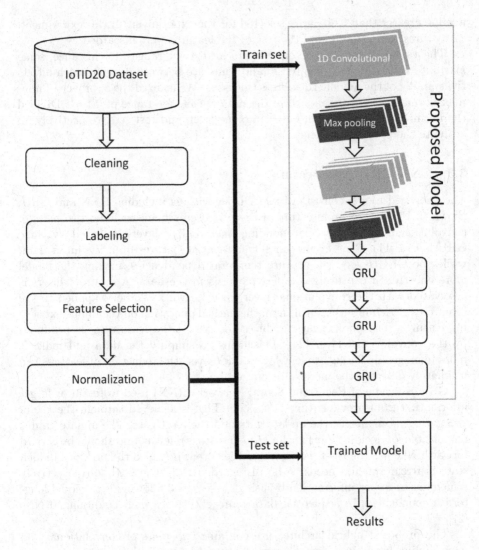

Fig. 2. The proposed CNN-GRU architecture

4   Examination of the Results

This section provides a brief discussion of evaluation metrics, performance evaluation of the proposed model, and comparisons with other ML and DL methods.

4.1   Evaluation Metrics

There are several evaluation metrics that are used to evaluate the effectiveness of the model. In this experiment, the utilized evaluation metrics are accuracy

(ACC), precision (PR), recall (RE), and F1-score. ACC is the proportion of accurately detected instances out of the total number of instances. PR is the proportion of true positive predictions out of all positive predictions. It measures the model's ability to avoid false positives. RE is the number of times that a prediction turned out to be right out of all the times that it was right. It measures the model's ability to detect all positive instances. The F1-score is the harmonic mean of precision and recall, which provides a single score that balances the trade-off between precision and recall.

4.2 Outcomes of the Proposed CNN-GRU

Experiments were performed for one and two CNN layers with different GRU layers to determine the optimal solution for attack classification. We conducted experiments on both binary class classification and multiclass classification to comprehensively evaluate the proposed model's performance across different scenarios. The comparison of the proposed approach in different layers is shown in Table 2 and Table 3 on binary and multi-class classification, respectively. The evaluation results show that the proposed approach performs well with two CNN and three GRU layers. The max pooling layers in CNN reduce the dimensionality of the convolutional layers' feature maps while preserving important features. This helps to avoid overfitting and improves the generalization ability of the model while the three-layer GRU model is capable of capturing long-term dependencies in the network data while avoiding the vanishing gradient problem. An adam optimizer was used with a cross-entropy loss function. The proposed model was trained with 50 epochs for both binary and multiclass classifications. The 5-fold cross-validation technique is used to validate the results of the model.

Table 2. A layer-wise performance comparison of CNN and GRU for binary classification.

Layers		Evaluation results			
CNN	GRU	ACC	PR	RE	F1-Score
1	1	0.9664	0.8940	0.8036	0.8420
2	1	0.9876	0.9813	0.9789	0.9798
2	2	0.9977	0.9915	0.9893	0.9904
2	**3**	**0.9989**	**0.9972**	**0.9915**	**0.9941**
2	4	0.9961	0.9899	0.9878	0.9892

Table 3. A layer-wise performance comparison of CNN and GRU for multiclass classification.

Layers		Results			
CNN	GRU	ACC	PR	RE	F1-Score
1	1	0.9454	0.9199	0.9288	0.9228
2	1	0.9781	0.9745	0.9645	0.9691
2	2	0.9834	0.9756	0.9764	0.9758
2	**3**	**0.9841**	**0.9753**	**0.9799**	**0.9775**
2	4	0.9783	0.9686	0.9724	0.9702

4.3 Comparisons with Other ML and DL Approaches

To assess the effectiveness of the proposed CNN-GRU, its results were compared to that of other Ml and DL models. Naive Bayes (NB), logistic regression (LR), and support vector machine (SVM) are common ML approaches. DNN, autoencoder (AE), deep belief network (DBN), GRU, and LSTM, are advanced DL algorithms. All of these approaches were tested within the same environment as the proposed method. All hyperparameters were the same for the DL methods. Table 4, Table 5 and Table 6 show a detailed comparison of the proposed approach with other methods. Analysis of the results shows that the proposed approach outperforms other methods. The proposed CNN-GRU achieved optimal results because the CNN is capable of recognizing spatial attributes in input data by using filters to convolve over the input. On the other hand, GRU is useful for processing sequential data by selectively updating and resetting the model's internal state using gating mechanisms.

Table 4. Results comparisons with other ML and DL methods on binary.

Models	ACC	PR	RE	F1-Score
NB	0.6521	0.5778	0.8103	0.6749
LR	0.9653	0.9028	0.7872	0.8339
SVM	0.9739	0.9194	0.8547	0.8839
DNN	0.9976	0.9978	0.9857	0.9917
AE	0.9969	0.9889	0.9881	0.9886
DBN	0.9964	0.9932	0.9801	0.9865
GRU	0.9954	0.9851	0.9802	0.9827
LSTM	0.9947	0.9938	0.9657	0.9792
Proposed CNN-GRU	**0.9989**	**0.9972**	**0.9915**	**0.9941**

Table 5. Results comparisons with other ML and DL methods on multiclass.

Model	ACC	PR	RE	F1-Score
NB	0.6778	0.6634	0.7387	0.6485
LR	0.8319	0.7733	0.7302	0.7316
SVM	0.8563	0.8451	0.7851	0.7889
DNN	0.9551	0.9345	0.9452	0.9372
AE	0.9649	0.9521	0.9446	0.9462
DBN	0.9594	0.9435	0.9554	0.9474
LSTM	0.9588	0.9546	0.9204	0.9358
GRU	0.9685	0.9581	0.9473	0.9524
Proposed CNN-GRU	**0.9841**	**0.9753**	**0.9799**	**0.9775**

Table 6. Average results comparisons with related papers.

Papers	ACC	PR	F1-Score
[28]	0.9050	0.8535	0.9035
[29]	0.8700	0.8700	0.8700
[30]	0.9868	–	–
[31]	0.9800	0.9840	0.9880
Proposed CNN-GRU	**0.9915**	**0.9862**	**0.9858**

5   Conclusion

This study proposed a hybrid scheme of CNN and GRU to improve the performance of existing systems. The proposed hybrid scheme integrates two CNN layers and three GRU layers. Experiments were performed on IoTID20 dataset for binary and multi-class classification. The results were compared with advanced DL and ML models to prove the effectiveness of the proposed scheme. With regards to binary and multi-class classification, the proposed approach demonstrated impressive accuracies of 99.89% and 98.41%, respectively, which is optimal than other models.

References

1. Asgharzadeh, H., Ghaffari, A., Masdari, M., Gharehchopogh, F.S.: Anomaly-based intrusion detection system in the internet of things using a convolutional neural network and multi-objective enhanced capuchin search algorithm. J. Parallel Distri. Comput. **175**, 1–21 (2023)
2. Anushiya, R., Lavanya, V.S.: A new deep-learning with swarm based feature selection for intelligent intrusion detection for the internet of things. Measur.: Sens. **26**, 100700 (2023)

3. Kandavalli, S.R., et al.: Application of sophisticated sensors to advance the monitoring of machining processes: analysis and holistic review. Int. J. Adv. Manuf. Technol. **125**, 1–26 (2023)
4. Rashid, M.M., et al.: A federated learning-based approach for improving intrusion detection in industrial internet of things networks. Network **3**(1), 158–179 (2023)
5. Farooq, O., Singh, P., Hedabou, M., Boulila, W., Benjdira, B.: Machine learning analytic-based two-staged data management framework for internet of things. Sensors **23**(5), 2427 (2023)
6. Al-Bahri, M., Yankovsky, A., Borodin, A., Kirichek, R.: Testbed for identify IoT-devices based on digital object architecture. In: Galinina, O., Andreev, S., Balandin, S., Koucheryavy, Y. (eds.) NEW2AN/ruSMART -2018. LNCS, vol. 11118, pp. 129–137. Springer, Cham (2018). https://doi.org/10.1007/978-3-030-01168-0_12
7. Zhang, Y., Li, P., Wang, X.: Intrusion detection for IoT based on improved genetic algorithm and deep belief network. IEEE Access **7**, 31711–31722 (2019)
8. Imane Laassar and Moulay Youssef Hadi: Intrusion detection systems for internet of thing based big data: a review. Int. J. Reconfigurable Embedded Syst. **12**(1), 87–96 (2023)
9. Ahmad, J., Shah, S.A., Latif, S., Ahmed, F., Zou, Z., Pitropakis, N.: DraNN_PSO: a deep random neural network with particle swarm optimization for intrusion detection in the industrial internet of things. J. King Saud Univ.-Comput. Inf. Sci. **34**(10), 8112–8121 (2022)
10. Wu, J., Dai, H., Wang, Y., Ye, K., Xu, C.: Heterogeneous domain adaptation for IoT intrusion detection: a geometric graph alignment approach. IEEE Internet Things J. (2023)
11. Gao, Z.J., Pansare, N., Jermaine, C.: Declarative parameterizations of user-defined functions for large-scale machine learning and optimization. IEEE Trans. Knowl. Data Eng. **31**(11), 2079–2092 (2018)
12. Khan, M.A., et al.: Voting classifier-based intrusion detection for IoT networks. In: Saeed, F., Al-Hadhrami, T., Mohammed, E., Al-Sarem, M. (eds.) Advances on Smart and Soft Computing. AISC, vol. 1399, pp. 313–328. Springer, Singapore (2022). https://doi.org/10.1007/978-981-16-5559-3_26
13. Vitorino, J., Praca, I., Maia, E.: Towards adversarial realism and robust learning for IoT intrusion detection and classification. Ann. Telecommun., 1–12 (2023)
14. Goodfellow, I., Bengio, Y., Courville, A.: Deep Learning, p. 800. The MIT Press, Cambridge (2016). ISBN: 0262035618. Genet. Program. Evolvable Mach. **19**(1-2), 305-307 (2018)
15. Mighan, S.N., Kahani, M.: A novel scalable intrusion detection system based on deep learning. Int. J. Inf. Secur. **20**, 387–403 (2021)
16. Khanday, S.A., Fatima, H., Rakesh, N.: Implementation of intrusion detection model for DDoS attacks in lightweight IoT networks. Expert Syst. Appl. **215**, 119330 (2023)
17. Al-Turaiki, I., Altwaijry, N.: A convolutional neural network for improved anomaly-based network intrusion detection. Big Data **9**(3), 233–252 (2021)
18. Dina, A.S., Siddique, A.B., Manivannan, D.: A deep learning approach for intrusion detection in internet of things using focal loss function. Internet Things **22**, 100699 (2023)
19. Abd Elaziz, M., Al-qaness, M.A., Dahou, A., Ibrahim, R.A., Abd El-Latif, A.A.: Intrusion detection approach for cloud and IoT environments using deep learning and capuchin search algorithm. Adv. Eng. Softw. **176**, 103402 (2023)

20. Ullah, S., et al.: A new intrusion detection system for the internet of things via deep convolutional neural network and feature engineering. Sensors **22**(10), 3607 (2022)
21. Awajan, A.: A novel deep learning-based intrusion detection system for IoT networks. Computers **12**(2), 34 (2023)
22. Ullah, S., et al.: HDL-IDS: a hybrid deep learning architecture for intrusion detection in the internet of vehicles. Sensors **22**(4), 1340 (2022)
23. Khan, M.A., et al.: A deep learning-based intrusion detection system for MQTT enabled IoT. Sensors **21**(21), 7016 (2021)
24. Riyaz, B., Ganapathy, S.: A deep learning approach for effective intrusion detection in wireless networks using CNN. Soft. Comput. **24**, 17265–17278 (2020)
25. Henry, A., Gautam, S.: Intelligent intrusion detection system using deep learning technique. In: Panda, S.K., Rout, R.R., Sadam, R.C., Rayanoothala, B.V.S., Li, KC., Buyya, R. (eds.) Computing, Communication and Learning. Communications in Computer and Information Science, vol. 1729, pp. 220–230. Springer, Cham (2023). https://doi.org/10.1007/978-3-031-21750-0_19
26. Henry, A., et al.: Composition of hybrid deep learning model and feature optimization for intrusion detection system. Sensors **23**(2), 890 (2023)
27. Wang, Z., Xie, X., Chen, L., Song, S., Wang, Z.: Intrusion detection and network information security based on deep learning algorithm in urban rail transit management system. IEEE Trans. Intell. Transp. Syst. **24**, 2135–2143 (2023)
28. Sarwar, A., Hasan, S., Khan, W.U., Ahmed, S., Marwat, S.N.K.: Design of an advance intrusion detection system for IoT networks. In 2022 2nd International Conference on Artificial Intelligence (ICAI), pp. 46–51. IEEE (2022)
29. Ullah, I., Mahmoud, Q.H.: A scheme for generating a dataset for anomalous activity detection in IoT networks. In: Goutte, C., Zhu, X. (eds.) Canadian AI 2020. LNCS (LNAI), vol. 12109, pp. 508–520. Springer, Cham (2020). https://doi.org/10.1007/978-3-030-47358-7_52
30. Jeyanthi, D.V., Indrani, B.: IoT based intrusion detection system for healthcare using RNNBILSTM deep learning strategy with custom features (2022)
31. Alkahtani, H., Aldhyani, T.H.H.: Intrusion detection system to advance internet of things infrastructure-based deep learning algorithms. Complexity **1–18**, 2021 (2021)

Identifying Discourse Markers in French Spoken Corpora: Using Machine Learning and Rule-Based Approaches

Abdelhalim Hafedh Dahou(✉) ⓘ

GESIS - Leibniz-Institute for the Social Sciences, Cologne, Germany
abdelhalim.dahou@gesis.org

Abstract. The objective of this work is to study the identification of French discourse markers (DM), in particular the polyfunctional occurrences such as '*attetion*', *bon, quoi, la preuve*. A number of words identified as DM, and traditionally considered as adverbs or interjections, are also, for instance, adjectives or nouns. For example *bon* can be a DM or an adjective, '*attetion*' can be a DM or a noun, etc. Hand annotation is in general robust but time consuming. The main difficulty with automatic identification is to take the context of the DM candidate correctly into account. To do that, a mechanisms based on rule-based and machine learning approaches was built, in order to reach an acceptable level of performance and reduce the expert effort. This study will provide a comprehensive use case of a machine learning algorithm, which has proved a good efficiency in dealing with such linguistic phenomena. In addition, an evaluation was done for the *Unitex* platform in order to determine the efficiency and drawbacks of this platform when dealing with such type of tasks.

Keywords: Discourse markers · DM · French corpora · KNN · Transfert learning · Machine Learning

1 Introduction

As known in the literature, discourse consists of collocated, ordered, and coherent clusters of sentences, instead of single and unrelated sentences. Discourse is coherent if among other things there are significant coherence relations between the utterances. For example, to justify the connection between utterances, some type of explanation could be required. One of the lexical phenomena that ensures coherence in discourse is the presence of a discourse marker (DM).

DM are linguistic expressions that have been proven to be effective for: (i) segmenting discourse into meaningful units and (ii) recognizing relationships between these units. Determining the meanings of discourse markers is frequently necessary for understanding the message (Petukhova and Bunt, 2009); (Heeman, Byron, and Allen, 1998); (Han et al., 2021).

A. Bennour et al. (Eds.): ISPR 2023, CCIS 1941, pp. 288–299, 2024.
https://doi.org/10.1007/978-3-031-46338-9_22

Some DM, called connectors (DMC), can express semantic relations between discourse segments. For example, *because* expresses cause/explanation, *but* expresses contrast, etc. Other DM, called particles (DMP), can express speakers' attitudes, their emotional or belief state, and play a role in the management of interaction, etc. The best known examples are emotional interjections like *oh* or *ah*, but there are many other attitudinal DM, like *bon, tu parles, quoi, hein, tu vois, écoute*, etc. in French, whose meaning is more difficult to describe.

The most difficult aspect of DM identification lies in the fact that the category of markers is fairly clearly a functional-pragmatic, and not a formal, morpho-syntactic one (Lamiroy and Swiggers, 1991); (Schiffrin, 1987). DM can come from a variety of distributional classes, and they frequently have formally identical counterparts that are not used as markers but do contribute to propositional content (whereas markers don't) (Fraser, 1999), as shown in the following examples, in (1) *bon* acts as a NON DM (adjective) and in (2) acts as a DM.

Nous avons passé/sommes restés un bon moment chez nos voisins.

We stayed with a good/long time at our neighbors.

- A. je vais te faire un super cadeau pour ta fête.
- B. bon j'ai hâte de voir ça
- A. I'll give you a great gift for your party.
- B. Well, I cannot wait to see that

Based on the few previous works, the problem of discriminating functions for particles is not new (Zufferey and Popescu-Belis, 2004) but has remained marginal, possibly because it is much less crucial for written language. Actually, in recent literature, the discourse markers which are used as pivots to predict text continuations (Wu et al., 2020) or learn discourse relations (Nie, Bennett, and Goodman, 2019); (Passonneau and Litman, 1997); (Zufferey, Degand, et al., 2012) are 'standard' DM like *because* or *but*, and not particles. A more liberal inventory is used by (Sileo et al., 2020) but is not intended to address the ambiguity problem. To our best knowledge, there is no systematic attempt to tackle this issue in recent work and based on that, we present this work. Our contribution to tackle the mentioned issue. The main contribution of this work is to study the identification of French discourse markers (DM) in spoken corpora with a focus on the polyfunctional occurrences such as '*attetion*', *bon, quoi, la preuve* and to build a mechanism based on rule-based and machine learning approaches that can identify DMs with an acceptable level of performance and reduce the expert effort. The study provides a comprehensive use case of a machine learning algorithm that can efficiently deal with such linguistic phenomena. Additionally, in this work, we evaluated the efficiency and drawbacks of the *Unitex* platform when dealing with such types of tasks.

This paper can be outlined as follows: A detailed methodology section will describe the processing pipeline of each mechanism. This will be followed by an experiment and results section, which will present the datasets used in this study, the model's configuration, and the results of each scenario. Prior to the conclusion, there will be a discussion section that will interpret the results and highlight the limitations of each mechanism. Finally, the paper will conclude by summarizing the work and providing a perspective on future research ideas.

2   Methodology

As mentioned in the above, the goal of this study is to identify polyfunctional DMs in French spoken corpora. In the literature, the identification process of those polyfunctional DMs is not dependent just on the DM expression. According to (Lamiroy and Swiggers, 1991), "the category of markers is fairly clearly a functional-pragmatic, and not a formal, morpho-syntactic one", which means that we need to take the context or the environment of those DMs in consideration to identify their function.

In this section, we describe the processing pipeline of each mechanism in details. Each mechanism will borrow some features from one of the two approaches: rule-based and machine learning. Investigating this topic from multiple perspectives yields more reliable, convincing results, as well as better research perspectives.

2.1   *Unitex* with Internal and External Ressources

This method is divided into two scenarios: (i) the first scenario uses *Unitex*'s[1] internal resources without any external intervention; (ii) the second scenario uses external resources, by passing *Unitex*'s resources such as dictionaries and linearized tagger, to estimate the effect of *Unitex*'s resources and explore their drawbacks.

Unitex with Internal Resources. This scenario used just *Unitex*'s resources (dictionary and linearized tagger) and local grammars (graphs) that describe syntactic rules sensitive to the left and right positive context for each DM. Figure 1 illustrate the workflow of this scenario.

Firstly, we read the text corpus by the *Unitex* platform and preprocess it. The preprocessing operations are: normalization of separators, normalization of non-ambiguous forms, splitting into sentences and tokens and applying dictionaries.

Secondly, construct the text automaton which is an effective and visual way of representing such ambiguities. Due to the many paths of tags because of the lexical ambiguity in the dictionaries, we apply a linearization process based on path scoring in order to select a single path and remove the others.

Thirdly, build a graph that represents our linguistic phenomena by using local grammars, in the form of visual graphs as shown in Fig. 2. In this method, we used the positive context where the polyfunctional DM acts as a DM and not as a descriptive content part such as noun, adjective or pronoun. We ran a series of trials to cover as many environments as possible for each DM. Analyzing each DM's behavior in the spoken corpus, as well as the various contexts that can arise in it, allows us to better define the positive context of each DM.

---

[1] https://Unitexgramlab.org/fr.

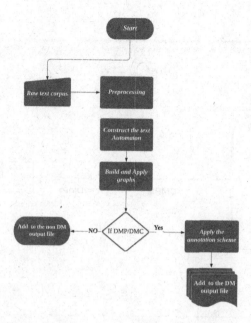

Fig. 1. *Unitex* with internal resources pipeline.

Finally, we compile the graph and apply it to search for all the sequences that match the pattern defined in the graph. For that, we asked *Unitex* (1) to consider all the matching sequences, (2) to base its exploration on the text automaton (high precision), (3) to keep the *DMP* tags in the result, and (4) to store the matching occurrences with a left and right context of 100 characters. *Unitex* can then produce an appropriate concordance table in an .html file.

***Unitex* with External Resources.** In this scenario we avoid using *Unitex*'s resources (dictionaries and linearized tagger) and use just an external resource to extract the syntactic information by the help of the *CamemBERT* model, as illustrated in Fig. 3. The rest is the same as mentioned in the first scenario. Seems the *CamemBERT* model generate different syntactic tags such as *CC* for coordinating conjunction, *ET* for foreign word, *NC* for common noun and *VINF* for infinitive verb, we need just to change the *Unitex*'s tags with *CamemBERT's* tags.

2.2 Syntactic and Lexical Patterns

In this section, we explain the second rule-based method which has less expert intervention compared to the previous rule-based method.

The idea here is to use the syntactic structure of the left and right context of the token in a negative way with lexical forms. This idea will impose a huge knowledge base to support different environments. Since the items under study are to be found in spoken language, our negative comparison point should be

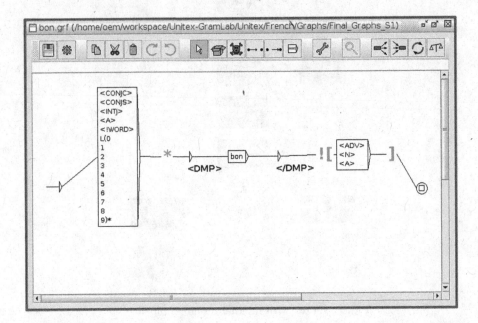

Fig. 2. Graph of the marker *bon* where the left/right boxe represents the left/right context and the exclamation mark means the negation of the box. (Codes: N = Noun, A = Adjective, V = Verb, ADV = Adverb, CONJC = Coordinating Conjunction, CONJS = Subordinating Conjunction, DET = Determiner, INTJ = Interjection, !WORD = non-(lexical word), L0, L1, etc. = speaker's number at the beginning of a speaking turn)

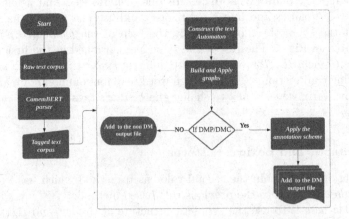

Fig. 3. *Unitex* with internal resources pipeline.

a big set of sentences in written language. If a potential DM occurs in such a set there is a great chance that it is not a real DM. For that, we used the *Le Monde* corpus in order to construct a knowledge base (patterns) due to (1) availability and access; (2) large number of words and segments; (3) compliance with the standard French written language, a fact which can help us to find various negative contexts for the studied DM.

The method consists of creating a list of patterns that holds the negative contexts of the possible DM item in the *Le Monde* corpus and applying this list to the spoken corpus, as shown in the Fig. 4. When the pattern and the DM item in the spoken corpus match, the DM item is identified as a NON DM; otherwise, as a DM.

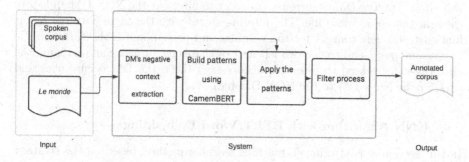

Fig. 4. Syntactic and lexical patterns pipeline process.

The process started with the extraction of all the environments of the 'DM'[2] that exist in the corpus by using the *Unitex* platform. The output is all the occurrences of the DM under study that exist in the *Le monde* corpus in HTML files.

After that, we used the *CamemBERT* in order to generate a tagged text that assigns a syntactic category to each token. Finally, we organize each DM into a pattern that includes three words with their syntactic categories from the left and right contexts as shown in Fig. 5.

Before applying the patterns, firstly, we converted the spoken corpus structure to the same structure of the patterns by assigning a syntactic category to each token. Secondly, an exception list for 'attetion' and 'bon ' is created which contains some multi words such as "alors 'attetion'", "mais bon", "enfin bon", "ah bon" and others. The goal here is to prevent spoken corpus occurrences from being classified incorrectly. Finally, we use the patterns to determine whether the occurrence is a DM or a NON DM in matching mode and add it to its corresponding output file.

---

[2] Let us recall that, given that *Le Monde* is written language, most forms of *bon* or other markers are in fact not the DM but their functional variants, like the adjective *bon*.

CL3	CL2	CL1	WORD	CR1	CR2	CR3	Example
peut , V	pas , ADV	faire , VINF	attention	tous , ADJ	ses , DET	enfants , NC	peut pas faire attention tous ses enfants
On , CLS-SU	doit , V	faire , VINF	attention	une , DET	dramatisation , P	de , P	On doit faire attention une dramatisation de
bon , ADJ	chat , NC	, , PONCT	bon	rat , NC	: , PONCT	plus , ADV	bon chat , bon rat : plus
ma , DET	tante , NC	a , V	bon	coeur , NC	, , PONCT	il , CLS-SUJ	ma tante a bon coeur , il
commencent , P	à , P	administrer ,	la preuve	que , CS	l' , DET	on , CLS	commencent à administrer la preuve que l' on
a , V	affirmé , VPP	avoir , VINF	la preuve	que , CS	cette , DET	opération , NC	a affirmé avoir la preuve que cette opération
tout , ADV	haut , ADJ	en , P	quoi	elle , CLS-SUJ	consiste , V	, , PONCT	tout haut en quoi elle consiste .
seule , ADJ	à , P	dire , VINF	quoi	que , CS	ce , PRO	soit , VS	seule à dire quoi que ce soit

Fig. 5. The pattern's extraction results that used to classify the DM and NON DM.

An additional step was conducted, because we discovered that the patterns in use categorize correctly the NON DM occurrences, but that the potential DM occurrences gave rise to a number of false positives. For that, a filter procedure was applied on the DM occurrences in order to filter out the NON DM and store them in the appropriate file. The filtering starts with the same patterns but is limited to the left context for the nominal and pronominal DMs ('*attetion*', *la preuve* and *quoi*) and to the two left and right tokens for *bon*. Summarizing, the patterns will be used to filter the occurrences (keep the DM occurrences and transfer the NON DM to the NON DM file).

2.3   KNN Algorithm with BERT Word Embedding

In this section, we introduce a machine learning method based on the strategy of K Nearest Neighbor (KNN) by using BERT word embedding to represent the DM under study. By the help of FlauBERT (Le et al., 2019) word embedding vectors, we are capable of capturing the context of the items in a discourse, the semantic and syntactic similarities, the relation with other words, etc. The concept is as follows: using a labeled dataset, one may estimate the class of new data by examining the majority class of nearby data (hence the name of the algorithm).

The creation of the labeled dataset is the initial phase. Due to the good performance attained and evaluated, we used the results supplied by method Sect. 2.2. We chose the most frequent environment to guarantee that we covered a majority of the different types of DMs environments that exist in the spoken corpora. In the end, we built two clusters for each DM, the first called DM and the second called NON DM as shown in Fig. 6.

Second, we fine-tuned the k (number of neighbors) hyper-parameter in this approach to obtain the ideal value in order to produce satisfactory results. Finally, we used the cosine similarity to compute the similarity between vectors in the classification process.

3   Experiments and Results

In this section, we outline the specifics of the spoken corpora utilized in the study, present the findings of each experiment, and summarize these results in conclusion.

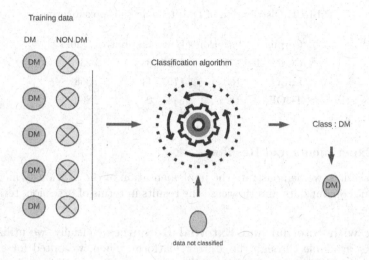

Fig. 6. The process flow of the K-Nearest Neighbor machine learning technique.

3.1   Dataset

In this study, we used 3 French corpora taken from different sources, covering diverse topics and writing styles. *CORPAIX* (Pallaud and Henry, 2004), is a corpus comprising 941624 tokens, 40241 segments, and 33 speakers that was produced in the 1970s with the goal of gathering typical spoken French data and converting it to a uniform format. *ESLO*[3] is a collection of recorded speech. The current version, as well as the one utilized for this study, is a concatenation of *ESLO 1* and *ESLO 2*, resulting in a 700-hour recording with 649081 tokens, 53772 segments, and 28 speakers. *TCOF*[4] (Traitement de Corpus Oraux en Français) is divided into two categories: (1) adult-child interaction recordings (children under the age of seven) and (2) adult interaction recordings. Both are of varying lengths, ranging from 5 to 45 min or more. We used 51 documents from the adult interaction recordings in our study, totaling 149292 tokens and 19527 segments.

After analyzing the data, we used a cleaning and normalization technique to ensure that all corpora were compatible with the parser and that no ambiguities remained. The distribution of each analyzed DM item in the three corpora is represented in the Table 1.

---

[3] http://eslo.huma-num.fr/.
[4] https://www.atilf.fr/ressources/tcof/.

Table 1. Distribution of DM items in the spoken corpus.

Corpus	'attetion'	Bon	La preuve	Quoi
CORPAIS	171	3867	8	2380
ESLO	163	1797	11	958
TCOF	18	545	0	862

3.2   Experiments and Results

In this section, we demonstrate the implementation of the previously mentioned techniques on our data and showcase the results in terms of precision, recall, and F1-score.

***Unitex* with Internal and External Resources.** Initially, we utilized the resources available through the *Unitex* platform. Then, we opted for external resources rather than internal. The experiment was conducted with two primary goals in mind: (1) to assess the efficiency of the method for discourse marker identification; and (2) to evaluate the performance and effectiveness of the resources offered by *Unitex*.

Table 2. Evaluation of *Unitex* with its internal resources on the identification of marker *bon* using the CORPAIX data.

Corpus	Precision	Recall	F1-score
CORPAIX	0.91	0.49	0.63

Table 3. Evaluation of *Unitex* with external resources on the identification of marker *bon* using the CORPAIX data.

Corpus	Precision	Recall	F1-score
CORPAIX	1	0.49	0.65

Tables 2 and 3 present the evaluation results obtained using the CORPAIX data for two different scenarios: one using only *Unitex* resources and the other using a combination of *Unitex* and external resources. The results reveal a significant difference in terms of precision, with the external resources outperforming *Unitex*'s internal resources by a substantial margin. The use of external resources in conjunction with *Unitex* results in a moderate improvement in performance compared to using *Unitex* resources alone in the CORPAIX corpus. On the other hand, the results from the two remaining corpora appear to be similar, with no change observed in the recall measure between the two scenarios.

Syntactic and Lexical Patterns. The experiment for this method is almost identical to the previous one. The final evaluation for each DM in different corpora is reported in Table 4. As we can observe, the identification of '*attetion*' in *CORPAIX* and *TCOF* data is slightly better than the identification in *ESLO*. This could be attributed to the data characteristics. Also, this method works well and is stable in terms of identification of *quoi*, *bon* and *la preuve*. Finally, based on the precision value of '*attetion*' over the three corpora, we can observe that the method's sensitivity over the positive context of '*attetion*' is very low.

Table 4. Results obtained for the categorization of DM in all corpora.

DM	Precision			Recall			F1-score		
CORPUS	COR.	ESLO	TCOF	COR.	ESLO	TCOF	COR.	ESLO	TCOF
'attetion'	0.62	0.31	0.67	0.94	1	0.67	0.74	0.47	0.67
bon	0.98	0.96	0.95	0.92	0.91	0.76	0.94	0.93	0.84
la preuve	0.8	-	-	1	-	-	0.88	-	-
quoi	0.95	0.93	0.94	0.88	0.84	0.85	0.91	0.88	0.89

KNN Algorithm with BERT Word Embedding. In this experiment, the training examples are vectors obtained from the FlauBERT large pre-trained model on French language, each with a class label (DM or non-DM). The training phase of the algorithm consists only in storing the feature vectors and class labels of the training samples (13 vectors for DM context and 13 for non-DM). The parameter k is set to 5.

Table 5. Results obtained for the categorization of DM in all corpora.

DM	Precision			Recall			F1-score		
CORPUS	COR.	ESLO	TCOF	COR.	ESLO	TCOF	COR.	ESLO	TCOF
'attetion'	0.88	0.89	1	0.88	0.88	1	0.88	0.88	0.66
bon	0.99	0.91	0.94	0.93	0.91	0.97	0.96	0.91	0.95
la preuve	-	-	-	-	-	-	-	-	-
quoi	0.65	0.67	0.63	0.96	0.99	0.98	0.78	0.8	0.77

The results of this experiment are reported in Table 5. The KNN algorithm performs better in *CORPAIX* and *ESLO* than in *TCOF* corpus. This is due to the *TCOF* corpus' characteristics and at the same time to the low generality of the training samples. For *la preuve*, we were unable to separate the data for training and testing due to the small amount of samples in the three corpora.

4   Discussion

In this section, we delve into the significance and pertinence of the results obtained from each methodology applied. The section also focuses on outlining and assessing our findings.

The first scenario from the first method, which exploited the *Unitex* platform with its own resources, had some problems in the identification of DM, as observed in Table 2. This difficulty is due, on the one hand, to the internal resources which affected the precision metric and, on the other side, to the low and noisy rules used in the graphs. The problem of precision was solved with the second scenario as illustrated in Table 3 by using the *CamemBERT* parser, which takes into account the word context before determining the final grammatical category. However, the problem of covering all DM still exists, because of the use of positive context. When compared to the first method, the second method performs well, leading us to believe that the negative context can be more helpful than the positive context, and that a larger knowledge base can make a significant difference when using a rule-based approach, because it provides us with more information.

The second technique demonstrates that DM identification cannot be accomplished only through the use of syntactic patterns. Using additional lexical information will assist the algorithm in distinguishing between DM and non-DM as we can see in Table 4.

Based on the outcomes of the KNN algorithm, we observed that using word embedding can identify correctly the function of '*attetion*' in different contexts, owing to the information contained in the embedding vector. As previously stated, word embedding can extract syntactic, lexical, and semantic information from the context, allowing us to determine whether the investigated occurrence is a DM or a descriptive content component.

In the case of the *TCOF* corpus, we may claim that the findings are influenced by the corpus' characteristics and by a lack of data in the training dataset. For the pre-trained model, the test was done for the classification of *bon* and '*attetion*'. Another method from deep learning approach was attempted, which involved fine-tune a pre-trained French language model (FlauBERT). However, due to the limited amount of data, the results were satisfactory in terms of precision but fell short in recall due to the small quantity of training data utilized. Despite the restricted diversity of context and imbalance in the training data, the results from the *bon* item were more favorable compared to those from the '*attetion*' item.

5   Conclusion

The research reported in this study concerns the grammatical functions of some polyfunctional DM such as '*attetion*', *bon, quoi, la preuve* in spoken dialogue. It applies rule-based and machine learning methods in order to detect the category of each occurrence in several contexts.

To conclude, the identification of the polyfunctional occurrences is still a challenge and needs a lot of analysis to determine the better way to deal with each DM because it is obvious that they have different patterns and behavior. The machine learning methods show their ability to detect correctly some DM, on the basis on relatively low resources. However, they need a lot of data to have a significant number of DM and environments in order to cover a lot of situations to adapt to different discourse genres and registers. The work is still continuing and our perspective is to contribute to this task, in particular by extending the analysis to other polyfunctional DM such as *tiens, tu parles, bref* and *bon Dieu*.

References

Fraser, B.: What are discourse markers? J. Pragmatics **31**(7), 931–952 (1999)

Han, H., et al.: Using connectives in implicit discourse relation recognition. In: 2021 IEEE 11th International Conference on Electronics Information and Emergency Communication (ICEIEC), pp. 186–190. IEEE (2021)

Heeman, P.A., Byron, D., Allen, J.F.: Identifying discourse markers in spoken dialog. Discourse **58298**(6163), 8278 (1998)

Lamiroy, B., Swiggers, P.: The status of imperatives as discourse signals. In: Fleischman, S., Waugh, L.R. (eds.) Routledge, London, pp. 120–146 (1991)

Le, H., et al.: Flaubert: unsupervised language model pre-training for French. arXiv preprint arXiv:1912.05372 abs/1909.05364 (2019). https://arxiv.org/abs/1912.05372

Nie, A., Bennett, E., Goodman, N.: DisSent: learning sentence representations from explicit discourse relations. In: Proceedings of the 57th Annual Meeting of the Association for Computational Linguistics, Florence, Italy, pp. 4497–4510. Association for Computational Linguistics (2019). https://aclanthology.org/P19-1442

Pallaud, B., Henry, S.: Amorces de mots et répétitions: des hésitations plus que des erreurs en fran, cais parlé. In: 7es Journées Internationales d'Analyse Statistique des Données Textuelles, pp. 848–858 (2004)

Passonneau, R.J., Litman, D.: Discourse segmentation by human and automated means. Comput. Linguist. **23**(1), 103–139 (1997)

Petukhova, V., Bunt, H.: Towards a multidimensional semantics of discourse markers in spoken dialogue. In: Proceedings of the Eight International Conference on Computational Semantics, pp. 157–168 (2009)

Schiffrin, D.: Discourse Markers, vol. 5. Cambridge University Press, Cambridge (1987)

Sileo, D., et al.: DiscSense: automated semantic analysis of discourse markers. In: Proceedings of the 12th Language Resources and Evaluation Conference, Marseille, France: European Language Resources Association, pp. 991–999 (2020). https://aclanthology.org/2020.lrec-1.125

Wu, X., et al.: TransSent: Towards Generation of Structured Sentences with Discourse Marker. CoRR abs/1909.05364 (2020). http://arxiv.org/abs/1909.05364

Zufferey, S., Popescu-Belis, A.: Towards automatic identification of discourse markers in dialogs: the case of like. In: Proceedings of the 5th SIGdial Workshop on Discourse and Dialogue at HLT-NAACL 2004, pp. 63–71. Association for Computational Linguistics (2004). https://aclanthology.org/W04-2313

Zufferey, S., Degand, L., et al.: Empirical validations of multilingual annotation schemes for discourse relations. In: Eighth Joint ISOACL/SIGSEM Workshop on Interoperable Semantic Annotation, pp. 77–84 (2012)

A Comparative Study of the Impact of Different First Order Optimizers on the Learning Process of UNet for Change Detection Task

Basma Dokkar[1]($\boxtimes$), Bouthaina Meddour[1], Khadra Bouanane[1], Mebarka Allaoui[1], and Mohamed Lamine Kherfi[2]

[1] Computer Science and IT Department, Laboratoire d'Intelligence Artificielle et Technologies de l'Information, LINATI, Kasdi Merbah University, Ouargla, Algeria
{Dokkar.Basma,Meddour.Bouthayna,Bouanane.khadra,
Allaoui.Mebarka}@univ-ouargla.dz
[2] The National Higher School of Artificial Intelligence, ENSIA, Algiers, Algeria
mohammedlamine.kherfi@ensia.edu.dz

Abstract. UNet is an encoder-decoder neural network that has been used to detect changes in remote-sensing images. This paper provides a comparative study on the performance of UNet when trained with different optimizers for the Change Detection task. Although several previous works aim to compare different UNet models for change detection, this paper is, as far as we know, the first work that investigates UNet regarding the optimization method that is used in the learning process. This can help designing of more efficient UNet models, especially with limited training resources. We compare five common gradient-based optimization techniques: Gradient descent with Momentum (Momentum GD), Nesterov Accelerated Gradient (NAG), Adaptive Gradient (AdaGrad), Root Mean Square Propagation optimizer (RMSProp), and the adaptive moment estimation optimizer (Adam). For this purpose, UNet is trained over 200 epochs using ONERA dataset for the optimization of the binary cross entropy. The model is assessed using three metrics: Accuracy, Precision, and F1-score. According to the obtained results, RMSProp, NAG, and AdaGrad reached the highest validation accuracies: $0.976, 0.978$, and 0.979 with $10^{-2}, 10^{-3}$ and 10^{-4} respectively. Adam was the fastest to converge and scored the lowest validation loss. Moreover, Adam scored the highest precision and F1-score across all learning rates, with 0.491 and 0.376 respectively. We also note that both Momentum-based algorithms and adaptive algorithms perform better with relatively small learning rate values.

Keywords: Comparative study · UNet · First order optimizer · Change Detection · Remote sensing satellite images

A. Bennour et al. (Eds.): ISPR 2023, CCIS 1941, pp. 300–315, 2024.
https://doi.org/10.1007/978-3-031-46338-9_23

1   Introduction

In the past decade, Deep Learning has successfully solved various challenging problems across different kinds of applications in many domains. Some of its models have reached the utmost level when applied to Computer Vision and Natural Language Processing tasks.

However, it is well-known that building a performant DL machine comes at an expensive cost in terms of memory and data consumption. It becomes then necessary when designing a new model, to select judiciously the three main components of DL: Data, network architecture, and the learning process.

For instance, the size and sparsity of the data or the model architecture [4,10,17,32,34] play an essential role in constructing a robust machine. Also, a careful choice of the cost function and the optimization method to perform the model's training is crucial for improving the learning process [1,15]. This motivates the elaboration of comparative studies that evaluate the performance and effectiveness of the models according to each of the three components.

In the context of the learning process, several first-order optimization methods have been adopted to overcome training issues arising from the particularity of the data on the one hand and that of the chosen architecture on the other. Gradient descent with Momentum [29] employs an exponentially decaying average of the gradients, which helps memorize the recent direction to accelerate the convergence. Nesterov's Momentum [25] as well, preserves the moving average but with pre-estimation of the position before performing an update. AdaGrad [11], unlike Momentum-based methods, treats all the gradients equally by summing their squares and dividing the learning rate at time step t by the corresponding square root of the sum. This causes the learning rate to decay continually but slowly to slow down the convergence as we get closer to the minimum hence avoiding overshooting. It should be mentioned that Nesterov addresses the overshooting problem by making more accurate steps than Momentum. RMSProp [35] uses a decaying average of squared gradients and adapts the learning rate accordingly by dividing each time step's gradient by the current square root average, which acts like normalization. Adam [16] combines the update rule of Momentum and RMSProp. A decaying average of gradients and another of their squares are used to adapt the learning rate.

The comparison of optimization methods in terms of speed of convergence has been the focus of several works in which, the performance of the convolutional neural network (CNN) has been assessed. Fatima [12] studied seven first-order methods: SGD, RMSProp, Adagrad, AdaDelta, Adam, Adamax, and Nadam by optimizing a simple sequential DL model using four different datasets. Dogo et al. [9] conducted a comparative study of the seven aforementioned methods using a basic CNN trained on three randomly selected data sets. Poojary and Pai [28] compared the performance of three gradient-based optimizers on well-known pre-trained models, namely ResNet-50 and Inception-V3, for transfer learning. Martenez et al. [22] considered eight optimizers to train ResNet on different points for vehicles on California roads for object recognition to improve the intelligent real-time sensor for autonomous driving.

A well-known variant of CNNs is UNet [31], which has been initially proposed for the segmentation of medical images and extended to other fields since then.

U-Net [31] is an encoder-decoder architecture. The encoder part consists of several blocks. Each block contains two successive convolutions of 3×3, doubling the number of channels in the feature map, a ReLU unit, and max pooling. In the decoder part, each block consists of an upsampling, a 2×2 up-convolution that halves the number of feature channels. Concatenation with the correspondingly cropped feature map from the contracting path and two 3×3 convolutions, each followed by a 3×3 convolution. Figure 1 illustrates a UNet architecture.

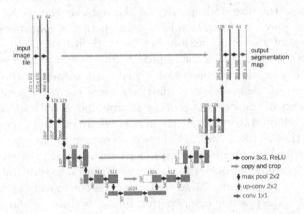

Fig. 1. UNet original architecture where blue boxes represent feature maps, white boxes copied feature maps with numbers of channels on top. Different operations are denoted with arrows of different colors [31] (Color figure online)

UNet is known to reach effective learning with few data images., which makes it suitable for the CD multi-spectral datasets like ONERA given the small number of images of some of these data-sets [8,18,36].

The use and performance of UNet architecture for detecting changes in Remote Sensing(RS) images have recently been investigated [2,6,19,24]. In [2], UNet and UNet++ have been trained and evaluated using high-resolution satellite images. The authors analyzed the effect of two loss functions on the performance of the models as well as the use of data augmentation and deep supervision techniques on the performance of the models. In [24], the generality and performance of UNet and four of its variants were assessed when trained and evaluated on datasets suffering from class imbalance and a small region of interest. Lv et al. [21] compared an existing DL-based approach along with a UNet-based approach proposed within the work and in [20], a residual UNet was evaluated with Sentinel-1 SAR CD dataset for urbanization.

As we can notice, no study on the performance of UNet architecture that involves the optimization method used in the learning process has been conducted for CD task. We believe that such a study will help in the design of more performant UNet models for CD, especially with limited resources.

In this work, we aim to investigate the impact of the choice of the optimization algorithm on the performance of UNet. For this purpose, we consider the first-order methods that are commonly used in Deep Learning and provide a comparative study of the five major gradient-based optimization algorithms. Namely, Momentum GD, NAG, AdaGrad, RMSProp, and Adam.

The rest of this paper is organized as follows: In Sect. 2, we give an explicit description of the adopted methodology. Section 3 is dedicated to the experimental settings and the discussion of the obtained results while Sect. 5 concludes and provides the future perspectives for this work.

2 Methodology

To investigate the performance of UNet when trained with different optimizers for CD in RS images, we train a UNet model by optimizing binary cross-entropy cost (BCE) using the five aforementioned gradient-based methods. Our methodology is detailed in the following subsections.

2.1 Change Detection Task(CD)

CD is the process of identifying differences in the state of an object or phenomenon by observing it at different times [33]. The process requires multitemporal images and certain criteria specified by the mask. In the case of Binary CD, the mask is a binary image where black regions correspond to areas with negligible change while white regions represent significant changes [30]. Figure 2 is a sample from Onera CD data set, where the change mask corresponds to the bi-temporal images. Therefore, the task is reduced to binary image segmentation.

Fig. 2. Beirut city. Sample from Onera Satellite Change Detection dataset created by Daudt et al. [7]

Difference image techniques are suitable for medium-resolution images like those of the dataset used in our experiments, based on the observation that the smaller the resolution, the less spatial-contextual information affects the resulting CD map [26]. DL-based approaches are summarized under three categories: feature-based, patch-based, and image-based [26]. The method we use falls under patch-based DL methods, where different images are calculated and then divided into smaller patches of 256×256 due to the large size of the images.

2.2 Optimization Algorithms

We compare five gradient-based methods that are among the most popular [13, 14]. The gradient descent algorithm consists in iteratively updating the target variable based on the partial derivative of the objective function with respect to this variable. Equation (1) represents the update rule:

$$w_{t+1} = w_t - \alpha \frac{\partial C}{\partial w_t} \tag{1}$$

where:

w_{t+1} and w_t:the weight parameter vector at time t and t+1, after the update.

α: the learning rate, a hyperparameter that controls the step size of the update.

C: and $\frac{\partial C}{\partial w_t}$: the cost function and its partial derivative w.r.t the weight parameter vector w_t.

Momentum. Momentum [29] introduced another term to gradient descent to accelerate the convergence by employing the concept of velocity. It has additional configurable hyper-parameters: Momentum β. Momentum defines an exponentially weighted moving average of the gradients for every iteration. Hence, the update depends on the newest history of the gradients. The average is defined as:

$$v_t = \beta v_{t-1} + (1 - \beta) \frac{\partial C}{\partial w_t} \tag{2}$$

where:

β: the momentum hyperparameter.

v_{t-1}: the velocity at time t-1.

$\frac{\partial C}{\partial w_t}$: the partial derivative of the cost function w.r.t the weight parameter vector.

β controls how much the update depends on past gradients. Momentum can reduce the oscillations resulting from insufficient or noisy gradient information, Hence moving mostly in the correct direction. The update rule is the following:

$$w_{t+1} = w_t - \alpha v_t \tag{3}$$

However, Momentum is highly prone to overshooting the minimum due to its large velocity. Hence, it will take a large step and does not change its direction even when a small step is sufficient [3].

NAG. The difference between NAG [25] and Momentum is that NAG estimates the future position of the variable and then corrects it by taking a Momentum step along with the previous Momentum, then a gradient step. This is particularly useful when we get very close to the minimum

$$v_t = \beta v_t + \alpha \frac{\partial C}{\partial (w_t - \beta v_t)} \tag{4}$$

$$w_{t+1} = w_t - v_t \tag{5}$$

In this case, the derivative itself depends on the history, not just the update. Nevertheless, both Momentum and NAG treat the gradients of all parameters equally regardless of the fact that they have different magnitudes.

AdaGrad. To reduce the oscillations ensuing from the different magnitudes which arise in gradient descent, Momentum and NAG, AdaGrad [11] normalizes the gradient of each parameter by the sum of squared gradients of all iterations preventing large magnitudes from dominating the direction of the step towards a certain parameter. Equation (6) is the sum of squared gradients, (7) is the adaptive learning rate, and (8) is the update rule.

$$v_t = \sum_{i=0}^{t}(\frac{\partial C}{\partial w_i})^2 \tag{6}$$

$$\alpha_t = \alpha_{t-1}/(\sqrt{v_t} + \epsilon) \tag{7}$$

$$w_t = w_{t-1} - \alpha_t \frac{\partial C}{\partial w_t} \tag{8}$$

However, the sum increases at every iteration without control of how much history should impact the step i.e. recent and old history are treated equally.

RMSProp. To overcome Adagrad's issue, RMSProp [35] uses the same concept of the weighted average as Momentum, But instead of gradients, it accumulates squared gradients for each weight. The square root of this latter acts as a normalization to the gradients' magnitudes, similar to AdaGrad. Equation (9) is the moving average of squared gradients, and Eq. (10) is the update rule.

$$v_t = \beta v_{t-1} + (1 - \beta)(\frac{\partial C}{\partial w_t})^2 \tag{9}$$

$$w_t = w_{t-1} - \frac{\alpha}{\sqrt{v_t} + \epsilon} \frac{\partial C}{\partial w_t} \tag{10}$$

Adam. Adam [16] combines the update rule of Momentum and RMSProp. Hence, two moving averages (Eqs. (11), (12)). Bias correction is then applied on the moments (Eqs. (13), (14)). Equation (15) is the update rule.

$$m_t = \beta_1 m_{t-1} + (1 - \beta_1)g_t \tag{11}$$

$$v_t = \beta_2 v_{t-1} + (1 - \beta_2)g_t^2 \tag{12}$$

$$\hat{m}_t = \frac{m_t}{1 - \beta_1^t} \tag{13}$$

$$\hat{v}_t = \frac{v_t}{1 - \beta_2^t} \tag{14}$$

$$w_t = w_{t-1} - \alpha \frac{\hat{m}_t}{\sqrt{\hat{v}_t} + \epsilon} \tag{15}$$

where $\beta_1, \beta_2 \in]0, 1]$ control the m_t the moving average of gradients, and v_t the moving average of squared gradients, $\hat{m}_t$ and $\hat{v}_t$ are the bias corrected m_t and v_t respectively.

2.3 Binary Cross Entropy(BCE)

BCE is a distribution-based loss function [15] i.e. it measures the difference between the probability distributions of the true labels and the predicted values. It is generally used for binary classification. As for our case of (binary segmentation), BCE was used for satellite images data as well as medical images data and obtained satisfactory results, either alone like in [23] or combined with other loss functions like in [5]. The BCE is formalized as follows:

$$L = ylog(\hat{y}) + (1 - y)log(1 - \hat{y}) \tag{16}$$

3 Experiments

3.1 Data Set

We conducted the experiments using ONERA, a CD data set consisting of 24 pairs of multi-spectral images taken from the Sentinel-2 satellites between 2015 and 2018 of different cities worldwide. The spatial resolution ranges between 10 m and 60m. It contains the corresponding masks, where the black regions represent unchanged areas and the white segments correspond to changes [7]. It is worth mentioning that to train standard deep neural network, the size of this data set is considered insufficient. But according to the authors [31], UNet can be trained efficiently using few images.

3.2 Experimental Setting

Preprocessing. The original images were split into 777 patches of 256×256 images for preprocessing. The image difference of the pairs was calculated according to the root mean squared error(RMSE) of the difference image and the ground truth. Hence, we combined average intensity(AI) with absolute distance(AD).

Training. Using one of the five optimizers, the model was trained for 200 epochs each time. We chose three different learning rates: $10^-2, 10^-3, and 10^-4$. All the training was done on Google Colab with the below specifications which was enough to meet our requirements.
GPU: NVIDIA T4 16 GB, 2560 CUDA cores, 585 MHz.
RAM: 12 GB, 3200 MHz memory clock.

Evaluation Metrics. We assess the performance of the model using three metrics: Accuracy (acc), Precision (prec), and F1-score (F1).

Table 1. The results of training UNet using the five optimization methods for 200 epochs.

learning rate	10^{-2}			10^{-3}			10^{-4}		
	Acc	prec	F1	Acc	prec	F1	Acc	precision	F1
Momentum	0.975	**0.382**	0.219	0.956	0.243	0.209	0.977	0.027	0.011
NAG	0.960	0.0002	0.0004	**0.978**	0.250	0.208	0.977	0.0006	0.0004
AdaGrad	0.971	0.290	0.200	0.972	0.201	0.198	**0.979**	0.002	0.004
RMSProp	**0.976**	0.274	0.249	0.975	0.435	**0.373**	0.973	0.350	**0.330**
Adam	0.971	0.330	**0.310**	0.977	**0.491**	**0.376**	0.972	**0.388**	**0.331**

4   Results and Discussion

This experiment aims to understand the behavior of UNet model when trained for the optimization of the BCE using the five optimizer with different learning rates. Table 1 provides us with the accuracy, precision, and F1-score of the model when trained over 200 epochs while Tables 2,3,4,5 and 6 illustrates the plots of the training and validation loss over epochs for the three learning rates values.

From Tables 2,3,4,5 and 6, we can notice that Momentum-based methods, i.e., Momentum and NAG, have more stable convergence concerning the learning rate than the rest of the methods. However, all of the optimizers stop significantly converging through the early epochs. On the other hand, at 10^{-3}, Adam, Ada-Grad, and RMSProp converge even further. These three optimizers eliminate the effect of different scales among the model's parameters. Therefore, they converge with more accurate steps than Momentum and NAG and are less prone to falling into local minima (empirically shown by Kingma and Lei Ba [16] that Adam outperforms other methods in cases of non-convexity).

Moreover, regardless of the learning rate value, AdaGrad, Momentum, and NAG have large loss values compared to Adam and RMSProp. We observe that Momentum and NAG have similar behavior while the rest of the methods behave in another way similarly. We can interpret these two different ways as follow:

- Momentum and NAG Arrive at a local minimum due to large steps biased toward large-scale parameters.
- AdaGrad did not converge because of vanishing gradients.

From Table 5, despite RMSProp scoring decent quantitative results, it is the most sensitive to changing the learning rate value. In contrast, Momentum and NAG are the least sensitive. These latter demonstrated an interesting behavior of similarity in the way that they converge, while they have different update mechanisms (2.2, 2.2). During optimization, Momentum and NAG start with large steps in the first epochs and then slow down, given that in the early iteration, these two algorithms have equivalent update rules (moving average equals 0). Accelerating learning seems to have no significant effect on NAG. Since the convergence has not improved, reinforcing the hypothesis that Momentum and

Table 2. SGD with Momentum: Training and Validation Loss.

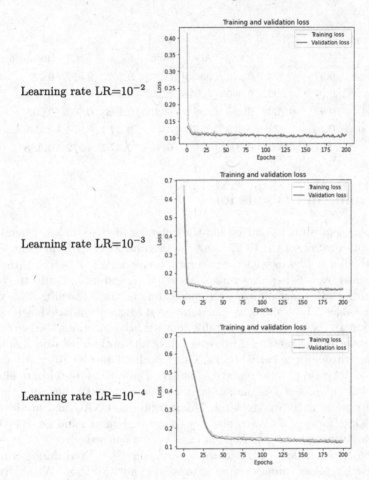

NAG arrive at local minima. At 10^{-4}, all methods' convergence slows remarkably.

Although the authors of [31] stated that UNet can be trained effectively with small data sets, this does not seem to be valid for all tasks and data patterns. With more exposure to the training data, the model can learn the patterns and relationships in the data thus make better predictions. However, over-fitting occurs when a model becomes too complex and starts to memorize the training data instead of learning the underlying patterns, resulting in poor performance on unseen data. Since ONERA is a small dataset, this is exactly what happens when training the UNET model on this dataset using different optimization

Table 3. NAG: Training and Validation Loss.

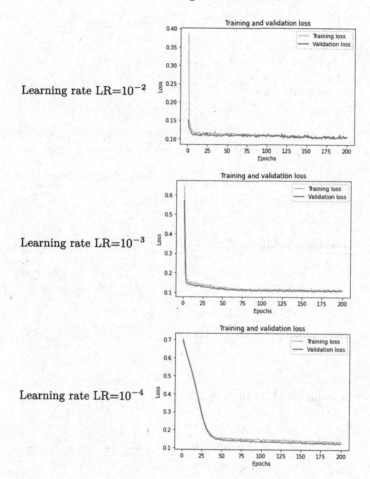

techniques. Even after we increased the dataset size using data augmentation techniques, it was insufficient to train the model well. The medium-resolution dataset(ONERA) has irregular and non-continuous change patterns which makes it harder for the model to generalize. Hence, the obtained quantitative results. This issue arises in similar tasks like crack detection [27] where the authors also attributed the difficulty of the task to the challenging patterns of cracks.

Table 4. AdaGrad: Training and Validation Loss.

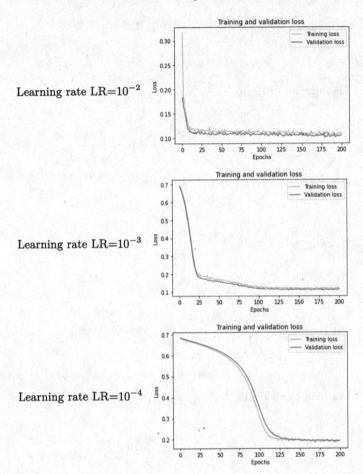

Learning rate LR=10^{-2}

Learning rate LR=10^{-3}

Learning rate LR=10^{-4}

Table 5. RMSProp: Training and Validation Loss.

Learning rate LR=10^{-2}

Learning rate LR=10^{-3}

Learning rate LR=10^{-4}

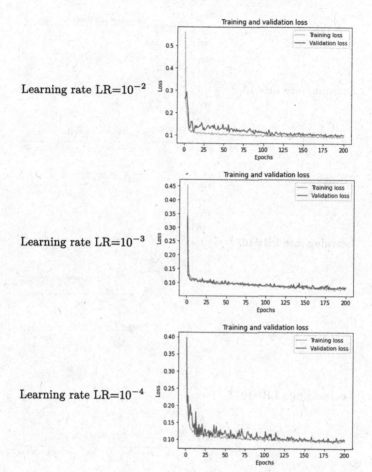

Table 6. Adam: Training and Validation Loss.

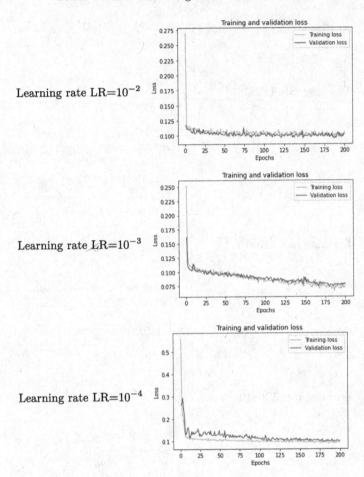

5   Conclusion

This paper presented a comparative study to understand the behavior of UNet model using five first-order optimization methods: Momentum GD, NAG, Ada-Grad, RMSProp, and Adam on the very specific case of binary CD in RS satellite images. The study showed that for different learning rates, different algorithms namely RMSProp, NAG, and AdaGrad outperformed the rest in terms of accuracy. On the other hand, Adam attained superior precision and F1-score and was the fastest to converge. The obtained results indicate that more can be explored with reference to the existing challenges: The small dataset volume, which potentially led to insufficient training as well as the irregular change patterns and the poor continuity. Our future work will focus on exploring potential improvements of the optimizers and the architecture against the proposed challenges.

References

1. Akbari, A., Awais, M., Bashar, M., Kittler, J.: How does loss function affect generalization performance of deep learning? Application to human age estimation. In: International Conference on Machine Learning, pp. 141–151. PMLR (2021)
2. Alexakis, E.B., Armenakis, C.: Evaluation of UNet and UNet++ architectures in high resolution image change detection applications. Int. Arch. Photogrammetry, Remote Sens. Spat. Inf. Sci. **43**, 1507–1514 (2020)
3. An, W., Wang, H., Sun, Q., Xu, J., Dai, Q., Zhang, L.: A PID controller approach for stochastic optimization of deep networks. In: Proceedings of the IEEE Conference on Computer Vision and Pattern Recognition, pp. 8522–8531 (2018)
4. Basha, S.S., Dubey, S.R., Pulabaigari, V., Mukherjee, S.: Impact of fully connected layers on performance of convolutional neural networks for image classification. Neurocomputing **378**, 112–119 (2020)
5. Brahmbhatt, P., Rajan, S.N.: Skin lesion segmentation using SegNet with binary crossentropy. In: Proceedings of the International Conference on Artificial Intelligence and Speech Technology (AIST2019), Delhi, India, pp. 14–15 (2019)
6. Daudt, R.C., Le Saux, B., Boulch, A.: Fully convolutional Siamese networks for change detection. In: 2018 25th IEEE International Conference on Image Processing (ICIP), pp. 4063–4067. IEEE (2018)
7. Daudt, R.C., Le Saux, B., Boulch, A., Gousseau, Y.: Urban change detection for multispectral earth observation using convolutional neural networks. In: IGARSS 2018–2018 IEEE International Geoscience and Remote Sensing Symposium, pp. 2115–2118. IEEE (2018)
8. Daudt, R.C., Lé Saux, B., Boulch, A., Gousseau, Y.: Multitask learning for large-scale semantic change detection. Comput. Vis. Image Underst. **187**, 102783 (2019)
9. Dogo, E., Afolabi, O., Nwulu, N., Twala, B., Aigbavboa, C.: A comparative analysis of gradient descent-based optimization algorithms on convolutional neural networks. In: 2018 International Conference on Computational Techniques, Electronics and Mechanical Systems (CTEMS), pp. 92–99. IEEE (2018)
10. Dong, S., Wang, P., Abbas, K.: A survey on deep learning and its applications. Comput. Sci. Rev. **40**, 100379 (2021). https://doi.org/10.1016/j.cosrev.2021.100379, https://www.sciencedirect.com/science/article/pii/S1574013721000198
11. Duchi, J., Hazan, E., Singer, Y.: Adaptive subgradient methods for online learning and stochastic optimization. J. Mach. Learn. Res. **12**(7) (2011)
12. Fatima, N.: Enhancing performance of a deep neural network: a comparative analysis of optimization algorithms. ADCAIJ: Adv. Distrib. Comput. Artif. Intell. J. **9**(2), 79–90 (2020)
13. Gaddam, D.K.R., Ansari, M.D., Vuppala, S., Gunjan, V.K., Sati, M.M.: A performance comparison of optimization algorithms on a generated dataset. In: Kumar, A., Senatore, S., Gunjan, V.K. (eds.) ICDSMLA 2020. LNEE, vol. 783, pp. 1407–1415. Springer, Singapore (2022). https://doi.org/10.1007/978-981-16-3690-5_135
14. Haji, S.H., Abdulazeez, A.M.: Comparison of optimization techniques based on gradient descent algorithm: a review. PalArch's J. Archaeol. Egypt/Egyptol. **18**(4), 2715–2743 (2021)
15. Jadon, S.: A survey of loss functions for semantic segmentation. In: 2020 IEEE Conference on Computational Intelligence in Bioinformatics and Computational Biology (CIBCB), pp. 1–7. IEEE (2020)
16. Kingma, D.P., Ba, J.: Adam: a method for stochastic optimization. arXiv preprint: arXiv:1412.6980 (2014)

17. Krishnan, R., Liang, D., Hoffman, M.: On the challenges of learning with inference networks on sparse, high-dimensional data. In: Proceedings of the Twenty-First International Conference on Artificial Intelligence and Statistics, pp. 143–151. PMLR (2018)

18. Leenstra, M., Marcos, D., Bovolo, F., Tuia, D.: Self-supervised pre-training enhances change detection in sentinel-2 imagery. In: Del Bimbo, A., et al. (eds.) ICPR 2021. LNCS, vol. 12667, pp. 578–590. Springer, Cham (2021). https://doi.org/10.1007/978-3-030-68787-8_42

19. Lei, T., Zhang, Y., Lv, Z., Li, S., Liu, S., Nandi, A.K.: Landslide inventory mapping from bitemporal images using deep convolutional neural networks. IEEE Geosci. Remote Sens. Lett. **16**(6), 982–986 (2019)

20. Li, L., Wang, C., Zhang, H., Zhang, B.: Residual UNet for urban building change detection with sentinel-1 SAR data. In: IGARSS 2019–2019 IEEE International Geoscience and Remote Sensing Symposium, pp. 1498–1501. IEEE (2019)

21. Lv, Z., Huang, H., Gao, L., Benediktsson, J.A., Zhao, M., Shi, C.: Simple multiscale UNet for change detection with heterogeneous remote sensing images. IEEE Geosci. Remote Sens. Lett. **19**, 1–5 (2022)

22. Martenez, F., Montiel, H., Martenez, F.: Comparative study of optimization algorithms on convolutional network for autonomous driving. Int. J. Electr. Comput. Eng. (2088–8708) **12**(6) (2022)

23. Mohanty, S.P., et al.: Deep learning for understanding satellite imagery: an experimental survey. Front. Artif. Intell. **3**, 534696 (2020)

24. Moustafa, M.S., Mohamed, S.A., Ahmed, S., Nasr, A.H.: Hyperspectral change detection based on modification of UNet neural networks. J. Appl. Remote Sens. **15**(2), 028505 (2021)

25. Nesterov, Y.: A method for unconstrained convex minimization problem with the rate of convergence o $(1/k^2)$. In: Doklady an USSR, vol. 269, pp. 543–547 (1983)

26. Peng, D., Zhang, Y., Guan, H.: End-to-end change detection for high resolution satellite images using improved UNet++. Remote Sens. **11**(11), 1382 (2019)

27. Polovnikov, V., Alekseev, D., Vinogradov, I., Lashkia, G.V.: DAUNet: deep augmented neural network for pavement crack segmentation. IEEE Access **9**, 125714–125723 (2021)

28. Poojary, R., Pai, A.: Comparative study of model optimization techniques in fine-tuned CNN models. In: 2019 International Conference on Electrical and Computing Technologies and Applications (ICECTA), pp. 1–4. IEEE (2019)

29. Qian, N.: On the momentum term in gradient descent learning algorithms. Neural Netw. **12**(1), 145–151 (1999)

30. Renza, D., Martinez, E., Arquero, A.: A new approach to change detection in multispectral images by means of ERGAS index. IEEE Geosci. Remote Sens. Lett. **10**(1), 76–80 (2012)

31. Ronneberger, O., Fischer, P., Brox, T.: U-Net: convolutional networks for biomedical image segmentation. In: Navab, N., Hornegger, J., Wells, W.M., Frangi, A.F. (eds.) MICCAI 2015. LNCS, vol. 9351, pp. 234–241. Springer, Cham (2015). https://doi.org/10.1007/978-3-319-24574-4_28

32. Sankararaman, K.A., De, S., Xu, Z., Huang, W.R., Goldstein, T.: The impact of neural network overparameterization on gradient confusion and stochastic gradient descent. In: International Conference on Machine Learning, pp. 8469–8479. PMLR (2020)

33. Singh, A.: Review article digital change detection techniques using remotely-sensed data. Int. J. Remote Sens. **10**(6), 989–1003 (1989)

34. Sun, S., Chen, W., Wang, L., Liu, X., Liu, T.Y.: On the depth of deep neural networks: a theoretical view. In: Proceedings of the AAAI Conference on Artificial Intelligence, vol. 30 (2016)
35. Tieleman, T., Hinton, G.: RMSPROP: divide the gradient by a running average of its recent magnitude. coursera: neural networks for machine learning. COURSERA Neural Netw. Mach. Learn. (2012)
36. Wang, Q., Yuan, Z., Du, Q., Li, X.: GETNET: a general end-to-end 2-D CNN framework for hyperspectral image change detection. IEEE Trans. Geosci. Remote Sens. **57**(1), 3–13 (2018)

Author Index

A. Bennour et al. (Eds.): ISPR 2023, CCIS 1941, pp. 317–318, 2024.
https://doi.org/10.1007/978-3-031-46338-9

Printed in the United States
by Baker & Taylor Publisher Services